COMPARATIVE
GENETICS *in*
Monkeys,
Apes and
Man

COMPARATIVE GENETICS *in* Monkeys, Apes and Man

Edited by

A. B. CHIARELLI

Institute of Anthropology,
Primatology Centre,
University of Turin, Italy

Proceedings of a symposium
on Comparative Genetics in
Primates and Human Heredity
held at Ernice, Sicily, July 1970

1971

 Academic Press · London and New York

ACADEMIC PRESS INC. (LONDON) LTD
24–28 Oval Road,
London, NW1 7DX

U.S. Edition published by
ACADEMIC PRESS INC.
111 Fifth Avenue,
New York, New York 10003

Library of Congress Catalog Card Number: 70-170750
ISBN: 0-12-172530-8

PRINTED IN GREAT BRITAIN BY
W. S. COWELL LTD IPSWICH

Contributors

H. BALNER, *Radiobiological Institute TNO, Rijswijk Z.H., The Netherlands*

J. BARNABAS, *Department of Anatomy, Wayne State University, Medical School, Detroit, Michigan, U.S.A.*

N. A. BARNICOT, *Department of Anthropology, University College London, England*

A. C. BERRY, *Department of Animal Genetics, University College London, England*

R. J. BERRY, *Royal Free Hospital School of Medicine, London, England*

A. O. CARBONARA, *Instituto Genetica Medica, University of Turin, Italy*

A. B. CHIARELLI, *Institute of Anthropology, Primatology Centre, University of Turin, Italy*

H. DERSJANT, *Radiobiological Institute TNO, Rijswijk Z.H., The Netherlands*

B. W. GABB, *Radiobiological Institute TNO, Rijswijk Z.H., The Netherlands*

M. GOODMAN, *Department of Anatomy, Wayne State University, Medical School, Detroit, Michigan, U.S.A.*

H. KALMUS, *Galton Laboratory, University College London, England*

A. L. KOEN, *Hawthorn-Plymouth Research Laboratory, Northville, Michigan, U.S.A.*

A. v. LEEUWEN, *Radiobiological Institute TNO, Rijswijk Z.H., The Netherlands*

G. WILLIAM MOORE, *Biomathematics Program, Institute of Statistics, North Carolina State University at Rayleigh, U.S.A.*

J. MOOR-JANKOWSKI, *Laboratory for Experimental Medicine and Surgery in Primates, New York University School of Medicine, New York, U.S.A.*

J. MAVALWALA, *Department of Anthropology, University of Toronto, Canada*

J. J. v. ROOD, *Radiobiological Institute TNO, Rijswijk Z.H, The Netherlands*

B. SULLIVAN, *Department of Biochemistry, Duke University Medical Center, Durham, North Carolina, U.S.A.*

W. v. VREESWIJK, *Radiobiological Institute TNO, Rijswijk, Z.H., The Netherlands*

M. L. WEISS, *Department of Anthropology, Wayne State University, Detroit, Michigan, U.S.A.*

A. S. WIENER, *Department of Forensic Medicine, New York University of Medicine, New York, U.S.A.*

Contents

A. B. CHIARELLI

Comparative Primate Genetics and Human Heredity

H. KALMUS

Epigenetic Polymorphism in the Primate Skeleton

A. CAROLINE BERRY AND R. J. BERRY

The Heredity of Dermatoglyphic Traits in Non-human Primates and Man

J. MAVALWALA

Phenylthiourea Testing in Primates

H. KALMUS

Blood Groups of Non-human Primates and Their Relationship to the Blood Groups of Man

ALEXANDER S. WIENER AND J. MOOR-JANKOWSKI

Leukocyte Groups of Non-human Primates; their Relation to Histocompatibility and to Human HL-A Antigens

H. BALNER, B. W. GABB, H. DERSJANT, W. v. VREESWIJK, A. v. LEEUWEN AND J. J. v. ROOD

Genetic Structure and Systematics of some Macaques and Men

MARK L. WEISS AND MORRIS GOODMAN

Evolving Primate Genes and Proteins

MORRIS GOODMAN, ANN L. KOEN, JOHN BARNABAS AND G. WILLIAM MOORE

Comparison of the Hemoglobins in Non-human Primates and their Importance in the Study of Human Hemoglobins

BOLLING SULLIVAN

Phylogenesis of Immunoglobulins in Primates

A. O. CARBONARA

Comparative Cytogenetics in Primates and its Relevance for Human Cytogenetics

A. B. CHIARELLI

Concluding Remarks   309

N. A. BARNICOT

Introduction

A. B. CHIARELLI

Institute of Anthropology, Primatology Centre, University of Turin, Italy

I am delighted to welcome you to Italy. I would like to say just a few words of thanks to the contributors and the participants who have accepted our invitation to this Round Table and who have shown their interest by coming to Erice, at the far end of Sicily.

I wish to make it clear at the outset that this is a Round Table and that we can dispense with the pretentious name "Symposium". This meeting was primarily organized as an informal discussion among a group of friends who want to focus attention on different aspects of the same problem, the main purpose being a stimulation and interchange of ideas.

Why a round table about Comparative Genetics in Primates and Human Heredity? The answer will be partly found in some personal recollection.

When I graduated in Anthropology, one of my main subjects was a study on "Comparative Genetics and Cytogenetics in Primates and their Impact on Human Evolution". Primate genetics was still in its infancy at that time. The chromosome number had been determined, imperfectly at that, for only a few species; mainly through the efforts of Darlington (1955), Painter (1922, 1924, 1925) and Matthey (1955). Hereditary traits were known only for some erythrocyte blood groups (Candela, 1940a, b; Butts, 1953; Dahr, 1939; Landsteiner and Miller, 1952; Voronoff, 1949; Wiener, 1952) and P.T.C. testing (Fisher *et al.*, 1939). I had planned some promising research work and I was confident that my hopes would be realized. I had, however, overlooked the state of affairs in the universities in Italy. My applications for hospitality or support were repeatedly rejected on financial grounds. After a short period in the Institute of Human Anatomy in Florence under the kind and understanding guidance of Professor Fazzari, I had the privilege of meeting Professor Adriano Buzzati Traverso of the Department of Genetics in Pavia. As is well known, his stimulating activity in biological research in Italy between 1958 and 1966 was unique. I worked in the Department of Genetics at the University of Pavia for 4 years and I owe practically all that I was able to accomplish in developing this research direction to Professor Traverso. I am sure that many of you have faced similar experiences in beginning research in this new field.

However, let us return to our meeting here. The programme of this

1

round table was originally prepared for the Third International Congress of Primatology at Atlanta 3 years ago. Many of you certainly remember our contact during 1967 and in early 1968. For several reasons, in part connected with the escalation of the Vietnam War during the spring of 1968, that meeting was cancelled. The idea of gathering in book form the knowledge provided by various scientists on such a specific topic as the "Comparative Genetics of Primates" was brought back to life by Mr. John Cruise of Academic Press, London. It is due to his enthusiasm that we are here together now.

What is the aim of this round table?

At a genetic level, the largest body of information is provided by *Drosophila*, the mouse, and Man. Human biology has essentially developed under the stimulus of clinical research based on a pathological–normal comparison. Comparative investigations with other species are restricted to the few tamed by man and commonly used in pharmacological research, for example the rat and the mouse. These species differ biologically from man to such a great extent, however, that any conclusions drawn from these comparisons are debatable, to say the least. Moreover, in genetic investigations, comparison between such distant species is meaningless. On the contrary, it may be anticipated that comparison with species more closely related to man will provide more significant information. This will be demonstrated by the data that will be discussed in the next few days.

Although this may seem quite obvious, a special feature is generally neglected, chiefly by primatologists. It should be stressed that the Primate group represents a zoological "unicum" from both a phylogenetic and taxonomic viewpoint. In fact, the Primates are a zoological group in which the existing species recapitulate, with a good degree of plausibility, the group's own evolutionary history.

It may be assumed that the Prosimians represent the living relics of the most remote Primate forms from which the entire stock of the upper Primates was derived. Catarrhine monkeys exemplify, both morphologically and chronologically, a following evolutionary stage. Anthropoid apes and, finally, Man himself are a still further evolved step in this pathway of differentiation. This particular situation should be carefully considered by geneticists who are thus given the opportunity to trace the evolution of some genes or gene complexes and to study their adaptation capacities.

Some of you will certainly disagree with the general idea of comparative genetics, especially when this concept is applied to the comparison of hereditary traits in Man and the non-human Primates from a theoretical and practical point of view. This is understandable but what is most important is that we are here together to discuss a problem. We are, therefore, looking forward to a fruitful dialogue.

References

Butts, D. C. (1953). Hemoagglutinogens of the chimpanzee. *Amer. J. Phys. Anthropol.* **11**, 213–224.

Candela, P. B. (1940a). New data on the serology of the Anthropoid Apes. *Amer. J. Phys. Anthropol.* **72**, 209–211.

Candela, P. B. (1940b). Serology of the Anthropoid Apes. *Amer. J. Phys. Anthropol.* **27**, 479–480.

Dahr, P. (1939). Uber Blutgruppen bei Anthropoiden. *Z. Morphol. Anthropol.* **38**, 38–45.

Darlington, C. D. and Haque, A. (1955). Chromosomes of Monkeys and men. *Nature* **175**, 32.

Fisher, R. A., Ford, E. B. and Huxley, S. (1939). Taste testing the Anthropoid Apres. *Nature* **1944**, 750.

Landsteiner, K. and Miller, C. P. (1952). Serological studies on the blood of the Primates. I. The differentiation of human and anthropoid bloods. *J. Exp. Med.* **42**, 841–852.

Matthey, R. (1955). Les chromosomes de Galago senegalensis. *Rev. Suisse Zool.* **67**, 190–197.

Painter, T. S. (1922). Studies in mammalian spermatogenesis. IV. The sex chromosomes of Monkeys. *J. Exp. Zool.* **39**, 433–463.

Painter, T. S. (1925). A comparative study of the chromosomes of mammals. *Amer. Natur.* **59**, 385–409.

Voronoff, S. (1949). Les groupes sanguins chez les singes (Doine, ed.).

Wiener, A. S. (1952). Blood group factors in anthropoid apes and monkeys. I. Studies on a chimpanzee. *Amer. J. Phys. Anthropol.* **10**, 372–375.

Comparative Primate Genetics and Human Heredity

H. KALMUS

Galton Laboratory, University College London, England

INTRODUCTION

The aims of comparative primate genetics are both speculative and practical. The latter are usually well-defined and merely serve to substitute primates for Man, when we cannot use humans for experimental or clinical procedures. Speculations—especially evolutionary speculations—based on genetical material suffer from very much the same conceptual and operational difficulties as speculations based on morphological or behavioural characteristics; as a consequence, hypotheses are either vague or inconclusive, and as genetics and in particular population genetics lends itself to mathematical treatment, the particularly insidious pitfalls of mathematical confabulation are added, which usually take the form of models or "trees".

In this introductory paper I shall try to (i) clarify and describe the main approaches of comparative genetics; (ii) to criticise pseudohistorical evolutionary speculations; (iii) to indicate the value of an unbiased survey of variability and (iv) give a few examples of the practical value of comparing hereditary, morbid conditions in Man and animal.

IDENTITY AND SIMILARITY

It is trivial but necessary to state that no two items in the material world can ever be identical; even if indistinguishable in any other way they differ in localization, time or context. To call two genes "the same" is therefore loose talk. All we can do is to discuss the similarities of genes and their various effects.

Comparisons

Comparative genetics is subject to similar constraints and errors as comparative morphology. It is true that genes are more fundamental than phenotypes and that it is more up to date to deal with chemistry and gene frequencies (Baglioni, 1967) than with bones and indices. On the other hand, compared with the morphologists, the geneticist is at a disadvantage because he is confined to studying the present; and when speculating about evolution he lacks fossil records. True, fossil morphs of gastropod shells,

5

which are probably genetical in nature, exist and the frequencies of them can be counted. As far as primate material is concerned, however, genetical fossil evidence is so far non-existent, the blood grouping of mummies having become rather dubious. Attempts at determining various kinds of haemoglobin, in particular that responsible for thalassemia in mummified human bodies, may be more successful (personal communication, B. Chiarelli) but even so are not likely to indicate any particular evolutionary pathways.

With the growth of genetics comparisons were extended to all the entities, which were at any particular time the subject of study, e.g. to characters, genes, chromosomes, gene frequencies and polymorphisms, and in these comparisons the concepts of homology and analogy, of convergence and parallel change were variously applied.

As early as 1927, Haldane speculated that there might exist a "more fundamental relation between genes than homology, namely chemical identity". At that time it was not realized that the carriers of genetical information were the ribonucleic acids, and Haldane therefore thought of protein identity. In practice even today comparisons of base differences between the DNAs and RNAs of higher organisms are not feasible and only overall measurements of the degree of binding when "hybridizing" the DNAs of two species, for instance the mouse and the Norwegian rat, have been attempted (McLaren and Walker, 1965).

At present the most fundamental kind of comparison possible is that of proteins, like the haemoglobins (Ingram, 1963), cytochromes, insulins and a number of enzymes (see Barnicot, 1969; Barnicot and Cohen, 1970). In many of these proteins the substitution of amino acids at particular points of the polypeptide chains is known. All this will be dealt with later on in the book.

Comparisons can also be made of chromosomes and linkage relationships, either separately or in conjunction. Comparing the size, shape or peculiar features, e.g. satellites of the chromosomes of related species, is just comparative morphology, but homology of chromosomal segments can also be inferred from the pairing of hybrid chromosomes.

In primates chromosomal comparisons between the species have so far only been possible for rather gross features. Thus it appears that all primates have an XX/XY sex mechanism (Mittwoch, 1967). On the other hand primates vary greatly in the chromosome numbers of species of many primate groups (Epozcue, 1969). The evolutionary possibilities of translocations and of centromere fusion or fission will be discussed by Chiarelli (this volume), as will be the probable loss of one pair of autosomes during anthropogenesis.

In not very closely related mammals similar genes seem to have similar localizations. Thus some G6PD characteristics of horse and donkey have been compared. These characters are as in Man sex linked in both species and in both possible hybrids the male offspring inherits its type from its mother (Trujillo *et al.*, 1965; Mathai *et al.*, 1966). It is perhaps worth mentioning that this character is also sex linked in two species of hares and their hybrids. I do not know of any report concerning non-human primates.

Also sex linked are the mutant genes for haemophilia A and B, both in Man and dog (Hutt *et al.*, 1948; Graham *et al.*, 1950; Mustard *et al.*, 1960). Searle (1968) has shown that the autosomal loci for the albino series and the pink-eyed dilute series are linked in several mammalian species.

The most numerous homologies of gene localization have been described in *Drosophila*. Sturtevant (1929) has shown that more than twenty phenotypically similar mutants of *Drosophila melanogaster* and *D. simulans* belong to the same linkage groups and produce similar deficiencies. On the other hand Spassky and Dobzhansky (1950) have shown that "yellow" and "white" are closely linked in some species but far apart in others.

In 1922, Vavilov formulated his "law of homologous series in variation", by which we understand the existence of similar multiple allelic series in different species. Examples for this are provided by the petal pigments in many species of flowers (Scott-Moncrief, 1939; Wagner and Mitchell, 1955) and by the coat colour polymorphisms in mammals (Searle, 1968). In the primates, genetical protein polymorphisms are the most widely discussed and among those the serological tissue and specificities.

The similarity of enzyme blocks in mutants of such widely separated organisms as *Neurospora* and Man contrasts sharply with the diversity of normal metabolic pathways in much more closely related species and the co-existence of this unity and diversity of biochemical constitution must be taken into account for any kind of evolutionary speculation.

Metabolic similarity can sometimes be demonstrated by transplantation experiments. Beadle and Ephrussi (1936) transplanted the eye discs of eye colour mutant between *Drosophila melanogaster* and *D. pseudoobscura*. Two of these mutants in each species behaved autonomously when transplanted into wild type hosts but two did not, developing instead wild eye colour. One of these pairs had formerly been called vermilion in both species but the other had been named cinabar in *D. melanogaster*, but orange in *D. pseudoobscura*. Modern methods of immuno repression and of cell hybridization promise similar insights into mammalian gene homologies.

Gross hereditary malformations such as polydactily, brachydactily, dwarfism and many others occur of course in many species and have also been found similar in their development in related species, while in other instances indistinguishable defects may be caused in the same species by mimicking genes. Thus homologies of phenotypically similar genes are usually problematical. While allelism of recessives within a group of related species may be tested by cross-breeding this is impossible between non-hybridizing species and altogether impossible for dominants.

Homologies between physiological mutants of different species are equally hazardous. It would, for instance, require extensive embryological and histological studies to compare in detail a newly discovered labyrinthine deficiency in a mammal with any of the numerous behavioural mutants of the mouse, such as "waltzing", "shaker", etc. Perhaps the taste polymorphism which I will deal with later in this book is a more straightforward case.

Genetical comparisons and evolutionary speculations based on behavioural traits of a non-pathological nature are the most dubious, especially when applied to rather distantly related species. Charles-Dominique and Martin (1970) recently tried to revise the traditional family tree of primates on the basis of behavioural traits. They believe, for instance, that dwarf lemurs and bush babies, two primitive types of primates, which formerly were widely separated have in fact until fairly recently had common ancestors; while the tree shrews which are usually put near the lemurs are in fact not very closely related to them and should perhaps not be included with the primates at all. They base the evolutionary proximity of bush babies and lemurs on such behavioural traits as locomotion, which is a characteristic mixture of running and hopping, on the habit of carrying the young in the mother's mouth, on similar nesting patterns and on the formation of female associations: however, all of these may be convergent behaviour patterns. On slightly safer ground are speculations where more detailed comparison of behavioural patterns can be supplemented by the comparison of morphological traits. Such studies have been pioneered by Lorenz (1950) and his school on such groups as the canines, geese, finches or tilapias. Studies on primates have been summarized in a volume edited by De Vore (1965).

PHYLOGENETICAL ANALYSIS

Evolutionary speculations have been proffered by most research workers who have concerned themselves with comparative genetics and this will certainly be evident in the following chapters of this book. These speculations are naturally based on the special field of interest of the various authors and it would be inappropriate to review them here. Instead I shall attempt to discuss in more general ways the ideas and fallacies of many such approaches, basing myself mainly on a paper by Cavalli-Sforza and Edwards (1967). To my mind this paper is perhaps less remarkable for its positive suggestions which are presented in the form of models (trees) than for the clear statement of the limitations of evolutionary speculations, in so far as those are based on genetical evidence lacking any depth of time. The models which Cavalli-Sforza and Edwards discuss are branching trees of which only the present branches are projected onto a "now" plane, which I would call "the present". In the simplest model the frequency in a number of related populations of two alleles is shown in this plane, where it is pierced by the individual branches. This model can be expanded for poly-allelic systems, the combination of various polymorphisms, continuous variation, and of overall combination in multi dimensional Euclidian space.

At this juncture it is necessary to pause and consider the properties of trees constructed in this manner and what they signify. It is clear that they do not represent a historical description of an evolutionary process, but only a system of similarities which may or may not approximate past events. Thus the lengths of the branches of the tree do not represent time—at whatever scale—but only some rather arbitrary degree of relationship. The number of branchings, as well as the position of the real present—and possible past—populations or species on the trees are hypothetical. Never-theless certain features of the model are of interest, mainly those concerning the form of trees "branching" and the numbers of possible trees.

From a formula connecting the number of compared populations with the number of possible tree forms it appears, for instance, that there exist for ten populations 34,459,425 tree forms if the first split of the tree is known and 2,027,025 if—as usual—it is unknown. The number of branchings irrespective of position is only 98, but this is of little value for constructing an evolutionary tree.

While computers may be programmed for dealing with such large num-bers Cavalli-Sforza and Edwards suggest that it is necessary to select "promising forms" of trees on extraneous (biological) criteria. Thus in-tuition enters through the back door.

Next it is necessary to discuss the forces, which can produce the model "trees" and to compare them with the forces, which we assume are instru-mental in real evolution. The tree models are the product of "random drift" in the gene frequencies of neatly separating populations and do not cater for mutation pressure, migration and, most important, variable selec-tion, which we consider to be the main evolutionary forces.

It remains to point out two more unreal assumptions of the model men-tioned by the authors themselves, namely the absence of loops (representing hybridization) and the non-representation of extinct populations which may outnumber those represented and may conceal convergence. In fact the origin of the various hypothetical trees is conceived in probabilistic terms. The simplest of these are based on the theory of Brownian motion mak-ing various constraining assumptions. A second method is based on the assumption of "minimum evolution", that is a process which results in the shortest overall length of the tree. While this notion is intuitively plausible, it is far removed from reality mainly because of the reversibility and repeatability of much genetical change. This is obvious in the case of gene equilibria but it even applies to morphological characters which are probably polygenetically determined. Thus Kurtèn (1963) has shown that a second molar, a structure unknown in the felidae since the Miocene, has reappeared in the Lynx as have other dental features. This, of course, is completely at variance with the most cherished principle of evolutionary paleontology namely Dolo's Law. However it may be that the reappearance of these features is not brought about by a repetition of the same gene combinations but rather by different combinations producing a similar phenotype presumably influenced by similar environmental demands. In any case evolution can not always be considered as unidirectional.

The construction of tree models on the basis of the "additive tree procedure" or on the principle of maximum likelihood does not remove these factual difficulties. Altogether I must subscribe to the opinion of Cavalli-Sforza and Edwards, that for many reasons in the absence of fossil data "the reconstruction of evolutionary trees is a type of inductive inference which is likely to be especially weak" and that "prolonged periods of selection peculiar to individual populations will not be detectable without data from the past and no method of phylogenetic analysis can alter the fact that any observed diversity can be explained by any evolutionary tree, provided we are willing to postulate the necessary selection".

For the primatologist wanting to use gene frequency data for evolutionary speculation Cavalli-Sforza and Edwards provide a working example describing the relationship between four human populations (Eskimo, Bantu, English and Korean) based on the A_1 A_2 B O, Rh (four sera), M N S s, Fy and Di systems. But they also point out various difficulties and absurdities in previous similar attempts.

The study of variability

The main argument against the sanguine use of similarity studies for evolutionary arguments is of course that different criteria produce different degrees of similarity, while in actual historical fact there was only one course of evolution. Why then should haematologists, serologists, biochemists and other geneticists pursue in future comparitive studies of interspecies variation ? I think the answer is, that such studies are of considerable interest, not for evolutionary speculation, but for the more modest aim of exploring the range of mutability, which can pass the sieve of selection and in a more general way the biochemical and other possibilities and potentialities of an animal group.

Comparative pathology

The practical importance of this approach becomes obvious, if one considers the comparative pathology of genes. This subject has been explored for many years by Grüneberg and his school and the results have been summarized in two books (1947 and 1963). From a comparative point of view the most important finding is that several hereditary syndromes in Man are closely parallel by the pleiotropic gene effects in other mammals. For example, the hereditary osteosclerosis of Man which goes by the name of Albers-Schönberg's disease is very similar to the grey lethal in the mouse, both being based on defective secondary bone absorption.

Chondrodystrophic dwarfism though dominant in Man and recessive in the rabbit are very similar in their development and morphology in both species. Many anaemias as well as eye conditions are hereditary in Man, the domestic animals and laboratory species. This catalogue could be continued at length and there is little doubt that new and striking similarities between the hereditary diseases of many mammalian species including Man are waiting to be discovered.

How much the study of primates will contribute to this field remains to be seen. The increasing interest in primate studies in many countries augurs well, provided of course that the excessive capture of primates for experimental and clinical purposes does not lead to the extinction of most non-human primates.

REFERENCES

Baglioni, C. (1967). Molecular Evolution in Man. *Proc. Third Int. Congr. Hum. Genet., Chicago*, 317–337.

Barnicot, N. A. (1969). Some biochemical and serological aspects of primate evolution. *Sci. Progr., Oxford* **57**, 459–493.

Barnicot, N. A. and Cohen, P. (1970). Red cell enzymes of primates (anthropoidea). *Biochem. Genet.* **4**, 41–57.

Beadle, G. W. and Ephrussi, B. (1936). The differentiation of eye pigments in Drosophila as studied by transplantation. *Genetics* **21**, 225–247.

Epozcue, J. (1969). Primates. *In* "Comparative Mammalian Cytogenetics". (Benirschke, K., ed.) Springer, Berlin, Heidelberg, New York.

Cavalli-Sforza, L. L. and Edwards, A. W. F. (1967). Phylogenetic analysis, models and estimation procedures. *Amer. J. Hum. Genet.* **19**, 233–257.

Charles-Dominique, P. and Martin, R. D. (1970). Evolution of Lorises and Lemurs. *Nature Lond.* **227**, 257–259.

De Vore, I. (ed.) (1965). Primate behaviour. "Field Studies of Monkeys and Apes". Holt, Rinehard and Winston, New York.

Graham, J. B., Buckwalter, J. A., Hartley, J. L. and Brinkhouse, K. H. (1950). Canine haemophilia. *J. Exp. Med.* **90**, 97–111.

Grüneberg, H. (1947). "Animal Genetics and Medicine". Hamish Hamilton, London.

Grüneberg, H. (1963). "The Pathology of Development". Blackwell, Oxford.

Haldane, J. B. S. (1927). The comparative genetics of colour in rodents and carnivora. *Biol. Rev.* **2**, 199–212.

Hutt, F. B., Rickard, G. C. and Field, R. A. (1948). Sex-linked haemophila in dogs. *J. Hered.* **39**, 3–9.

Ingram, V. M. (1963). "The Haemoglobins in Genetics and Evolution". Columbia University Press, New York and London.

Kurtèn, B. (1963). Return of a lost structure in the evolution of the felid dentition. *Soc. Sci. Fenn. Commentat. Biol.* **XXV**, 4.

Lorenz, K. (1950). The comparative method of studying innate behaviour patterns. *Symp. Soc. Exp. Biol. IV. Animal Behav.* 221–254.

Mathai, C. K., Ohno, S. and Beutler, E. (1966). Sex linkage of the glucose-6-phosphate-dehydrogenase gene in Equidae. *Nature, Lond.* **210**, 115–116.

McLaren, Anne and Walker, P. M. B. (1965). Genetic discrimination by means of DNA/DNA binding. *Genet. Res.* **6**, 230–247.

Mittwoch, Ursula (1967). "Sex Chromosomes". Academic Press, New York and London.

Mustard, J. F., Boswell, H. C., Robinson, G. A., Hoeksema, T. D. and Downie, H. G. (1960). Canine haemophilia B (Christmas disease). *Brit. J. Haematol.* **6**, 259–266.

Ohno, S., Poole, J. and Gustavson, I. (1965). Sex linkage of erythrocyte-glucose-6-phosphate-dehydrogenase in two species of wild hares. *Science, N.Y.* **150**, 1737–1738.

Scott Moncrieff, R. (1939). The genetics and biochemistry of flower colour variation. *Ergeb. Enzymforsch.* **8**, 277–306.

Searle, A. B. (1968). "Comparative Genetics of Coat Colour in Mammals". Academic Press, New York and London.

Spassky, B. and Dobzhansky, T. (1950). Comparative genetics of *Drosophila willistoni*. *Heredity* **4**, 201–215.

Sturtevant, A. H. (1929). The genetics of *Drosophila. Carnegie Inst. Wash. Publ.* **399**, 1–62.

Trujillo, J. M., Walden, B., O'Neil, P. and Austall, H. B. (1965). Sex linkage of glucose-6-phosphate dehydrogenase in horse and donkey. *Science, N.Y.* **148**, 1603–1604.

Vavilov, N. I. (1922). The law of homologous series in variation. *J. Genet.* **12**, 47–89.

Wagner, R. P. and Mitchell, H. K. (1955). "Genetics and Metabolism". John Wiley, New York.

Epigenetic Polymorphism in the Primate Skeleton

A. CAROLINE BERRY AND R. J. BERRY

Department of Animal Genetics, University College, London, England;
Royal Free Hospital School of Medicine, London, England

The inherited variation existing in any animal population can be used to explore either the developmental processes and abnormalities therein (King, 1959; Edwards, 1960), or the genetical forces acting thereon (Ford, 1964; Berry, 1971). However, it is well to realize that many of the traits usually studied by population geneticists cannot be shown to be more than mere markers or indicators of some adaptive or physiological process. Even when the character in question is determined by a single gene and it is possible to observe an effect of the gene closely related to the putative primary gene product (as with studies of isozymes or blood groups), it is rare to be able to separate the adaptive importance of the character from its heuristic role as a pleiotropic, linked or relic manifestation of another process (e.g. Berry and Davis, 1970; Berry and Murphy, 1970).

Hence it follows that there is a distinction between the importance of a character to an animal, and its importance to a worker interested in studying the animal. We have laboured this point since it is basic to an understanding of the biology of the apparently entirely trivial variation described in this paper. We believe that in some situations minor skeletal variation may be more informative about individual and population processes and structure than more traditional genetical traits.

SKELETAL VERSUS BIOCHEMICAL CHARACTERIZATION

Forty years ago, ecological studies (particularly of non-human primates) were effectively non-existent, and population studies were based largely upon osteometric investigations (e.g. Morant, 1925; Schultz, 1926). These suffered from the twin problems that the environmental contribution (age, sex, seasonal, dietary, etc.) to the total variation was difficult to determine from the information available, while the data collected (on skeletal means and indices) were virtually impossible to interpret in terms of meaningful population and genetical parameters. Consequently, when immunological and later, electrophoretic techniques of characterizing individuals were developed, it is not surprising that there was a reaction against the uncertainties of osteometry. However, there are considerable disadvantages in working with what are now traditional biochemical methods: firstly, material has to be collected and preserved in a fresh condition, and this can be very difficult in the field; secondly, a considerable amount of labour is involved in gaining information about allele frequencies at even a fairly small number of gene loci. Clearly, there is much to be gained if genetical information can be extracted from the relatively imperishable parts of an animal (i.e. skin and skeleton).

This paper describes certain skeletal characteristics in the primate skeleton which can be used as genetical indicators because of what is known about their inheritance, and because they are relatively uninfluenced by environmental imponderables. These characters are the shape of certain skull sutures, the presence (or absence) of accessory foramina, the existence of sutural bones, and the like. They have been known as long as skeletal series have been looked at (e.g. Montagu, 1937; Straus, 1937). Sometimes their hereditary basis has been recorded (e.g. Schultz, 1923, 1960), but more often they are noted as mere anatomical curiosities or irrelevancies. However, many of them can be used as genetical markers and, because of their multifactorial determination, they can characterize a large part of the genome. (It is perhaps worth pointing out that the skeleton "records" variation in many systems—such as the place or extent of branching of a blood vessel or nerve fibre—and hence skeletal variation in the sense used in this paper reflects more than variation in genes affecting the skeleton, e.g. Froud, 1959.)

THE INHERITANCE OF MINOR SKELETAL VARIANTS

The mode of inheritance of minor skeletal variants was elucidated in laboratory mice twenty years ago by Grüneberg and his co-workers (summarized Grüneberg, 1963). The basic observation was that the frequency of any variant is constant in an inbred strain, although it can be altered by a mutational event (Deol et al., 1957; Grewal, 1962a). Crosses between strains showed that a variant is determined by a number of genes acting additively, although a developmental threshold leads to one of two (sometimes more) phenotypic alternatives, rather than a continuously distributed characteristic such as height or number of fingerprint ridges. The paradox between the mode of inheritance and the phenotypic expression of the character led Grüneberg (1952) to call this sort of variation "quasi-continuous". Berry and Searle (1963) used the term "epigenetic" for the same thing, to emphasize the contrast between the Mendelian nature of the underlying variable, and the ontogenetic nature of the phenotypic distinction. More recently, we have tended to use the older expression "non-metrical", as being non-committal about determination.

One of the most intensively studied epigenetic variants in mice is third molar agenesis. Eighteen per cent of CBA mice lack one or both of their lower third molars. Grüneberg (1951) showed that what is inherited in these mice is the size of the tooth itself: if the tooth germ falls below a certain critical size, the tooth fails to develop. There are apparently genes which control tooth size as such (shown by the increased variance of third molar size in the F_2 of a cross between two strains); the factors which determine whether a tooth will develop or not in an inbred strain are the environmental factors connected with maternal physiology which cannot be excluded even in a laboratory bred isogenic stock (Fig. 1). For example, tooth loss is commonest in large litters where the size of the young is small at birth; third molar size can be increased by fostering inbred young onto outbred females, whose lactational performance is better than that of their natural mothers; conversely, feeding the parents on deficient (unbalanced) diets leads to a decrease in tooth size (presumably by interfering with lactation).

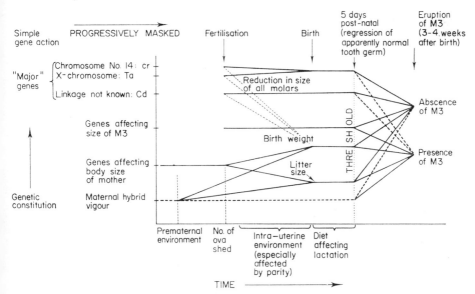

FIG. 1. Interaction of different genetical and environmental factors in the determination of third molar (M3) loss in the mouse. Based on the work of Deol and Truslove (1957), Grewal (1962b), Grüneberg (1951, 1965), and Searle (1954a, b, 1957). From Berry, 1968.

The physiological threshold which determines whether or not a tooth will develop has not been identified. Grüneberg originally suggested that all three molars might be competing for a limited amount of dental lamina, but this was not supported by the anatomical study of Grewal (1962b), who showed that the germs of third molars which fail to erupt are present and normal at first. On the sixth day after birth the germs appear smaller than normal, and development is arrested at the "cap" stage, followed by regression.

Grahnén (1956) has argued that third molar loss in mouse and Man could have a similar causation. There are indeed many similarities. For example, marked racial differences exist in the incidences of hypodontia, while Keene (1965) has shown that in Man parity and birth weight at least affect the chance of not developing third molars. Consequently, there would seem to be good grounds for accepting that third molar loss in Man is a complex affair, and at least analogous to the state of affairs in the mouse.

It is perhaps worth emphasizing that the *probability* of third molar loss is an inherited character in the mouse, despite the environmental factors which influence its manifestation. Loss of the tooth is a "satellite character" to small size of the tooth. There are about fifty other traits in the mouse inherited in the same way (Deol, 1955; Berry, 1963; Berry and Searle, 1963). It is encouraging that Howe and Parsons (1967) found that, although the manifestation of a few variants is affected by maternal factors (see also Searle, 1954a; Deol and Truslove, 1957), when information from a large number of variants (25 in Howe and Parson's work) is combined, there is no overall significant effect of the environment. In other words, the characters are as genetically characteristic of a population as are single gene characters such as blood groups, despite the environmental influence on the position of the threshold.

GENETICAL HOMOLOGY AND ANATOMICAL ANALOGY

Minor skeletal variants occur in all mammals studied. In a few examples in Man, a little is known about their transmission in families. Although the genetical interpretation of the evidence by different authors is wide and vague (Berry, 1968), there can be no doubt that the genetical determination is similar to that in the mouse. Perhaps more compelling are population studies which show that variant frequencies are similar in related populations, and that differences in frequency can be used as measures of genetical distinctiveness (e.g. Berry, 1969a, b). It is, of course, impossible to prove that similar alleles at the same loci are influencing epigenetic variation in different species, but for intra-specific studies this does not matter. This means that, although it is logically incorrect to claim that apparently anatomically identical variants are genetically the same in Man and the higher apes, we have no hesitation in asserting that they are equivalent from the point of view of their use as genetical markers (Searle, 1968; Robinson, 1970; and below).

Most of our work has been done with mice (R.J.B.) or men (A.C.B.). Our acquaintance with non-human primates is limited. We have scored a number of primate skull series for the same characters that we have previously used as genetical markers in Man (Berry and Berry, 1967, 1971; Berry *et al.*, 1967), and report here on the differences in incidence and problems in morphology of this comparative study.

Material Studied

We have restricted ourselves to examination of a single series each from the anthropoid genera *Pan* and *Pongo*. For *Hylobates* we have material from three sub-species of *H. lar* Illiger; for *Gorilla* we have scored four series. All the specimens we scored are in the British Museum (Natural History) with the exception of most of the western gorillas which are in the Powell-Cotton Museum, Birchington, Kent (Groves and Napier, 1966).

Hylobates lar. We have followed the recent revision of Groves (1970) and divided our material into *Hylobates lar lar* and *H.l. entelloides* from Thailand and the Malayan peninsula (39 specimens) and *H.l. muelleri* from Borneo and Sarawak (14 specimens).

Pan troglodytes Oken. 50 specimens.

Pongo pygmaeus L. 44 specimens.

Gorilla gorilla Geoffroy. Of the western, Lowland gorilla:
  52 specimens from the Batouri area;
  24 specimens from the Yaoundé area;
  26 specimens from the coastal plain of Gabon.

Of the eastern gorilla, *Gorilla gorilla beringei*: 9 specimens from Ruanda.

Other workers have listed a few epigenetic variants in a number of other species, but the only intensive study of these variants in other primate species that we know is by R. W. Thorington (unpublished, and Balan and Thorington, 1969) on the South American monkey species *Aotus trivirgatus* Humboldt and *Saimiri sciureus* Voigt, basing himself largely on our work in Man (Berry and Berry, 1967).

For comparison with the anthropoid series, we also include here three human series:

1. 182 Egyptian skulls from Sedment, Qurneh and Qau of Middle Kingdom date (ca. 2200 to 1800 B.C.), housed in the Duckworth Laboratory of the University of Cambridge (Berry *et al.*, 1967).

2. 56 skulls of Ashanti from Ghana, now in the British Museum (Natural History) (Berry and Berry, 1967).

3. 50 skulls from a mediaeval churchyard in Oslo, Norway, now in the Anatomical Institute of the University of Oslo (Berry and Berry, 1971).

We classify our human skulls for 30 variants. Most of these are the "classical" variants described by Wood-Jones (1930–31) and Brothwell (1963) but five of them (nos. 1, 14, 17, 22 and 25 below) were gleaned from "Gray's Anatomy". We have found it impossible to use some of the published variants (e.g. presence or absence of styloid processes) because of the condition of many of the skulls in museums.

Because we undertook the work reported here essentially as a comparative study with Man, we have attempted to classify the same variants in non-humans as humans. Accordingly we have based ourselves on the same criteria that we used in our human studies. The following descriptions are based on those in our 1967 paper (where illustrations of all the variants in Man are given), with notes on difficulties or amendments which it was necessary to make for using the variants in non-humans. We did not use the lower jaw in our previous work and we give here descriptions of three mandibular variants. With these exceptions we have not actively searched for variants confined to non-humans, although we include below a mention of several we noted in passing. Balan and Thorington (1969) list a number of variants in *Aotus* whose counterpart we have not looked for in Man. We have confined ourselves to skull variants for no reason other than the availability of material (cf. Grüneberg, 1952; Anderson, 1968).

Highest nuchal line present

In Man there are two, and sometimes three, nuchal lines forming well-marked ridges running horizontally across the occipital bones. In other primates, the degree of development of both anterio-posterior and lateral crests is very much more marked, but is considerably influenced by both age and sex. We have not used this character in non-humans.

Ossicle at the lambda (Fig. 2)

A bone may occur at the junction of the sagittal and lambdoid sutures (the position of the posterior fontanelle). We have described this as an ossicle, and have made no attempt to distinguish between a sutural bone in this position, and a "true" interparietal or Inca bone formed from the membraneous part of the occiput.

Lambdoid ossicles present (Fig. 2)

One or more ossicles may occur in the lambdoid suture. Up to about twelve distinct bones may be present on either side. Rare in non-human primates.

Parietal foramen present (Fig. 2)

This pierces the parietal bone near the sagittal suture a short distance in front of the lambda. It transmits a small emissary vein, and sometimes a branch of the occipital artery.

Bregmatic bone present

A sutural bone (the bregmatic or interfrontal) may occur at the junction of the sagittal suture with the coronal one (the position of the anterior fontanelle).

Receding bregma (Fig. 2)

In non-human primates (especially *Hylobates*) the sutural junction at the bregma may be displaced posteriorly giving a V-shaped pattern.

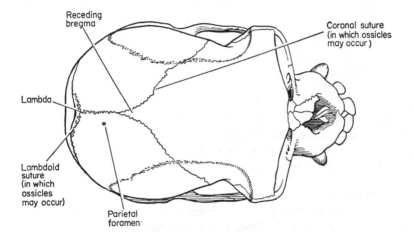

FIG. 2. Skull of *Hylobates lar* viewed from above, to show some of the variants described in the text. Where possible both expressions (e.g. "presence" and "absence") of a variant are shown, but only one alternative has been labelled.

Metopism

The medio-frontal suture disappears within the first two years of life in humans. In a few individuals it persists throughout life: this condition is known as metopism.

Coronal ossicles present (Fig. 2)

Ossicles are sometimes found in the coronal suture.

Epipteric bone present

A sutural bone (the epipteric bone or pterion ossicle) may be inserted between the anterior inferior angle of the parietal bone and the greater wing of the sphenoid. When large it may also articulate with the squamous part of the temporal bone.

Fronto-temporal articulation

Normally the frontal bone is separated from the squamous part of the temporal bone by the greater wing of the sphenoid and the anterior inferior angle of the parietal bone. Commonly in non-humans, but only occasionally in Man, the frontal and temporal bones are in direct contact, forming a fronto-temporal articulation.

Parietal notch bone present

The parietal notch is that part of the parietal bone that protrudes between the squamous and the mastoid portions of the temporal bone. It may form a separate ossicle which is known as the parietal notch bone. We did not observe this in any of our non-human material.

Ossicle at asterion

The junction of the posterior inferior angle of the parietal bone with the occipital bone and mastoid portion of the temporal bone is known as the asterion. A sutural bone may occur at this junction.

Auditory torus present

Rarely a bony ridge or torus is found on the floor of the external auditory meatus. Not observed in any of our non-human material.

Foramen of Huschke present

This is a foramen occurring in the floor of the external auditory meatus. It is always present in young children but only occasionally does it persist after the fifth year in humans. It is most easily scored from the inferior aspect of the tympanic part of the temporal bone. The floor of the auditory canal is much more substantial in non-human primates than in Man, and we did not find any variation in this region.

Mastoid foramen exsutural

Mastoid foramen absent

When present, the mastoid foramen usually lies in the suture between the mastoid part of the temporal bone and the occipital bone. Less frequently it lies exsuturally, piercing the mastoid part of the temporal bone, or, more rarely, the occipital bone.

Posterior condylar canal patent (Fig. 3)

The posterior condylar canal usually pierces the condylar fossa which lies immediately posterior to the occipital condyle. Sometimes it ends blindly in the bone, and has only been scored as patent when a seeker can be passed through it. In non-human primates the canal appears to be developed only in *Pan*. However small "pits" are frequently found in the fossa, and these have been scored as a variant.

Condylar facet double

Occasionally the articular surface of the occipital condyle is divided into two distinct facets. Not seen in our non-human material.

Precondylar tubercle present (Fig. 3)

Occasionally a bony tubercle lies immediately anterior and medial to the occipital condyle. A centrally placed tubercle has been regarded as two fused tubercles.

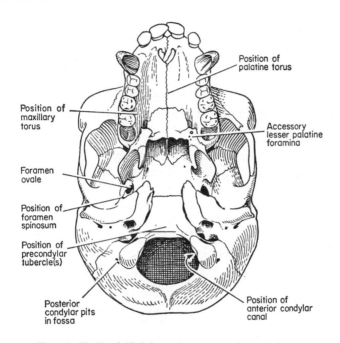

Position of
palatine torus

Position of
maxillary
torus

Accessory
lesser palatine
foramina

Foramen
ovale

Position of
foramen
spinosum

Position of
precondylar
tubercle(s)

Posterior
condylar pits
in fossa

Position of
anterior condylar
canal

FIG. 3. Skull of *Hylobates lar* viewed from below.

Anterior condylar canal multiple (Fig. 3)

This canal (foramen hypoglossi) pierces the anterior part of the occipital condyle and transmits the hypoglossal nerve. Embryologically the nerve originates from several segments. In Man the canal may be divided for part or all of its length; in non-human primates the canal may be represented by up to four channels.

Foramen ovale incomplete (Fig. 3)

Rarely the postero-lateral wall of the foramen ovale is incomplete.

Foramen spinosum open or absent (Fig. 3)

The posterior wall of the foramen spinosum is sometimes deficient. The same variant occurs in *Gorilla*, but in the smaller primates the foramen is frequently absent, and in this case we have scored presence or absence.

Accessory lesser palatine foramen present (Fig. 3)

The lesser palatine foramina lie on both sides of the posterior border of the hard palate immediately posterior to the greater palatine foramen, and transmit the lesser palatine nerves. When more than one (there may be three or four) foramen is present, it has been scored as accessory.

Palatine torus present (Fig. 3)

A bony ridge may run longitudinally down the mid-line of the hard palate. This is the palatine torus. It is common in north European humans, but otherwise rare.

Maxillary torus present (Fig. 3)

The maxillary torus is a bony ridge running along the lingual aspects of the roots of the molar teeth. It was not seen in this study.

Zygomatico-facial foramen single or absent (Fig. 4)

This is a small foramen which pierces the zygomatic bone opposite the junction of the infraorbital and lateral margins of the orbit. It transmits a nerve and small artery, and may be single, multiple or absent. Absence is rare in non-humans and we have scored the variant as single or absent versus multiple.

Supraorbital foramen complete (Fig. 4)

The supraorbital foramen transmits the supraorbital vessels and nerve. It is frequently incomplete (or open). The feature is much less marked in non-humans, and has been scored as presence or absence of a supraorbital "notch".

Frontal foramen present (Fig. 4)

A well-defined secondary foramen in the vicinity of (usually lateral to) the supraorbital foramen has been scored as a frontal foramen. Frequently a cluster of tiny foramina are present, but these have been ignored. However, scoring was inevitably somewhat arbitrary in a few borderline cases.

Anterior ethmoid foramen exsutural

The anterior ethmoid foramen pierces the medial wall of the orbit. It normally lies on the suture between the medial edge of the orbital plates of the frontal and ethmoid bones, but it occasionally emerges above the suture.

Posterior ethmoid foramen absent

The posterior ethmoid foramen lies just behind the anterior ethmoid foramen on the same suture line. Its absence can only be scored satisfactorily in well-preserved skulls. Not seen in non-humans.

Accessory infraorbital foramen present (Fig. 4)

A second foramen may lie immediately adjacent to the infraorbital foramen. In non-humans, up to six foramina are sometimes found.

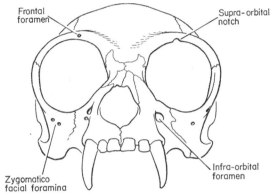

FIG. 4. Frontal aspect of a skull of *Hylobates lar*.

Accessory mental foramina present (Fig. 5)

The mental foramen may have one or more smaller accessory foramina near to it.

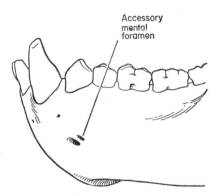

FIG. 5. Left external view of a mandible of *Gorilla gorilla*.

Anterior internal foramen of mandible single or absent (Fig. 6)

A number of small foramina are usually present on the internal aspect of

B

the mandible close to the symphysis menti. These may be reduced to absence or presence of a single foramen.

Anterior external foramen of mandible absent (Fig. 6)

There are usually a number of small foramina scattered on the external anterior aspect of the mandible. Rarely these are absent.

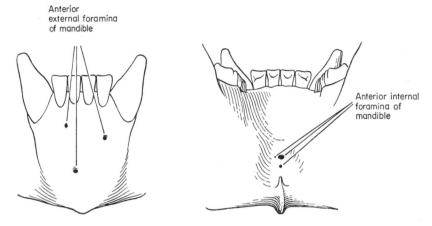

FIG. 6. External (left) and internal (right) views of the mandible of *Gorilla gorilla* in the region of the symphysis menti.

All our material was classified by one of us (A.C.B.). Following our previous practice, we have combined data for males and females, basing this on the lack of sex differences in variant incidence in mice and men. However, since there are greater osteological differences between the sexes in non-humans than in Man, we tested for homogeneity between the sexes in some of our gorilla material (Table I). Two variants which do not occur in Man (crest development and projections on the posterior border of the palate) proved to be greatly influenced by both sex and age, and we have eliminated them from further consideration. Two other of the 32 comparisons were "significantly" different at the 5% level. In one of these, bregmatic bones were found in four males only, none in females; in the other, three epipteric bones were found in males, none in females. These differences were so small that we have assumed that they could be due to chance, and that there are no marked sex differences in the manifestation of the variants in Table I.

In mice, there is no age correlation with any variant. With humans, we avoided using immature skulls. However a number of variants become extremely difficult to score in adult non-human primates (especially gorillas) because of the obliteration of sutures. Hence we have not restricted ourselves to adult skulls.

Interpretation of Variant Incidences

The results of our scoring are set out in Table I. For non-midline variants the variant incidence is based on the frequency of occurrence of the variant, i.e. the percentage of *sides* in which the variant occurs. Some of the frequencies are based on less than the total number of specimens (or sides) because of damage to specimens; more important is that some variants could not always be scored because of obliteration of sutures, etc. (see above).

Having assembled the data, the problem is to interpret them. There are two dimensions: inter-specific and intra-specific.

Inter-specific

As already noted, it is impossible to know whether an anatomically identical variant in two species is genetically the same. Since the basic pattern of development is so similar in all primates, it is in fact highly likely that (say) metopism in Man and gibbon is aetiologically the same thing. However, with a multifactorially determined character, there can be no guarantee that a variant is controlled by the same alleles in two individuals which are members of populations unable to breed together (Berry, 1969b). Consequently it is meaningless to attempt any quantitative measures of phylogenetic separation of different species. The fact that a broadly similar spectrum of variants occur in all the primate species examined encourages one to use them with confidence for intra-specific comparison, but the most that inter-specific comparisons can show are the degrees of developmental stability (or canalization) in different parts of the skeleton in the various species.

Intra-specific

In comparisons between populations which breed—or can breed—together, the position is entirely different. The differences between such populations are theoretically expressable in numbers of allele substitutions and frequency changes. In other words it is in principle possible to measure the genetical differences between populations in numerical terms (McLaren and Walker, 1970). It is our contention that the differences in variant incidence in two populations can be used as a measure—albeit an arbitrary one—of genetical divergence.

There are a number of ways of calculating population differences on the basis of epigenetic variants. Grüneberg and his co-workers expressed them in terms of the number of "significant" differences between variants. In

TABLE I. Percentage incidences of epigenetic skull variants

	Homo sapiens			Pan troglodytes	Pongo pygmaeus	Gorilla gorilla						Hylobates lar	
						G.g. gorilla			G.g. beringei	♀♀	♂♂		
	Middle Kingdom Egypt	Ashanti (Ghana)	Mediaeval Oslo			Batouri	Coastal	Yaoundé				Hylobates lar entelloides + H.l. lar	Hylobates lar muelleri
2. Ossicle at the lambda	12·6	12·5	14·0	12·8	2·4	15·0	14·3	40·0	0	12·5	20·0	2·7	20·0
3. Lambdoid ossicle present	32·3	25·9	40·0	2·1	0	2·9	0	0	10·0	2·3	7·1	9·9	0
4. Parietal foramen present	51·8	59·2	64·0	12·0	7·3	36·5	16·7	0	43·8	34·4	36·2	4·5	10·0
5. Bregmatic bone present	0	0	4·0	3·8	0	8·0	0	0	12·5	0	12·5	90·9	40·0
Receding bregma	—	—	—	0	0	2·0	7·1	37·5	0	0	3·1	0	60·0
6. Metopism	3·3	0	20·0	2·6	0	2·0	0	12·5	0	3·1	0	0	0
7. Coronal ossicle present	1·7	0	3·0	0	0	2·0	0	0	0	0	0	0	30·0
8. Epipteric bones present	14·1	6·2	18·0	0	4·1	2·0	2·6	0	0	0	4·7	0	0
9. Fronto-temporal articulation	2·3	9·8	1·0	96·9	33·8	99·0	100	100	87·6	100	98·4	20·6	50·0
10. Parietal notch bone present	9·4	6·2	12·0	0	0	0	0	0	0			0	0
11. Ossicle at asterion	11·7	14·3	6·0	16·9	13·3	5·0	0	11·8	7·1	4·7	9·4	2·7	0
12. Auditory tori present	0	0	0	0	0	0	0	0	0	0	0	0	0
13. Foramen of Huschke present	17·5	30·4	12·1	0	0	0	0	0	0	0	0	0	0
14. Mastoid foramen exsutural	42·1	36·9	48·0	21·0	48·8	67·7	58·8	70·0	72·2	67·0	70·2	34·7	55·0
15. Mastoid foramen absent	5·3	15·3	4·0	64·0	11·2	7·8	8·7	2·5	0	10·9	6·2	19·4	25·0
16. Posterior condylar canal patent	22·4	33·9	24·0	11·0	0	0	0	0	0	0	0	0	0
Pits in post. condylar fossa	—	—	—	68·0	—	—	—	—	44·4	—	—	4·1	0

17. Condylar facet double	0·9	0·9	1·0	0	0	0	0	0	0	0	0	0	0
18. Precondylar tubercle present	7·2	1·8	4·0	0	3·8	0	0	0	0	0	0	0	0
19. Anterior condylar canal double	17·7	11·6	12·0	32·0	71·2	53·2	53·2	68·9	72·2	47·0	57·8	25·7	33·3
Anterior condylar canal treble	—	—	—	9·0	2·5	13·6	4·4	4·4	0	9·4	17·2	0	0
20. Foramen ovale open	0·9	3·6	4·0	14·0	68·7	0	0	2·2	0	0	0	30·8	22·7
21. Foramen spinosum pres.	—	—	—	88·0	7·1	—	75·0	—	100	—	—	5·1	0
Foramen spinosum incomplete	16·0	7·1	8·1	—	—	4·4	—	2·4	—	12·2	3·6	—	—
22. Acc. lesser palatine for. present	72·3	41·0	56·5	41·0	4·6	78·9	74·0	70·9	11·1	84·5	75·0	16·9	38·1
23. Palatine torus present	0·6	0	38·0	6·0	2·4	0	0	0	0	0	0	0	0
24. Maxillary torus present	0	0	12·6	0	0	0	0	0	0	0	0	0	0
25. Zygomatico-facial foramen single or absent	15·9	18·7	9·0	27·3	2·3	28·0	44·0	23·4	77·7	42·2	25·0	2·6	38·1
26. Supra-orbital notch present	83·4	88·3	79·0	56·0	24·7	57·2	90·5	71·0	83·4	48·4	52·2	51·3	68·2
27. Frontal foramen present	51·1	30·4	58·0	11·0	21·2	67·2	69·2	84·3	88·8	70·3	70·3	18·4	13·6
28. Anterior ethmoid for. ex-sutural	18·3	15·4	29·3	21·1	22·1	22·2	20·0	4·6	50·0	21·9	20·3	12·9	33·3
29. Posterior ethmoid for. absent	2·5	0	0	0	0	0	0	0	0	0	0	0	0
30. Infra-orbital for. double	5·9	6·4	5·1	45·0	40·7	50·0	66·6	53·3	72·2	51·7	53·0	21·8	19·0
treble				20·0	36·0	7·7	2·1	4·4	27·8	7·8	3·1	1·3	0
quadruple +:				11·0	18·6	0	0	0	0	0	0	0	0
Accessory mental foramina present	—	—	—	7·1	44·6	22·6	27·0	27·9	83·3	18·8	31·2	0	0
Anterior int. for. of mandible single or absent	—	—	—	8·0	2·7	33·4	50·0	47·6	33·3	34·3	40·0	40·0	50·0
Anterior ext. for. of mandible absent	—	—	—	68·0	5·4	45·1	16·7	31·8	33·3	40·6	40·6	43·3	28·6

Numbering of variants refers to list in Berry and Berry (1967) this, they have been followed by Balan and Thorington. Laughlin and Jorgensen (1956) and Brothwell (1958) used Penrose's "size and shape" statistic in their analyses of human data. We have used a method devised by C. A. B. Smith and first used by Grewal (1962a). In this a single "measure of divergence" is calculated between each pair of populations. The advantage of the Smith method is that the computations involved are extremely simple, and the statistic is related to χ^2. Moreover the answers obtained by the Smith and Penrose methods are similar.

The measure of divergence between any two populations 1 and 2 for any character is defined as:

$$(\theta_1 - \theta_2)^2 - (1/n_1 + 1/n_2)$$

where θ is the angular transformation in radians of the percentage frequency of the character in the population (see Appendix) and n is the number of animals scored. (Professor Smith suggests that the statistic is improved by taking the square root of the measure of divergence. This ensures that, to a first approximation, the distance between two populations 1 and 3 is not greater than the sum of the distances between 1 and 2, and 2 and 3.)

This computation has the property that since θ has the variance $1/n$, $\theta_1 - \theta_2$ has variance $1/n_1 + 1/n_2 = V$, and where there is no real difference between the large populations from which the two samples are drawn, the observed $\theta_1 - \theta_2 = D$ will be a nearly normal deviate with mean zero and variance V. Thus $(\theta_1 - \theta_2)^2/V$ will be approximately distributed as χ^2 with one degree of freedom; and it will be significant at, for example, the 5% probability level if it is greater than $3V$, and at the 1% level if it is greater than $6V$.

The variance of D^2 will be approximately

$$4D^2 \times \text{variance of } D$$
$$= 4D^2(1/n_1 + 1/n_2).$$

An estimate of the mean measure of divergence between two populations is the simple arithmetical mean of the measures for individual characters. This is permissible because in mice and men where it has been tested (Truslove, 1961; Berry and Berry, 1967; see also Herzog, 1968), the correlation, or covariance, of different variants in the same skeleton is very low and can be treated as statistically independent. In genetical terms it must mean that the variants classified are the pleiotropic manifestations of many independent developmental processes, and that differences in the "spectrum" of epigenetic variation between individuals reveal variation at a large number of gene loci. Comparisons between the human populations included in Table I are set out in Table II to show the sort of results which are obtained.

TABLE II. Measures of divergence (x 100) between three human populations

	Egypt (Middle Kingdom)	Ashanti (Ghana)	Oslo (Mediaeval)
Sample size	182	56	50
Egypt (Middle Kingdom)		5·75 *1·34*	9·36 *1·78*
Ashanti (Ghana)			21·28 *3·28*

The italicized figures are estimates of the standard errors of the measures of divergence.

THE GENETICAL VALIDITY OF THE MEASURE OF DIVERGENCE

The gene loci controlling skeletal size do not seem to act in the same way as those determining epigenetic variants. This is the conclusion that comes from the finding that there is a strong negative correlation between distance statistics calculated from metrical and from epigenetic results in two sets of data (one on Indian rats: Berry and Smith, unpublished, and one on ancient Egyptian humans: Berry *et al.*, 1967). Although this is unexpected, it is supported by a longitudinal study of a closed population of house mice on an island (Berry, 1970; Berry and Murphy, 1970). Here it was found that the variance of epigenetic characters decreased during the winter (at a time of heavy mortality and no breeding) and increased during the summer, whereas the variance of skeletal measurements changed reciprocally, increasing during the winter and decreasing during the summer. During the same periods heterozygotes at one isozymic locus increased above expectation during the summer, while another locus showed heterosis during the winter only.

The question then arises as to whether metrical or epigenetic measures give the most useful estimates of genetical separation. Berry and Smith (unpublished) analysed data from sixteen populations of rats (*Rattus rattus*) collected on or near the Malabar Coast in south India (see Grüneberg *et al.*, 1966). They were able to formulate certain expectations about the magnitude of distance measurements from the nature of the populations sampled. For example, one rat "population" was trapped almost entirely in a cashew nut factory which had a large and probably fairly isolated rat population. In such a situation, founder effects play a large part (Grüneberg, 1961;

Berry, 1964, 1970), and this population would be expected to be distinct from its neighbours. It was distinct on the epigenetic measure of divergence but had very little distinctiveness on the metrical coefficient of racial likeness. Similarly, two samples from the same village and arbitrarily separated into two groups were found to be rather distinct on the metrical measure, but not on the epigenetic one. Such examples could be multiplied from these data.

Another completely unrelated study by Rees (1969) on American white-tailed deer (*Odocoileus virginianus*) came to the same conclusion about the superiority of the epigenetic distance. Rees correlated distance statistics with geographical origin for ten samples of deer skulls collected from a more or less continuous breeding range in north America. He found the correlation of measure of divergence with geographical distance was 0·71, but with Mahalanobis's D^2 it was only 0·37.

Finally, one of us (A.C.B.) has found that immunological data indicate a small degree of heterogeneity among the human population of Norway, but that differences are shown up much better by epigenetic variant data.

POPULATION STUDIES

If the measure of divergence can be used as an indicator of genetical difference, it is clearly a powerful tool in the dissection of the ecological and evolutionary processes acting in populations. For example, it has made possible an analysis of colonization in field mice (*Apodemus sylvaticus*) in the North Atlantic islands (Berry, 1969b); of migration between breeding colonies of the grey seal (*Halichoerus grypus*) (Berry, 1969a); of adaptation and selection in mice (Berry, 1970); and of the correlation between biological and cultural data in human societies (Berry *et al.*, 1967).

There is no doubt that similar methods can be applied in non-human primate studies, even from the pilot results reported here. For example, *Hylobates lar muelleri* of Borneo and Sarawak is clearly distinct from *H.l. lar* and *H.l. entelloides* from Thailand and the Malayan peninsula (measure of divergence: 7·76 ± 2·93). A more extensive study of differentiation in *Hylobates lar* using similar techniques would be of considerable interest in view of the striking polymorphisms and taxonomic difficulties in the species (Frisch, 1963; Groves, 1970).

The data of Balan and Thorington (1969) on *Aotus trivirgatus* Illiger (Table III) provide an opportunity to test the separateness of taxonomic sub-species. Corbet (1970) has argued that the naming of sub-species often derives from ignorance about the extent of local and clinal variation within a species range, while Berry (1969b) showed that the fifteen sub-species of *Apodemus sylvaticus* described on the Scottish islands are effectively the

TABLE III. Percentage incidences of epigenetic skull variants in sub-species of *Aotus trivirgatus* (from Balan and Thorington, 1969)

	grisei-membra	lemur-inus	micro-don	vocifer-ans	tri-virgatus	Total
2. Ossicle at the lambda						1
3. Lambdoid ossicles present	16	0	0	27	8	10
4. Parietal foramen present	48	0	18	14	6	17
5. Bregmatic bone present						6
Common parietal-sphenoid suture	49	43	40	41	81	51
14. Mastoid foramen exsutural	15	7	16	0	2	8
15. Mastoid foramen absent						0·8
16. Post. condylar canal patent	100	92	86	82	73	87
19. Ant. condylar canal double						12
Pterygoid-sphenoid fusion						40
Ant. alisphenoid for. pres.						3
21. Foramen spinosum d.						2
Superior zygomatic notch pres.	34	21	13	55	21	29
Superior zygomatic for. pres.	1	7	50	9	17	17
25. No. of zygomatico-facial for. 1						30
2						54
3						13
4						1
26. Supra-orbital for. incomplete	54	14	23	9	29	26
Optic foramen incomplete						0·8
Common lacrimal-nasal suture	33	11	26	36	31	27
For. rotundum incomplete	9	7	0	0	4	4
Lateral orbital for. pres.	20	0	45	73	8	29
Acc. post. orbital foramen	0	0	0	0	14	3
Mylohyoid groove 1A	6	0	29	9	10	11
1B	38	21	34	41	50	37
2A	53	61	37	50	37	48
2B	1	0	0	0	0	0·2
3A	0	0	0	0	0	0
3B	2	18	0	0	4	5
30. Acc. infra-orbital for. 1	30	72	53	5	14	35
2	49	25	37	55	36	40
3	20	4	11	37	36	21
4	0	0	0	5	16	4
Mental for. single	85	86	69	59	39	67

Numbering of variants refers to list in Berry and Berry (1967) chance consequence of a small number of founders establishing each island race. Any subsequent adaptation will be chiefly dependent upon the initially available genetical variation, and will affect the original "instant sub-speciation" only in a small way. Haffer (1967) and Mayr (1969) have interpreted the avifauna of northern South America in a similar way: that the present diversity is a result of restriction of species ranges during the Pleistocene followed by expansion in recent times. Clearly this example is relevant to the *Aotus* situation, where the sub-species are comparatively distinct (Fig. 7).

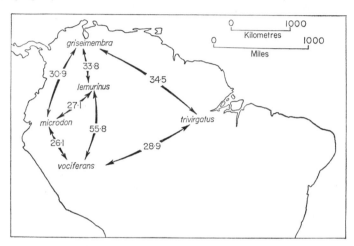

Fig. 7. Measures of divergence (× 100) between geographically separated sub-species of the night monkey, *Aotus trivirgatus*. Data of Balan and Thorington, 1969.

The farther one probes back into evolutionary time, the more speculation is involved. In contrast, our three arbitrarily and fortuitously chosen series of the western gorilla suggest an ecological situation which is open to test. The sample from Yaoundé is more like the geographically remote coastal sample than the nearby one from the Batouri district (Fig. 8). Indeed Batouri is more similar to coastal than to Yaoundé. The probable explanation for this is that the Yaoundé specimens come from the Sanaga river area, and are separated from the Batouri ones by a range of mountains. The Batouri sample is from an area which drains east and south into the Congo Basin. The inference is that there is gene-flow between Yaoundé and coastal, and between Batouri and coastal via the Congo Basin, but that Yaoundé and Batouri are reproductively isolated from each other. This genetical and geographical antithesis is thus exactly the same as that in the Eskimo settlements in Greenland studied by Laughlin (Laughlin and

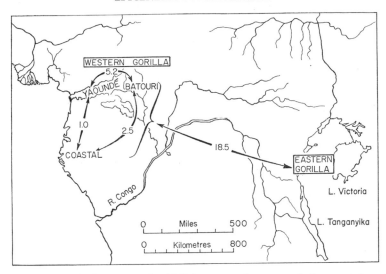

FIG. 8. Measure of divergence ($\times 100$) between three population samples of the western gorilla (*Gorilla gorilla gorilla*). Note that Yaoundé and Batouri are on opposite sides of a watershed draining into the Congo Basin to the east, and directly into the Atlantic Ocean to the west. The geographically isolated eastern gorilla (*G. g. beringei*) is much more distinct from the western sub-species, than are sub-samples of the western form from each other.

Jorgensen 1956; Laughlin, 1963; Berry, 1968), where relatively close geographical populations are separated by the Greenland ice-cap, and are genetically related only through a chain of coastal settlements. This is not a completely far-fetched suggestion in gorilla and African terms, particularly since individual movement over the Batouri-Yaoundé watershed does not necessarily mean that the two populations breed together (Anderson, 1965, 1970; Selander *et al.*, 1969; Kettlewell *et al.*, 1969; Berry, 1969a, 1971).

VARIANCE OF EPIGENETIC CHARACTERS

We have concentrated in this essay on the value of epigenetic variants merely as genetical markers. We believe that this is their main relevance in studies of non-human primates at the present time. However, most interest elsewhere has been concentrated on the aetiological implications of their occurrence. In 1934, Wright (basing himself on his work on polydactylism in guinea pigs) showed that if the underlying variable which determines the threshold is assumed to be normally distributed, it is possible to assign to it an arbitrary mean and variance in terms of the incidence of the character in question. Falconer (1960, 1965) has pointed out that this means that the heritability of a character, and hence its recurrence risk, can be calculated. Edwards (1969) has developed a more complicated model, based on the assumption that overstepping the threshold is exponential rather than linear. Either model can explain morbidity data for the common congenital malformations (W.H.O., 1970), and thus provide circumstantial support for the premises on which the model is based.

The next stage is to identify the action of the individual genes contributing to the "continuous" variable, since "multifactorial" inheritance is merely an expression of ignorance about the contributions of individual elements (Thoday, 1961; Spickett and Thoday, 1966). This could be important, since the effect of individual deleterious genes can—theoretically— be corrected (e.g. Berry, 1969c). The only disease where the multifactorial basis has been analysed with any success is congenital dislocation of the hip (Fig. 9) and even here we are far from understanding the biochemical mechanisms involved in morphogenesis.

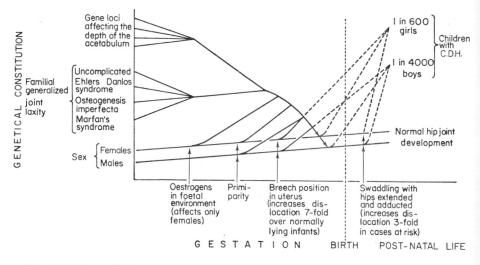

FIG. 9. Epigenetic interactions in the causation of congenital dislocation of the hip in man. Based on Carter and Wilkinson, 1964. Carter, 1966.

As far as the more trivial variants described in this essay are concerned, they are unlikely to have any great pathogenetic significance (although Torgersen, 1951, has argued that the sutural pattern of the skull is significant because "the close topographical relationship between the sutures and the meningeal membranes indicates that the physiological effect of the genes determining the situation of the sutures, the rate of obliteration and also the brain–skull correlation, is effected by these membranes"). Their importance is as indicators of developmental stability and genetical change, and as models of inherited disease processes (Searle, 1959; Berry, 1969c).

EPIGENETIC POLYMORPHISM

The skull variants described here can be regarded as falling within the classical definition of polymorphism of Ford (1940): "the occurrence together in the same habitat of two or more discontinuous forms, or 'phases', of a species in such proportions that the rarest of them cannot be maintained merely by recurrent mutation". It has been objected in the past that this is false, since the variants are merely pheno-deviants (Lerner, 1954), developmental "noise" (Waddington, 1957), or examples of fluctuating asymmetry (the inability of organisms to develop along precisely determined paths) (Ludwig, 1932; Van Valen, 1962). There is undoubtedly some truth in these terms when they are used in their less pejorative senses. Nevertheless, the weight of evidence is now such as to make it certain that "epigenetic morphs" can be valuable genetical markers in population studies, and there is little point in the context in forsaking pragmatics for polemics.

It would be stupid to claim that epigenetic variants can give a complete genetical characterization of any animal or population. Their great defect is that they can only be scored in dead animals, and sufficient material may be impossible to obtain. Furthermore, there is undoubtedly a subjective element in the scoring of some variants, but this can be minimised by consultation between workers in the field. On the credit side, scoring epigenetic variants is quicker and easier than measuring bones, they are little affected by age or sex, and the results obtained are easily understandable and computed.

Nature is a rag merchant, who works up every shred and ort and end into new creations; like a good chemist whom I found the other day in his laboratory, converting his old shirts into pure white sugar.

EMERSON: Conduct of Life.

This quotation brings us full cycle: we began by pointing out that adaptive relevance is not a necessary criterion for a character to be useful for a geneticist. The traits we have been concerned with in this essay may not unfairly be described as "phenotypic non-events", even though the common congenital malformations are epigenetic variants in the sense we use the term. Nevertheless, it is our contention that we can learn much of significance about populations from studying minor morphological variation.

Acknowledgements

We should like to express our thanks to Dr Richard W. Thorington, Jnr., of the Smithsonian Institute for sending us his unpublished data on *Aotus*. We are grateful to Dr John Napier for introducing us to non-human primates, to Professor B. Chiarelli for suggesting that we undertake the work reported here, to Professor C. A. B. Smith for his statistical help, to Dr G. B. Corbet of the British Museum (Natural History) and Mr L. R. Barton, Curator of the Powell-Cotton collection for access to the specimens in their care, and especially to Mr A. J. Lee who drew all the figures.

REFERENCES

Anderson, J. E. (1968). Skeletal "anomalies" as genetic indicators. *In* "The Skeletal Biology of Earlier Human Populations" (D. R. Brothwell, ed.), pp. 135–147. Pergamon, London.

Anderson, P. K. (1965). The role of breeding structure in evolutionary processes of *Mus musculus* populations. Symposium on "Mutation in Populations", pp. 17–21, Prague.

Anderson, P. K. (1970). Ecological structure and gene flow in small mammals. *Symp. zool. Soc. Lond.* **26**, 299–325.

Balan, B. B. and Thorington, R. W. (1969). Discrete variation in *Aotus* skulls. Paper read at Annual Meeting of Am. Soc. Mammalogists.

Berry, A. C. and Berry, R. J. (1967). Epigenetic variation in the human cranium. *J. Anat.* **101**, 361–379.

Berry, A. C. and Berry, R. J. (1971). Origins and relationships of the ancient Egyptians. *In* "The Population Biology of Ancient Egypt" (D. R. Brothwell and B. Chiarelli, eds). *J. hum. Evolution* (in the press).

Berry, A. C., Berry, R. J. and Ucko, P. J. (1967). Genetical change in ancient Egypt. *Man.* **2**, 551–568.

Berry, R. J. (1963). Epigenetic polymorphism in wild populations of *Mus musculus. Genet. Res.* **4**, 193–220.

Berry, R. J. (1964). The evolution of an island population of the house mouse. *Evolution, Lancaster, Pa.* **18**, 468–483.

Berry, R. J. (1968). The biology of non-metrical variation in mice and men. *In* "The Skeletal Biology of Earlier Human Populations" (D. R. Brothwell, ed.), pp. 103–133. Pergamon, London.

Berry, R. J. (1969a). Non-metrical variation in two Scottish colonies of the Grey seal. *J. Zool., Lond.* **157**, 11–18.

Berry, R. J. (1969b). History in the evolution of *Apodemus sylvaticus* (Mammalia) at one edge of its range. *J. Zool., Lond.* **159**, 311–328.

Berry, R. J. (1969c). Genetical factors in the aetiology of multiple sclerosis. *Acta neurol. Scand.* **45**, 459–483.

Berry, R. J. (1970). Covert and overt variation, as exemplified by British mouse populations. *Symp. zool. Soc. Lond.* **26**, 3–26.

Berry, R. J. (1971). Conservation aspects of the genetical constitution of populations. *In* "The Scientific Management of Animal and Plant Communities for Conservation" (E. A. G. Duffey, ed.), pp. 177–206. Blackwell, Oxford.

Berry, R. J. and Davis, P. E. (1970). Polymorphism and behaviour in the Arctic Skua (*Stercorarius parasiticus* (L.)). *Proc. R. Soc. B*, **175**, 255–267.

Berry, R. J. and Murphy, H. M. (1970). Biochemical genetics of an island population of the house mouse. *Proc. R. Soc. B*, **176**, 87–103.

Berry, R. J. and Searle, A. G. (1963). Epigenetic polymorphism of the rodent skeleton. *Proc. zool. Soc. Lond.* **140**, 577–615.

Brothwell, D. R. (1958). The use of non-metrical characters of the skull in differentiating populations. *J. Dt. Ges. Anthrop*, **6**, 103–112.

Brothwell, D. R. (1963). "Digging up Bones". British Museum (Natural History), London.

Carter, C. O. (1966). The inheritance of common congenital malformations. *Prog. med. Genet.* **4**, 59–84.

Carter, C. O. and Wilkinson, J. A. (1964). Genetic and environmental factors in the etiology of congenital dislocation of the hip. *Clin. Orthopaedics,* **33**, 119–128.

Corbet, G. B. (1970). Patterns of subspecific variation. *Symp. zool. Soc. Lond.* **26**, 105–116.

Deol, M. S. (1955). Genetical studies on the skeleton of the mouse. XIV. Minor variations of the skull. *J. Genet.* **53**, 498–514.

Deol, M. S., Grüneberg, H., Searle, A. G. and Truslove, G. M. (1957). Genetical differentiation involving morphological characters in an inbred strain of mice. I. A British branch of the C57BL strain. *J. Morph.* **100**, 345–376.

Deol, M. S. and Truslove, G. M. (1957). Genetical studies on the skeleton of the mouse. XX. Maternal physiology and variation in the skeleton of C57BL mice. *J. Genet.* **55**, 288–312.

Edwards, J. H. (1960). The simulation of Mendelism. *Acta genet.* **10**, 63–70.

Edwards, J. H. (1969). Familial predisposition in man. *Brit. med. Bull,* **25**, 58–64.

Falconer, D. S. (1960). "An Introduction to Quantitative Genetics". Oliver & Boyd, Edinburgh and London.

Falconer, D. S. (1965). The inheritance of liability to certain diseases, estimated from the incidence among relatives. *Ann. hum. Genet.* **29**, 51–76.

Frisch, J. E. (1963). Dental variability in a population of gibbons (*Hylobates lar*). *In* "Dental Anthropology" (D. R. Brothwell, ed.), pp. 15–28. Pergamon, London.

Ford, E. B. (1940). Polymorphism and taxonomy. *In* "The New Systematics" (J. Huxley, ed.), pp. 493–513. Clarendon Press, Oxford.

Ford, E. B. (1964). "Ecological Genetics". Methuen, London.

Froud, M. D. (1959). Studies on the arterial system of three inbred strains of mice. *J. Morph.* **104**, 441–478.

Grahnén, H. (1956). Hypodontia in the permanent dentition. A clinical and genetical investigation. *Odont. Revy.* **7**, Suppl. 3.

Grewal, M. S. (1962a). The rate of genetic divergence in the C57BL strain of mice. *Genet. Res.* **3**, 226–237.

Grewal, M. S. (1962b). The development of an inherited tooth defect in the mouse. *J. Embryol. exp. Morph.* **10**, 202–211.

Groves, C. P. (1970). Taxonomic and individual variation in gibbons. *Symp. zool. Soc. Lond.* **26**, 127–134.

Groves, C. P. and Napier, J. R. (1966). Skulls and skeletons of *Gorilla* in British collections. *J. Zool., Lond.* **148**, 153–161.

Grüneberg, H. (1951). The genetics of a tooth defect in the mouse. *Proc. R. Soc. B*, **138**, 437–451.

Grüneberg, H. (1952). Genetical studies on the skeleton of the mouse. IV. Quasi-continuous variations. *J. Genet.* **51**, 95–114.

Grüneberg, H. (1961). Evidence for genetic drift in Indian rats (*Rattus rattus* L.). *Evolution, Lancaster, Pa*, **15**, 259–262.

Grüneberg, H. (1963). "The Pathology of Development". Blackwell, Oxford.

Grüneberg, H. (1965). Genes and genotypes affecting the teeth of the mouse. *J. Embryol. exp. Morph.* **14**, 137–159.

Grüneberg, H., Bains, G. S., Berry, R. J., Riles, L. E., Smith, C. A. B. and Weiss, R. A. (1966). A search for genetic effects of high natural radioactivity in South India. *Spec. Rep. Ser. med. Res. Coun.* No. 307, 1–59.

Haffer, J. (1967). Zoogeographical notes on the "non-forest" lowland bird faunas of northwestern South America. *Hornero*, **10**, 315–333.

Herzog, K. P. (1968). Associations between discontinuous cranial traits. *Am. J. phys. Anthrop.* **29**, 397–404.

Howe, W. L. and Parsons, P. A. (1967). Genotype and environment in the determination of minor skeletal variants and body weight in mice. *J. Embryol. exp. Morph.* **17**, 283–292.

Keene, H. J. (1965). The relationship of maternal age, parity and birth weight to hypodontia in naval recruits. *Am. J. phys. Anthrop.* **23**, 330.

Kettlewell, H. B. D., Berry, R. J., Cadbury, C. J. and Phillips, G. C. (1969). Differences in behaviour, dominance and survival within a cline. *Heredity*, **24**, 15–25.

King, J. C. (1959). Differences between populations in embryonic developmental rates. *Am. Nat.* **93**, 171–180.

Laughlin, W. S. (1963). Eskimos and Aleuts: their origins and evolution. *Science, N.Y.* **142**, 633–645.

Laughlin, W. S. and Jorgensen, J. B. (1956). Isolate variation in Greenlandic Eskimo crania. *Acta genet.* **6**, 3–12.

Lerner, I. M. (1954). "Genetic Homeostasis". Oliver & Boyd, Edinburgh and London.

Ludwig, W. (1932). "Das Rechts-Links Problem in Tierreich und beim Men-schen". Springer, Berlin.

McLaren, A. and Walker, P. M. B. (1970). Rodent DNA: comparisons between species. *Evolution, Lancaster, Pa* **24**, 199–206.

Mayr, E. (1969). Bird speciation in the tropics. *Biol. J. Linn. Soc.* **1**, 1–17.

Montagu, M. F. A. (1937). The medio-frontal suture and the problem of meto-pism in the primates. *J.R. anthrop. Inst., Lond.* **67**, 157–201.

Morant, G. M. (1925). A study of Egyptian craniology from prehistoric to Roman times. *Biometrika* **17**, 1–52.

Rees, J. W. (1969). Morphological variation in the cranium and mandible of the white-tailed deer (*Odocoileus virginianus*): a comparative study of geographical and four biological distances. *J. Morph.* **128**, 95–112.

Robinson, R. (1970). Homologous mutants in mammalian coat colour variation. *Symp. zool. Soc. Lond.* **26**, 251–269.

Schultz, A. H. (1923). Bregmatic fontanelle in mammals. *J. Mammal.* **4**, 65–77.

Schultz, A. H. (1926). Studies on the variability of platyrrhine monkeys. *J. Mammal.* **7**, 286–305.

Schultz, A. H. (1960). Age changes and variability in the skulls and teeth of the central American monkeys *Alouatta*, *Cebus* and *Ateles*. *Proc. zool. Soc., Lond.* **133**, 337–390.

Searle, A. G. (1954a). Genetical studies on the skeleton of the mouse. IX. Causes of skeletal variation within pure lines. *J. Genet.* **52**, 68–102.

Searle, A. G. (1954b). Genetical studies on the skeleton of the mouse. XI. The influence of diet on variation within pure lines. *J. Genet.* **52**, 413–424.

Searle, A. G. (1957). Delayed hybrid vigour in mammals. *Proc. Int. Genet. Symp.* 1956 (Tokyo and Kyoto), 386–389.

Searle, A. G. (1959). The incidence of anencephaly in a polytypic population. *Ann. hum. Genet.* **23**, 279–288.

Searle, A. G. (1968). "Comparative Genetics of Coat Colour in Mammals". Logos and Academic, London.

Selander, R. K., Hunt, W. G. and Yang, S. Y. (1969). Protein polymorphism and genic heterozygosity in two European subspecies of the house mouse. *Evolution, Lancaster, Pa*, **23**, 379–390.

Spickett, S. G. and Thoday, J. M. (1966). Regular responses to selection. 3. Interaction between linked polygenes. *Genet. Res.* **2**, 96–121.

Straus, W. L. (1937). Cervical ribs in the woolly monkey. *J. Mammal.* **18**, 241–242.

Thoday, J. M. (1961). The location of polygenes. *Nature, Lond.* **191**, 368–370.

Truslove, G. M. (1961). Genetical studies on the skeleton of the mouse. XXX. A search for correlations between some minor variants. *Genet. Res.* **2**, 431–438.

Van Valen, L. (1962). A study of fluctuating asymmetry. *Evolution, Lancaster, Pa*, **16**, 125–142.

Waddington, C. H. (1957). "The Strategy of the Genes". Allen & Unwin, London.

W.H.O. (1970). "Genetic Factors in Congenital Malformations". *Wld Hlth Org. tcchn. Rep. Ser.* No. 438.

Wood-Jones, F. (1930–31). The non-metrical morphological characters of the skull as criteria for racial diagnosis. I, II, III. *J. Anat.* **65**, 179–195; 368–378; 438–445.

Wright, S. (1934). The results of crosses between inbred strains of guinea pigs differing in numbers of digits. *Genetics*, **19**, 537–551.

READ VALUES OVER 50% NEGATIVELY

APPENDIX

Table of angular transformations in radians

%	0	·1	·2	·3	·4	·5	·6	·7	·8	·9		%
0	1·571	1·508	1·481	1·461	1·444	1·429	1·416	1·403	1·392	1·381	1·371	99
1	1·371	1·361	1·351	1·342	1·333	1·325	1·317	1·309	1·302	1·294	1·287	98
2	1·287	1·280	1·273	1·266	1·260	1·253	1·247	1·241	1·235	1·229	1·223	97
3	1·223	1·217	1·211	1·205	1·200	1·194	1·189	1·184	1·178	1·173	1·168	96
4	1·168	1·163	1·158	1·153	1·148	1·143	1·139	1·134	1·129	1·124	1·120	95
5	1·120	1·115	1·111	1·106	1·102	1·097	1·093	1·089	1·084	1·080	1·076	94
6	1·076	1·072	1·068	1·063	1·059	1·055	1·051	1·047	1·043	1·039	1·035	93
7	1·035	1·031	1·028	1·024	1·020	1·016	1·012	1·008	1·005	1·001	·997	92
8	·997	·994	·990	·986	·983	·979	·976	·972	·968	·965	·961	91
9	·961	·958	·954	·951	·948	·944	·941	·937	·934	·931	·927	90
10	·927	·924	·921	·917	·914	·911	·908	·904	·901	·898	·895	89
11	·895	·892	·888	·885	·882	·879	·876	·873	·870	·866	·863	88
12	·863	·860	·857	·854	·851	·848	·845	·842	·839	·836	·833	87
13	·833	·830	·827	·824	·821	·818	·815	·813	·810	·807	·804	86
14	·804	·801	·798	·795	·792	·790	·787	·784	·781	·778	·775	85
15	·775	·773	·770	·767	·764	·762	·759	·756	·753	·751	·748	84
16	·748	·745	·742	·740	·737	·734	·732	·729	·726	·724	·721	83
17	·721	·718	·716	·713	·710	·708	·705	·702	·700	·697	·694	82
18	·694	·692	·689	·687	·684	·682	·679	·676	·674	·671	·669	81
19	·669	·666	·664	·661	·659	·656	·654	·651	·648	·646	·644	80
20	·644	·641	·639	·636	·634	·631	·629	·626	·624	·621	·619	79
21	·619	·616	·614	·611	·609	·606	·604	·602	·600	·597	·594	78
22	·594	·592	·590	·587	·585	·582	·580	·578	·575	·573	·570	77
23	·570	·568	·566	·563	·561	·559	·556	·554	·552	·549	·547	76
24	·547	·544	·542	·540	·538	·535	·533	·531	·528	·526	·524	75

%	·0	·1	·2	·3	·4	·5	·6	·7	·8	·9	0	%
74	·501	·503	·505	·508	·510	·512	·514	·517	·519	·521	·524	25
73	·478	·480	·482	·485	·487	·489	·492	·494	·496	·498	·501	26
72	·456	·458	·460	·462	·465	·467	·469	·471	·474	·476	·478	27
71	·433	·436	·438	·440	·442	·444	·447	·449	·451	·453	·456	28
70	·412	·414	·416	·418	·420	·422	·425	·427	·429	·431	·433	29
69	·390	·392	·394	·396	·399	·401	·403	·405	·407	·409	·412	30
68	·368	·370	·373	·375	·377	·379	·381	·383	·385	·388	·390	31
67	·347	·349	·351	·353	·355	·358	·360	·362	·364	·366	·368	32
66	·326	·328	·330	·332	·334	·336	·338	·341	·343	·345	·347	33
65	·305	·307	·309	·311	·313	·315	·317	·319	·322	·324	·326	34
64	·284	·286	·288	·290	·292	·294	·296	·298	·300	·303	·305	35
63	·263	·265	·267	·269	·271	·273	·276	·278	·280	·282	·284	36
62	·242	·244	·247	·249	·251	·253	·255	·257	·259	·261	·263	37
61	·222	·224	·226	·228	·230	·232	·234	·236	·238	·240	·242	38
60	·201	·203	·205	·208	·210	·212	·214	·216	·218	·220	·222	39
59	·181	·183	·185	·187	·189	·191	·193	·195	·197	·199	·201	40
58	·161	·163	·165	·167	·169	·171	·173	·175	·177	·179	·181	41
57	·140	·142	·144	·147	·149	·151	·153	·155	·157	·159	·161	42
56	·120	·122	·124	·126	·128	·130	·132	·134	·136	·138	·140	43
55	·100	·102	·104	·106	·108	·110	·112	·114	·116	·118	·120	44
54	·080	·082	·084	·086	·088	·090	·092	·094	·096	·098	·100	45
53	·060	·062	·064	·066	·068	·070	·072	·074	·076	·078	·080	46
52	·040	·042	·044	·046	·048	·050	·052	·054	·056	·058	·060	47
51	·020	·022	·024	·026	·028	·030	·032	·034	·036	·038	·040	48
50	·000	·002	·004	·006	·008	·010	·012	·014	·016	·018	·020	49

READ VALUES OVER 50% NEGATIVELY

The Heredity of Dermatoglyphic Traits in Non-human Primates and Man

J. MAVALWALA

Department of Anthropology, University of Toronto, Canada

INTRODUCTION

There is ample evidence now available to show that the ancient Chinese were aware of the use of fingerprints for identification purposes. Prints of the thumb or finger on pottery or on seals were found to have been an established practise in ninth century India, in seventh century Japan, and in fourteenth century Persia. A historical review is presented by Cummins and Midlo (1961) and Ökrös (1965). While finger prints, and in some cases palm prints, have a long history of use as validators for signatures, the skin ridges and the papillary patterns they form are first known to have been extensively studied by M. Malpighius (1686). Studies began to appear directed at elucidating the morphology of skin ridges and the various types of patterns they formed. Purkinje (1823) listed nine types of patterns in a scale of increasing complexity (see Fig. 1).

Galton (1892) set up a simpler classification and Galton's arch, loop and whorl terms are still the most commonly used classification today (see Fig. 2).

Galton conducted studies in the heredity of finger prints and also printed the same individuals several years apart to convince sceptics that papillary patterns do not change throughout the lifetime of an individual. But finger prints did not come into their own as an identification medium until Bertillon's anthropometric method (Bertillon, 1882) fell into disuse. As finger prints were internationally accepted (i) to be unique in each individual, and in fact on each finger (palms, soles and toe prints were later included) and (ii) as unchanging throughout the lifetime of the individual except for size, a surge of interest was aroused and finger prints immediately raised in one's mind an association with crime. A considerable amount of work was done to classify the prints, particularly finger prints, in an effort to expedite the filing and retrieval procedures in police headquarters. The United States Department of Justice, Federal Bureau of Investigation (1960) issued an excellent 200 page booklet that typifies the "classify–file–identify" approach.

While the enthusiasm with which the identification aspects of patterns formed by papillary ridges on the human skin did a great deal to attract attention to the extreme variability of this phenotype, it also distracted the majority of investigators from further probing into the nature of the genotype.

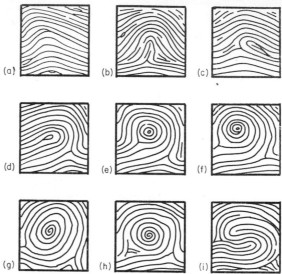

FIG. 1. The nine pattern types postulated by Purkinje are (a) transverse curves; (b) the central longitudinal stria; (c) the oblique stripe; (d) the oblique loop; (e) the almond whorl; (f) the spiral whorl; (g) the ellipse; (h) the circle and (i) the double whorl.

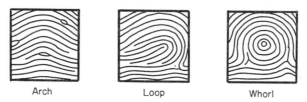

Arch   Loop   Whorl

FIG. 2. The three patterns postulated by F. Galton.

The delay in studies of the genotype also stems from the historical sequence of the development of scientific knowledge. While papillary ridge patterns were known to man for many centuries, Mendel's laws of inheritance were only rediscovered in 1900, even though scholars such as Maupertius had described the inheritance of polydactly in Man in Berlin in the 1750's, the Talmud clearly understood the implications of the heritability of haemophilia in establishing marriage and circumcision rules for certain members of affected families, and Horner described the inheritance of X-linked recessive colour blindness in 1876. McKusick (1969) reviews the development of interest in heredity processes in Man. Papillary patterns have, because of the variability of the phenotype, been less involved in studies by geneticists than other human phenotypes.

The term dermatoglyphics was coined by Cummins and Midlo (1926) who later wrote the excellent review of dermatoglyphics in 1943, revised in 1961 (Cummins and Midlo, 1961).

With the knowledge that dermatoglyphics were not only useful in identification and varied in pattern frequencies in different populations, but that they could be useful as a diagnostic aid in medicine (Alter, 1967), a resurgence of interest was created.

However, McKusick (1968) could report only two examples of Mendelian inheritance in dermatoglyphics, since focal dermal hypoplasia does not fall into the category of dermatoglyphics. The first instance was "absence of finger prints" as a dominant trait, better termed hypoplastic papillary patterns and the second condition was suggested by Walker (1941). She postulated that radial loops on the right index finger are inherited via X-linkage. Like the other studies discussed later in this chapter, suggested inheritance of specific patterns has not been proven.

A genetic mechanism has been worked out for finger-ridge counts (Holt, 1952). Holt established their multifactorial inheritance as being due to polygenes with additive effect.

A considerable number of studies are now appearing on dermatoglyphic variation in individuals with various anomalies, especially chromosomal, and it is hoped that the studies of the effects of chromosomal deletions will yield genetic information.

Some work has been done on the phenotypic variability of dermatoglyphics on non-human primates but no inheritance data is available.

Genetic studies in dermatoglyphics have focused on specific parts of the phenotype. The total dermatoglyphics of the individual should involve comprehension of *all* papillary ridge formations on the fingers, palms, soles and toes and include minor variations of ridges, minutiae, but as yet no model has been devised that can state the total dermatoglyphics of an individual in a fashion suitable to forms of genetic analysis available at present.

Attempts have been made to show that because of a high correlation between some features only the feature which proved to be more amenable to analysis should be studied. Finger and toe prints are positively correlated (Newman, 1936; Bali, 1964). Also, because toe prints are far more difficult to print than finger prints, a larger body of literature has accumulated on finger prints and relatively few studies are available on toe prints. Because of availability of data the bulk of analysis has been done with finger and palm prints. The last few years have seen an increase of interest in the prints of the sole, after it was pointed out by Dr Norma Ford Walker (1957) that the hallucal area of the sole showed altered dermatoglyphics in Down's Syndrome patients. But the studies are designed to distinguish whether sole dermatoglyphics of specific patients are altered significantly enough to serve as a diagnostic aid.

Figures 3 and 4 illustrate the general areas of the hand and foot that have undergone the scrutiny of workers in dermatoglyphics. While Cummins, the originator of the word dermatoglyphics, points out that flexion creases should not, strictly speaking, be included in the definition of dermato-glyphics, they are discussed here since they do appear on the palmar surface and appear to be significantly altered by some chromosomal aberrations.

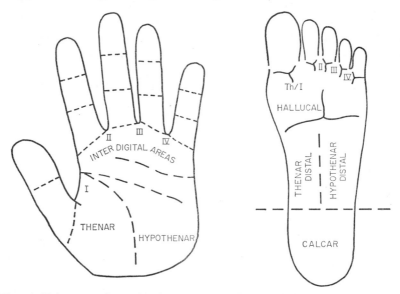

FIG. 3. Palmar configurational areas.   FIG. 4. Sole configurational areas.

FINGER PATTERNS

By far the largest body of data is available on prints of the terminal phalanges of the fingers. The proximal and medial sectors, while bearing ridged skin, display so little variation that they have not aroused much interest, even though Ploetz-Radmann (1937) classified their variation into twelve types depending on the slanting of the ridges. Whipple (1904) commented briefly on this region and MacArthur (1938) felt that the ridge formations on the middle and proximal phalanges were of diagnostic value in zygosity determination of twins.

The terminal phalange or the "ball" of the finger, however, provides a large volume of data and an analysis, both qualitative and quantitative.

Qualitative studies of finger prints have been heavily influenced by classificatory systems set up primarily for identification. While Sir Francis Galton (1892) was interested in the heredity of finger prints his main theme was to prove their utility for identification by demonstrating that patterns did not change during the life of an individual, except, of course, for gross size. Henry (1937), Bridges (1948) and Cherrill (1954), among many other workers, have been responsible for classificatory systems of finger prints. The system most widely used is described in detail by Cummins and Midlo (1961) and a Memorandum on Dermatoglyphic Nomenclature (Penrose, 1968) was issued by the Ciba Foundation. This was an attempt at international standardization. Some idea of the variety of ways in which the patterns can be classified is gained by a perusal of Practical Fingerprinting (Bridges, 1948) which describes the Henry System, the Battley system, the Vucetich system, and *forty-seven* other systems used in identification bureaus across the world. Genetic studies are not as diverse but unfortunately too often err on the side of simple classifications and lump all variations into arches, loops and whorls. Reports of chromosomal anomalies with references to dermatoglyphics have been known to achieve the epitomy of brevity. Workers in dermatoglyphics all have memories of reading an interesting paper expecting momentarily to come upon a description of the dermatoglyphics only to be stumped by "unusual dermatoglyphics were observed". The heredity or non-heredity of finger print patterns was debated first as a question of heredity versus non-heredity. The question now relates to the exact mode of inheritance.

The following workers, among others, postulated the inheritance of finger print patterns: Bonnevie (1924, 1929, 1931); Elderton (1920); Essen-Möller (1941); Galton (1892); Geipel (1937); Grüneberg (1928); Heindl (1927); Monga (1954); Poll (1914); Inez Whipple (1904); de Wilde (1953) and Wilder (1902, 1904, 1908, 1919). Other workers denied the influence of heredity. Forgeot (1892) failed to find any evidence of heredity when he studied several 3-generation pedigrees. Senet (1906), using 5-generation pedigrees, also refuted the influence of heredity, and Stockis (1908) agreed with this. Böhmer and Harren (1939) studied 100 families and claimed that finger print patterns were not inherited. The "non-heredity" workers were arguing on the assumption that children should show *exactly* the same patterns as their parents whereas the "heredity" workers argued that a *similarity* of patterns was evidence enough.

Bonnevie (1931) postulated three "factors", V, R, and U involved in genetically determining finger patterns. Factor V was responsible for the thickness of the epidermis under the pattern area, and U and R were

FINGER PRINTS

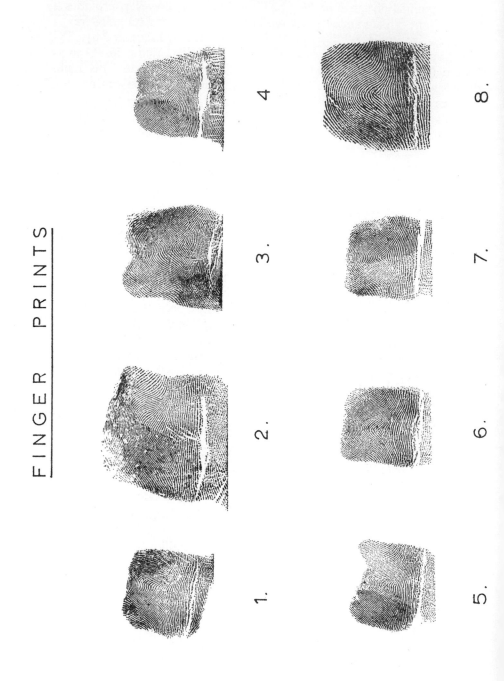

1.   2.   3.   4.

5.   6.   7.   8.

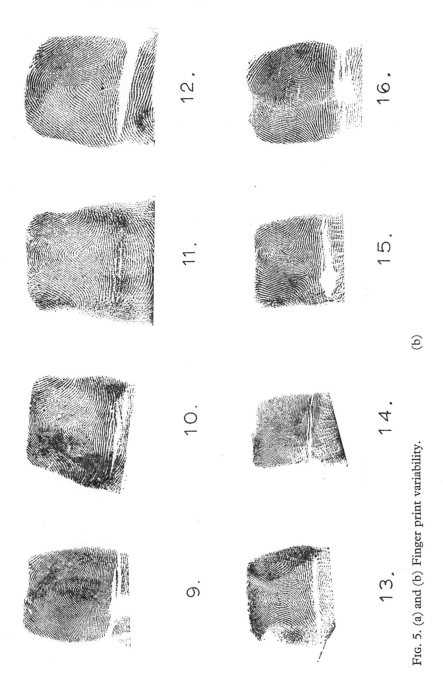

FIG. 5. (a) and (b) Finger print variability.

(b)

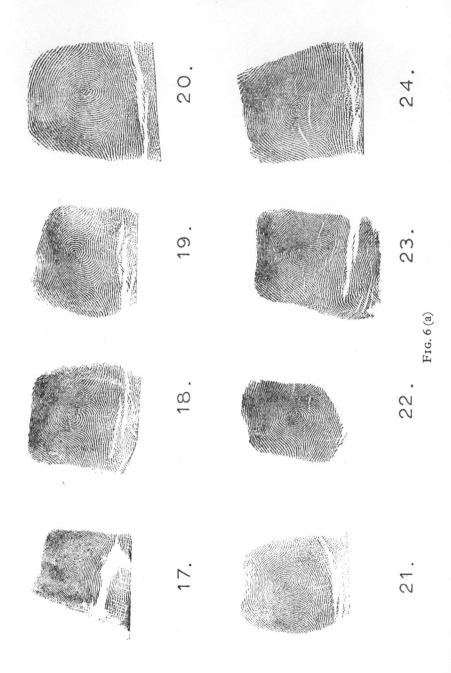

Fig. 6 (a)

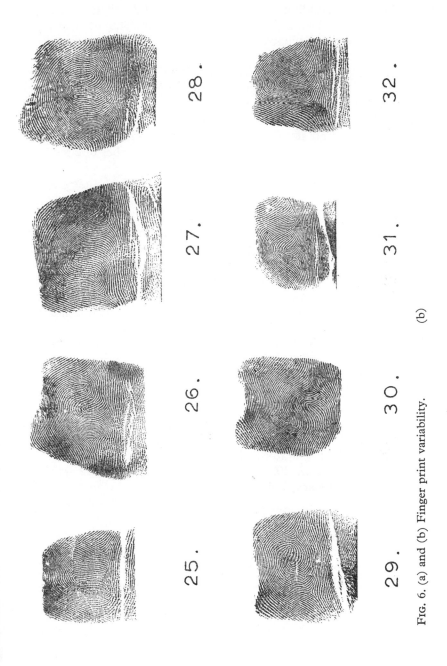

FIG. 6. (a) and (b) Finger print variability.

(b)

responsible for the padding or cushioning on the ulnar digits I, II and III and the radial digits IV and V. This view is too simplistic today in view of our knowledge of the sequence of events from DNA to the finished phenotypic product.

All qualitative studies, while demonstrating the importance of the role of heredity in the phenotypic expression of finger patterns, have proved inconclusive. It is tempting to claim that all our efforts have not unearthed the answer because we do not understand multifactorial inheritance mechanisms enough and all attempts to tabulate the continuous variation observed in patterns has been arbitrary and a meagre number of classes have been used.

Some idea of the difficulty of examining the heredity mechanisms underlying the formation of dermatoglyphic patterns can be gained by a brief look at the variation encountered in the phenotype. Figures 5 and 6 are finger prints put together in a series to show the gradual increase in geometric complexity. Workers in genetics divide this continuous variation arbitrarily into as few as three classes, following Galton (1892) or into as many as 95 classes or types (Ökrös, 1965). Ökrös found that in actual practise he could eliminate some rare types and use a working classification of 60 types. Reports containing data subdivided into three major types only, may be grossly underestimating the significance of the phenotypic variation encountered and those workers who try to take into account every aspect of every variation may be floundering in a plethora of genetically unimportant detail. No international agreement or standardization is in sight at present in spite of many attempts, the latest being the 1967 Ciba Foundation Symposium in London (Penrose, 1968). With pattern variation being studied in terms of heredity as far back as 1892 , no major understanding has been achieved (Di Bacco, 1965; Bat-Miriam and Guttman, 1961; Becker, 1954; Elderton, 1920; Grüneberg, 1928; Gupta, 1953; Kramp, 1954; Lamy et al., 1957; Lazaro et al., 1966; Monga, 1954; Müller, 1930; Rife, 1952; Walker, 1941).

PALMAR PATTERNS

The human palm displays specific configurational areas that are considered worthy of study by dermatoglyphic workers (see Fig. 3). These are the IInd, IIIrd and IVth interdigital areas, the Hypothenar area, and the Thenar area which is considered as a unit with the Ist interdigital area. Meyer-Heydenhagen (1934) analysed palmar patterns on the palms of twins and felt that the similarity of this feature was an excellent means of zygosity diagnosis.

Weninger (1935, 1947) states that palmar patterns were definitely not inherited via simple single factor inheritance and Weinand (1937) and Czik and Malan (1938) support this view.

Studies conducted to demonstrate "dominant" or "recessive" inheritance of some particular pattern have proved inconclusive (Bansal and Rife, 1962; Fang, 1950b).

The only non-human primates on whom dermatoglyphics are available in significant numbers are 56 *Macaca fuscata fuscata* of Takasakiyama and 49 *Macaca fuscata yakui* of Yakushima, Japan (Iwamoto, 1964). No inheritance is postulated. The study limits itself to morphology.

Brehme in a series of studies has examined the palms of various Primates and analysed pattern distribution and ridge direction utilizing the Main Line Index devised by Cummins and Midlo (1961). He reports on pattern frequencies for 61 *Presbytis* and 13 *Pygathrix* (Brehme, 1968a), on 526 individuals from the genus *Cercopithecus* and the genus *Erythrocebus* (Brehme, 1968b) where he feels the dermatoglyphic data justifies separating these two genera. Dermatoglyphics of twenty-one individuals of *Rhinopithecus* and ten *Simias* (Brehme, 1967) are similar to other langur species within the Colobinae.

Twenty-six *Black Saki* monkeys, *Chiropotes satanas* displayed high variability (Brehme, 1965a) in pattern intensity.

Brehme (1956b) also reports on palmar patterns of 150 *Nasalis larvatus*.

These large numbers prove that intra- and inter-species variation exists among the Primates but at this stage no further deductions can be made.

QUANTITATIVE STUDIES OF RIDGE COUNTS

A series of additions to knowledge have led to an understanding of the quantitative genetics of dermatoglyphics as compared to the qualitative aspects. In 1889, Galton proposed a statistic, the correlation coefficient, now widely used. In 1918 R. A. Fisher demonstrated how familial correlation values could be used to demonstrate Mendelian inheritance. For a detailed discussion of the technique see Holt (1968, p. 38).

Finger Ridge Counts

The ridge count (Fig. 7), is the total number of ridges actually touching a fine line drawn from the triradius to the core of a pattern. This was first used by Bonnevie (1924) and subsequently by many workers. Good descriptions of ridge counts and the various rules for obtaining them are to be found in Cummins and Midlo (1961) and Holt (1968).

Sarah Holt of the Galton Laboratory working over a period of many years (1949–1968) elegantly used the finger ridge count to demonstrate that ridge counts (indicative of pattern size) were inherited by many genes functioning in an additive fashion. A very complete exposition of this work is available (Holt, 1968). Penrose (1949) studied the effect of various modes of inheritance using correlations expected between various relatives.

RIDGE COUNT = 6

FIG. 7. The Finger Ridge Count.

Palmar Ridge Counts

The mechanism of counting ridges to evaluate size of a configuration and also to make the data amenable to quantitative analysis has been used in the interdigital areas of the palm (Fang, 1950a, b; Pons, 1959, 1964; Mitra *et al.*, 1966) (Fig. 8). Fang's studies indicated that about 80% of the effect was due to recessive heredity but Pons' data does not agree with these

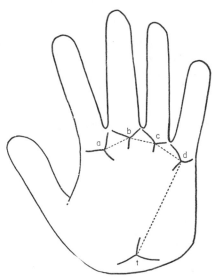

FIG. 8. Palmar Ridge Counts.

findings and Mitra *et al.* felt no exact inheritance mechanism could be postulated at this time. Glanville (1965), using ridge counts of patterns in the IInd, IIIrd and IVth interdigital areas, came to the conclusion that additive genes were responsible for the variation but that "an appreciable amount of variation of non-genetic origin was observed".

Other studies report on various ridge count values for populations. The *t–d* count has been reported by Berg (1968) for a British sample of 300 normals, 160 Down's Syndrome patients, 31 Turners and 27 Klinefelters. He reports that the *t–d* count correlates negatively with the *atd* angle in all groups, and only discriminates well between normals and Down's patients. *a–b* Counts are available on Bavarian populations (Baitsch and Schwarzfischer, 1959), Punjabis of India (Seth, 1963), Andhra Pradesh, India (Datta, 1961) and on Bhutanese (Bhasin, 1966). *a–d* Counts are available on Punjabis (Seth, 1963) and Bhasin (1966) reports on Bhutan giving values for *a–b*, *b–c*, *c–d* and *a–d* counts. No genetic conclusions can be drawn from these studies so far.

THE *atd* ANGLE

A triradius, *t*, is normally situated medially near the proximal border of the palm. Cummins (1939) was one of the first workers in dermatoglyphics to point out that in the palms of Down's patients the ridges were characteristically altered and that this triradius *t* was elevated distally. He classified the triradius as *t'* if elevated and if it was situated practically at the

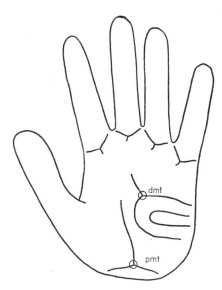

Fig. 9. The Proximal, and Distal, medial Triradii on the Palm.

C

middle of the palm it was designated t'' (see Cummins and Midlo, 1961, p. 99). Studies were later carried out by Penrose examining the elevation of the triradius t, or the axial triradius as it came to be called, in Down's patients as compared to normals (Penrose, 1949, 1954, 1961, 1963a, b, 1965a, b). Geipel also examined the effects of an extra chromosome in the G group (Geipel, 1961, 1963) on the position of the axial triradius on the human palm. There is no doubt that among other things an extra chromosome in the G group elevates the axial triradius. It may be appropriate to note the difference between a proximal medial triradius and a distal medial triradius. A PMT is a triradius present at the proximal border and may be concurrent with a DMT. A DMT is usually associated with a hypothenar pattern (Fig. 9). In Fig. 9 some workers would consider the palm to possess an elevated axial triradius whereas it should be distinguished from the palm in which there is only a PMT and it is distally situated.

THE FLEXION CREASES

The palm of primates including Man displays three major flexion creases. Shuttleworth published a paper on Mongolian Imbecility in 1909 and it was followed by a comment by Langdon-Down (1909) that mongol palms displayed only two flexion creases. Crookshank (1924) pointed out that the fusion of the two lateral flexion creases into one was a "mongoloid" trait. Further examination of this feature was made by Würth (1937), Portius (1937), Tillner (1953, 1958), Weninger and Navratil (1957) and Penrose (1949). It is now clearly established that the "Simian Crease" is significantly more frequent in Down's Syndrome and has been observed on the palms of other chromosomally anomalous patients, notably Trisomy 18, 5p− and 18q−.

Sarker (1961) describes the "Simian Crease" and reviews studies of this feature on the palms of the anthropoid apes. This so-called "Simian Crease" appears only on the palms of the Orangutan according to Crookshank (1924). Studies by Duckworth (1915), Sonntag (1924) and Wood Jones (1929) show that various forms of transitional flexion creases are to be found on the palms of the Orangutan, Gorilla and Chimpanzee. This variation is also illustrated by Biegert (1963) and Schultz (1968).

Studies are available showing the varying frequencies of different flexion creases in different populations of Man (Kutsuna, 1929; Walter, 1956; Leiber, 1960). Lestrange (1966, 1969) developed a classification of five types of flexion creases and found that there is a significant sex difference but that right and left hands are interdependent.

The phenotypic variation of the flexion creases is not as complex as finger print patterns but the types most commonly observed on the human palm can be subdivided into many types (Fig. 10). This classification is being used at the University of Toronto on current studies on samples of various chromosomal anomalies and normal populations. The term Single Transverse Flexion Crease (STFC) is preferred to other terms since the feature is not the same as observed on the hands of normal non-human primates. Types 1 and 2 would be considered normal, all the others forms of a STFC.

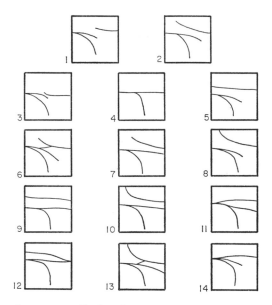

FIG. 10. Types of transverse flexion creases.

TOE AND SOLE PRINTS

Toe prints have been studied by very few workers (Bali, 1964, 1968) and like finger print patterns, the inheritance mechanisms are unclear.

Geipel (1952) examined patterns on the soles of twins, and Lehmann (1954) examined a pedigree with multiple malformations. While heredity was postulated no clear genetic model was unearthed. The hallucal area of the sole (Fig. 4) has drawn attention from medical geneticists (Alter, 1967; Smith, 1964; Hirsch, 1964; Walker, 1958). Brismar (1965) found a high mother–child correlation using hallucal pattern direction. Datta using a plantar main-line index to determine the transversality of the ridges across the sole (Datta, 1962, 1964, 1966) found significant values in sib pair and parent–child correlations and postulates polygenic inheritance with additive effect.

MINUTIAE

The dermal ridges sometimes fork or show small gaps or form little features similar to a road around a traffic island. These small features are termed minutiae.

Grüneberg (1928) and Steffens (1965) both studied the minutiae of identical twins and concluded that they were not inherited. Recently, however, Okajima (1966, 1967a, b, 1970) has studied the minutiae on twin data in the calcar area of the sole and on the fingers and palms. His studies lead him to believe that the frequency of forks in minutiae is at least partially controlled by heredity.

HYPOPLASTIC PAPILLARY PATTERNS

Furuya (1961) reports on 19 cases from 7 families. The skin ridges, instead of forming smooth linear lines are often broken up into what Abel (1936) called "Gestörte Papillarmuster", or a series of short broken lines. The breaks can be so severe as to totally obliterate any recognizable pattern.

Goddard (1950) reported a Japanese whose ridges were so broken that he presented no patterns on all ten fingers. The same man was examined further by Furuhata and Kuwashima (1950). Further data was reported by Matukura (1954) and Furuhata et al. (1957) suggesting simple Mendelian dominant inheritance. Further evidence (Okamoto, 1958, 1959; Baird, 1964) supports the evidence of dominant inheritance.

A large body of literature (Alter, 1967) also shows that this kind of broken ridges, referred to in the medical literature as dermal dysplasia, is present on some parts of the fingers or palms, more frequently among mentally retarded individuals than normals. Matukura (1953) reports from studies on 3-month-old foetuses that the dysplasia is apparent while the dermal tissue on the finger prints is still in the early stages of development.

The dysplasia varies in intensity and location and Furuya recognizes two forms.

1. A total dot-and-short-ridge pattern, generalized over the entire dermatoglyphic area, inherited as an autosomal dominant trait, with 77·8% penetrance.

2. A partial dot-and-short-ridge pattern, inherited as an autosomal dominant trait with 62·5% penetrance.

He feels that these two forms are distinctly separate phenotypes, inherited independently.

EVIDENCE FROM MEDICAL GENETICS

A rapidly growing body of data is now becoming available showing that dermatoglyphics are affected by factors affecting growth in the first 3–4 months of foetal life. Chromosomal aberrations affect dermatoglyphics (Alter, 1967; Thompson, 1965; Taylor, 1968; Holt, 1968; Dallapiccola, 1968) as do toxic factors such as thalidomide (Davies and Smallpiece, 1963) and rubella (Achs et al., 1966).

We hope that dermatoglyphic data of patients with chromosomal deletions will furnish some genetic leads, but the material is usually hampered by lack of data from parents or siblings.

While sometimes dramatic changes are seen as the result of a deletion, for example 18q– (Mavalwala et al., 1970) equally important dermatoglyphic changes are also observed caused by rubella. Schiono et al. (1969) report that the parents of children with Waardenburg's Syndrome had dermatoglyphics similar to those found in their affected children.

CONCLUSION

If the heredity mechanisms of dermatoglyphics are stages of development controlled by separate factors emanating from as diverse sites as chromosome 5, to the G group and the sex chromosomes, and the final phenotype discussed here is the result of the influence of non-genetic factors, then the difficult task that lies ahead is to determine exactly what pathways are involved, what do they actually dictate, in how much detail, and how much of the dictation can we assume will be affected by non-genetic factors.

Data accumulated from patients with chromosomal deletions along with extensive family data is one avenue of approach. If extensive family pedigrees are to be studied our methods of phenotype recognition will have to be revised since a large number of such studies have yielded inconclusive results.

Man has been intrigued by the inheritance of dermatoglyphics for nearly a century now and it is highly likely that with the pace of science today the inheritance mechanisms underlying as variable a trait as dermatoglyphics will be clearly understood within the next one hundred years. It is unlikely that the answer will be a simple pathway directly from the DNA code to the patterns as we see them. It is more probable that the pathways will lead to the formation of a particular morphology of the hand or foot which in turn will limit the expression of the final ridge configurations. Overlying this mechanism must be a multitude of environmental factors, only very few of which are known to us today.

While the scholars who have laboured over dermatoglyphics in the past have not achieved a major breakthrough, they have pointed out the way for further studies by workers in the future, and a final understanding of the heredity of dermatoglyphics will be, in part, their victory also.

REFERENCES

Abel, W. (1936). Z. Morphol. Anthropol. 36, 1–37.
Achs, R., Harper, R. G. and Siegel, M. (1966). New England J. Med. 274, 148–150.
Alter, M. (1967). Medicine (Balt.) 46, 35–56.
Baird, H. W. (1964). J. Pediat. 64, 621–631.
Baitsch, H. and Schwarzfischer, F. (1959). Homo 10, 226–236.
Bali, R. S. (1968). Z. Morphol. Anthropol. 59, 244–272.
Bali, R. S. (1964). "Toe Dermatoglyphics–Inheritance and Comparison with the Finger and Ball Patterns". Delhi.
Bansal, P. and Rife, D. C. (1962). Acta Genet. Med. (Roma) 11, 29.
Bat-Miriam, M. and Guttman, A. L. (1961). Int. Congr. Hum. Genet. 3, 1481–1483.
Becker, E. (1954). Z. Menschl. Vererb. u. Konstit. Lehre 32, 106–115.
Berg, J. M. (1968). Hum. Biol. 40, 375–385.
Bertillon, J. (1882). "Identification anthropometrique". Annales de demographie internationale. Paris.
Bhasin, M. K. (1966). Int. Sym. Der. Abstracts, Delhi.
Biegert, J. (1963). In "Classification and Human Evolution" (S. L. Washburn, ed.), Viking Fund Publs. Anthrop. No. 37. Aldine Publishing Co., Chicago.
Böhmer, K. and Harren, F. (1939). Deut. Z. Gesamte Gericht Med. 32, 73–82.
Bonnevie, K. (1924). J. Genet. 15, 1.
Bonnevie, K. (1929). Z. Indukt. Abstamm. Vererbungsl. 50, 219.
Bonnevie, K. (1931). Z. Indukt. Abstamm. Vererbungsl. 59, 1.
Brehme, H. (1965a). Z. Morph. Anthropol. 56, 206–216.
Brehme, H. (1965b). Homo Supplement, 244–248.
Brehme, H. (1967). Anthropol. Anz. 30, 149–161.
Brehme, H. (1968a). Mitt. Anthrop. Ges. Wien 98, 1–14.
Brehme, H. (1968b). Folia primat. 9, 41–67.
Bridges, B. C. (1948). "Practical Fingerprinting". Funk and Wagnalls, New York.
Brismar, B. (1965). A. Ge. Me. Ge. 14, 86–92.
Cherrill, F. R. (1954). "The Finger Print System at Scotland Yard". H.M. Stationery Office, London.
Crookshank, F. G. (1924). "The Mongol in Our Midst". New York. E. P. Dutton.
Cummins, H. (1926). Am. J. Phys. Anthropol. 9, 471–502.
Cummins, H. (1939). Anat. Rec. 73, 407.
Cummins, H. and Midlo, C. (1961). "Fingerprints, Palms and Soles. An Introduction to Dermatoglyphics". Dover Publications, New York.
Czik, L. and Malan, M. (1938). Z. Konst. Lehre 21, 186.

Dallapiccola, B. (1968). "I dermatoglifi della Mano". Zambon S.p.A. Milano-Vicenza.

Datta, P. K. (1961). *Man* **63**, 189–190.

Datta, P. K. (1962). *Anthropologist* **9**, 1–5.

Datta, P. K. (1964). *Acta Genet. Med. (Roma)* **13**, 400–405.

Datta, P. K. (1966). *Acta Genet. (Basel)* **16**, 89–94.

Davies, P. and Smallpiece, V. (1963). *Devel. Med. Child. Neurol.* **5**, 491–496.

Di Bacco, M. (1965). *A. Ge. Me. Ge.* **14**, 305–316.

Duckworth, W. L. H. (1915). "Morphology and Anthropology". Cambridge.

Elderton, E. M. (1920). *Biometrika* **12**, 57–91.

Essen-Möller, E. (1941). *Hereditas* **27**, 1–50.

Fang, T. C. (1950a). "The inheritance of the *a—b* ridge-count on the human palm, with a note on its relation to mongolism". Ph.D. thesis, University of London.

Fang, T. C. (1950b). *J. Ment. Sci.* **96**, 780.

Federal Bureau of Investigation. (1960). "The Science of Fingerprints". The United States Department of Justice.

Fisher, R. A. (1918). *Trans. Roy. Soc. Edinb.* **52**, 399.

Forgeot (1892). *Arch. de l'anthrop. crim.* **6**, 387–404.

Furuhata, T., Furuya, Y., Tanaka, T. and Nakajima, H. (1957). *Proc. Japan Acad.* **33**, 410–413.

Furuhata, T. and Kuwashima, N. (1950). *Proc. Japan Acad.* **26**, 41–43.

Furuya, Y. (1961). *Jap. J. Human Genetics* **6**, 102–117.

Galton, F. (1892). "Finger Prints". Macmillan & Co., London.

Geipel, G. (1937). *Z. Morphol. Anthropol.* **36**, 330–361.

Geipel, G. (1952). *Z. Morphol. Anthropol.* **44**, 70.

Geipel, G. (1961). *Z. Morphol. Anthropol.* **51**, 333–338.

Geipel, G. (1963). *Z. Morphol. Anthropol.* **54**, 71–81.

Glanville, E. V. (1965). *A. Ge. Me. Ge.* **14**, 295–304.

Goddard, C. H. (1950). *Finger Print and Identification Magazine* **31**, 4–5.

Grüneberg, H. (1928). *Z. indukt. Abstamm. u. Verb. Lehre* **46**, 285–310.

Gupta, S. P. (1953). *The Eastern Anthropologist* **7**, 47–49.

Heindl, R. (1927). "System und Praxis der Daktyloskopie". De Gruyter & Co., Berlin.

Henry, E. R. (1937). "Classification and Uses of Finger Prints". H.M. Stationery Office, London.

Hirsch, W. (1964). *Humangenetik* **1**, 246.

Holt, S. (1952). *Ann. Eugen., Lond.* **17**, 140.

Holt, S. (1968). "The Genetics of Dermal Ridges". Charles C. Thomas, Springfield, Illinois.

Iwamoto, M. (1964). *Primates* **5**, 53–73.

Kramp, P. (1954). *Homo* **3**, 175–177.

Kutsuna, M. (1929). *J. Anthrop. Soc., Tokyo* **44**, 471–489.

Lamy, M., Frezal, J., Grouchy, J. de and Kelley, J. (1957). *Ann. Hum. Genet., Lond.* **21**, 374–396.

Langdon-Down, R. L. (1909). *Brit. Med. J.* **12**, 665.

Lazaro, C., Kolski, R., Olaizola, L. and Scvortzoff, E. (1966). *A. Ge. Me. Ge.* **15**, 307–313.

Lehmann, W. (1954). *Z. Morphol. Anthropol.* **46**, 189.

Leiber, B. (1960). *Z. menschl. Vererb. u. Konstit. Lehre.* **35**, 205.

Lestrange, M. Th. de (1966). *Bull. et. Mem. Soc. Anthrop., Paris* **10**, 103–118.

Lestrange, M. Th. de (1969). *Bull. et. Mem. Soc. Anthrop., Paris* **5**, 251–267.

MacArthur, J. W. (1938). *Hum. Biol.* **10**, 12–35.

Malpighius, M. (1686). "De Externo Tactus Organo". London.

Matukura, T. (1953). *Jap. J. Leg. Med.* **7**, 290–292.

Matukura, T. (1954). *Jap. J. Leg. Med.* **8**, 443–444.

Mavalwala, J., Wilson, M. G. and Parker, C. E. (1970). *Amer. J. Phys. Anthropol.* **32**, 443–449.

McKusick, V. A. (1969). "Human Genetics". Prentice-Hall, New Jersey.

McKusick, V. A. (1968). "Mendelian Inheritance in Man". The John Hopkins Press, Baltimore.

Meyer-Heydenhagen, G. (1934). *Z. Morphol. Anthropol.* **33**, 1.

Mitra, A. K., Chattopadhyay, P. K., Dashsharma, P. and Bardhan, A. (1966). *Humangenetik* **2**, 25–27.

Monga, I. P. S. (1954). *The Anthropologist* **1**, 39–49.

Müller, B. (1930). *Deutsch. Z. Ges. Gericht. Med.* **17**, 407.

Newman, M. T. (1936). *Hum. Biol.* **8**, 531–552.

Okajima, M. (1966). *Z. Morphol. Anthropol.* **58**, 1–37.

Okajima, M. (1967a). *Am. J. Hum. Genet.* **19**, 660–673.

Okajima, M. (1967b). *Jap. J. Leg. Med.* **21**, 536–537.

Okajima, M. (1970). *Am. J. Phys. Anthropol.* **32**, 41–48.

Okamoto, K. (1958). *Finger Print and Identification Magazine* **39**, 15–18.

Okamoto, K. (1959). *Hanzai to Kanshiki* **55**, 4–7.

Ökrös, S. (1965). "The Heredity of Papillary Patterns". Akademiai Kiado, Budapest.

Penrose, L. S. (1949). *Hereditas suppl. vol.* 412.

Penrose, L. S. (1954). *Ann. Hum. Genet.* **19**, 10.

Penrose, L. S. (1963a). *Nature, Lond.* **197**, 933.

Penrose, L. S. (1963b). *Ann. Hum. Genet.* **27**, 183.

Penrose, L. S. (1965a). *Proc. XI Int. Congr. Genet.* **3**, 973.

Penrose, L. S. (1965b). *Nature, Lond.* **205**, 544.

Penrose, L. S. (1968). *Birth Defects OAS, National Foundation—March of Dimes* **4**, 1–13.

Ploetz-Radmann, M. (1937). *Z. Morphol. Anthropol.* **36**, 281–310.

Poll, H. (1914). *Z. Ethnol.* **46**, 87–105.

Pons, J. (1959). *Amer. J. Hum. Genet.* **11**, 252.

Pons, J. (1964). *Ann. Hum. Genet.* **27**, 273.

Portius, W. (1937). *Z. Morphol. Anthropol.* **36**, 382–390.

Purkinje, J. E. (1823). "Commentatio de examine physiologico organi visus et systematis cutannei". Československà Akademie, Bratislava.

Rife, D. C. (1952). *Hum. Biol.* **24**, 53–56.

Sarker, S. S. (1961). *Z. Morphol. Anthrop.* **51**, 212–219.

Schiono, H., Bandou, K. and Kadowaki, J. (1969). *Tohoku J. exd. Med.* **99**, 45–50.

Schultz, A. H. (1968). *In* "Perspectives in Human Evolution" (S. L. Washburn and P. C. Jay, eds). Holt, Rinehart and Winston.

Senet (1906). "La dactiloscopie y la Herencia". Bericht für den Kongress für Kriminologie in Turin.

Seth, P. K. (1963). *Man* **63**, 189–190.

Shuttleworth, G. E. (1909). *Brit. Med. J.* **12**, 665.

Smith, G. F. (1964). *J. Ment. Defic. Res.* **8**, 125.

Sonntag, C. F. (1924). "The Morphology and Evolution of the Apes and Man". J. Bale and Co., London.

Steffens, C. (1965). *Anthropol. Anz.* **29**, 243–249.

Stockis (1908). *In* "The Heredity of Papillary Patterns" (S. Ökrös, ed). Akademiai Kiado, Budapest.

Taylor, A. (1968). *J. Med. Genet.* **5**, 227–252.

Thompson, H. (1965). *Am. J. Med. Sci.* **250**, 718–735.

Tillner, I. (1953). *Z. Menschl. Vererb. u. Konstit. Lehre* **32**, 56–67.

Tillner, I. (1958). *Anthropol. Anz.* **22**, 260–276.

Walker, J. F. (1941). *J. Hered.* **32**, 279–280.

Walker, N. F. (1957). *J. Pediat.* **50**, 19.

Walker, N. F. (1958). *Pediat. Clin. N. Amer.* **5**, 531–543.

Walter, H. (1956). *Homo* **8**, 26–34.

Weinand, H. (1937). *Z. Morphol. Anthrop.* **36**, 418.

Weninger, M. (1935). *Mitt. Anthrop. Ges. Wien* **55**, 182.

Weninger, M. (1947). *Mitt. öst. Ges. Anthrop.* **73**, 55.

Weninger, M. and Navratil, L. (1957). *Mitt. Anthrop. Ges. Wien* **87**, 1–21.

Whipple, I. L. (1904). *A. Morphol. Anthrop.* **7**, 261–368.

Wilde, A. G. de (1953). "De Grondslagen der Overerving van het Vingerpatroon". The Hague.

Wilder, H. H. (1902). *Amer. J. Anat.* **1**, 423–441.

Wilder, H. H. (1904). *Amer. J. Anat.* **3**, 387–472.

Wilder, H. H. (1908). *Anat. Anz.* **32**, 193–200.

Wilder, H. H. (1919). *J. Hered.* **10**, 410–420.

Wood Jones, F. (1929). "Man's Place among the Mammals". Edward Arnold, London.

Würth, A. (1937). *Z. Morph. Anthrop.* **36**, 187–214.

Phenylthiourea Testing in Primates

H. KALMUS

Galton Laboratory, University College London, England

INTRODUCTION

Among the more tractable entities which can be considered by comparative geneticists, are the "homologous series of variation" (Vailov, 1922) or what we now call similar genetical polymorphisms in related species. Such polymorphisms can be observed at many levels of phenotypical organization in proteins and pigments, tissue and organ diversity, in form or function.

The taster polymorphism which is the subject of this chapter, can at present only be explored for operational purposes by behavioural investigations, although it has metabolic and probably hormonal connotations. The peculiar difficulties arising from this situation will be now described.

THE TASTER POLYMORPHISM IN MAN

"Taste blindness" against bitter substances containing a $N-C = S$ group was first described by Fox (1932). It has since been independently discovered several times and an extensive literature of specialist and review papers on the subject exists. At present it appears that the taster polymorphism in Man is largely determined by an autosomal gene so that one homozygote (TT) and the heterozygote (Tt) are highly sensitive to solutions of any of those substances while the other homozygotes (tt) are insensitive (non-tasters). In European and Indian populations between one-fifth to one-third are non-tasters, while amongst the Africans and Amerindians and others very few non-tasters occur. The $N-C = S$ substances have antithyroid (goitrogenic) properties and some of them occur in cruciferous vegetables. Statistical correlations of taster status and thyroid diseases have been established in several populations, but the causal relation between clinical manifestation and taster status is by no means clear (Kalmus, 1961).

PHENYLTHIOUREA (PTC) TESTS IN APES

During the International Congress of Genetics at Edinburgh in 1939, Fisher *et al.* (1939) taste-tested some chimpanzees, orang outangs, gorillas

and gibbons in the Zoological Gardens of Edinburgh, London and Whip-
snade and found that 20 out of 27 chimpanzees rejected sugar solutions
containing various concentrations of Phenylthiourea. The remaining seven
chimpanzees accepted the solutions and the animals were declared "tasters"
and "non-tasters" respectively. Both reactions were observed among the
orangs and gibbons, but of the two tested gorillas, both were apparently
tasters.

From these slender observations the three authors drew the following
far-reaching conclusions:

"The existence of a dimorphism in the taste test, parallel to that known
in man, in at least two of the manlike apes, is a most remarkable fact. Its
significance is emphasized by the circumstance that, certainly in the
chimpanzee, the proportion of tasters to non-tasters is nearly the same
as in human populations, which, as Boyd has shown, vary somewhat
among themselves. Without the conditions of stable equilibrium it is
scarcely conceivable that the gene-ratio should have remained the same
over the million or more generations which have elapsed since the separation of the anthropoid and hominid stocks. The remarkable inference
follows that over this period the heterozygotes for this apparently value-
less character have enjoyed a selective advantage over both the homozygotes, and this, both in the lineage of the evolving chimpanzees and
in that of evolving man. Wherein the selective advantages lie, it would at
present be useless to conjecture, but of the existence of a stably balanced
and enduring dimorphism determined by this gene there can be no room
for doubt."

The Fisher *et al.* communication (1939) was the starting point for all
following enquiries into the problem of taste polymorphism in apes and
monkeys. Their findings and conclusions also got into several textbooks
and were quoted as examples of a long-lasting polymorphism.

However, doubts were also expressed. Barnicot pointed out (in a lecture)
that the proportion of tasters to non-tasters in chimpanzees may be similar
to Europeans, but that Africans are very rarely non-tasters.

More seriously, doubts arose about the factual findings, and in particular
their interpretation, as soon as other experimentalists tried to verify and
to expand the original reports. Some of these are expounded in a paper by
the author (Kalmus, 1969) in which the methodology of taste testing in
primates and most of the results obtained until then are also described.

Methods

Humans have been tested by putting PTC crystals or paper squares
impregnated with PTC solution on their tongues, which gives quite in-

accurate results, or by letting them taste or drink small quantities of graded solutions. A particular procedure described by Harris and Kalmus (1949) is now almost universally used for humans (Weiner and Lowrie, 1969). For animals, infants and mental defectives, however, other less efficient methods must be used. Solutions have been tried for primates (Chiarelli, 1959), and fruit or sugar cane sprinkled with PTC crystals or soaked in PTC solutions have been used as well as ice cream into which certain amounts of PTC solution had been incorporated. These presentations are all satisfactory. The real difficulties stem from a quite different source; the flexibility, ambiguity and capriciousness of primate behaviour.

THE CRITERIA FOR TASTE SENSATION IN PRIMATES

Some reactions of apes and monkeys to strong taste stimuli are so similar to those of humans, and in particular unhibited humans (i.e. infants or certain defectives) that anthropomorphic inferences concerning their sensations are perfectly acceptable. Grimacing, spitting, even tantrums of a young chimpanzee immediately after tasting a bitter morsel or sipping of a bitter liquid can be accepted as signs of rejection and by implication as signs of a taste sensation and so can the refusal of accepting a further sample after such an experience. Other reactions are more difficult to interpret and may often go unnoticed. Thus the author has observed, the orang outangs after placidly munching some PTC or quinine containing ice cream, would quietly take what is left out of their mouth and put it on the ground —occasionally to try and taste it again. This was never seen to happen with ice cream without a bitter admixture. The behaviour is reminiscent of the reactions of some mongoloid imbeciles, who when asked would happily agree that a strong PTC-solution was "tasting nice" but could afterwards be observed to wipe their tongues with a handkerchief. The assessment of overt behavioural reactions to weaker stimuli (solutions or impregnated morsels) is that even changes of prediliction can influence the actions of the animals and the measures based on these actions. Thus a liquid may very well "have a taste" for a monkey without this being detected in a whole series of experiments.

The usual indirect methods, by which preference is inferred from the relative consumption of bitter drink or food as compared with the controls, suffer from many difficulties, as expounded previously by the author (Kalmus, 1969).

In these circumstances it is tempting to try objective methods, which will in most circumstances indicate whether a stimulus has or has not triggered a peripheral transducer—in our case some taste buds on the tongue, even if this does not tell us much about the centrally generated sensation. Thus if a certain PTC solution put on the tongue of one individual—a taster— produces an action potential from the chorda tympani, while the same procedure does not have this result when applied to another individual—a non-taster—we might conclude that the difference between the two is peripheral and specific; such a conclusion would be made even more cred- ible if a quinine solution elicits positive electro-physiological reactions in both individuals. Experiments of this kind have incidentally been per- formed during operations on patients suffering from otosclerosis, the PTC thresholds of whom were known by the ordinary method of taste-testing (Zotterman, personal communication) but have so far not been performed on primates. Consequently we are at present still restricted to behavioural methods.

RESULTS AND CONCLUSIONS

The primates taste-tested since the appearance of the paper by Fisher *et al.* were inmates of several Zoological Gardens and primate centres in Western Europe and North America, and a few animals in Africa. Many of the individuals and small family groups were investigated by all the investigators working during this period, namely Chiarelli (1959, 1963), Eaton and Gavan (1965) and Kalmus (1969). The results obtained by these authors mostly agree but it is difficult to be sure as to which individuals had been repeatedly tested. Therefore numbers from the various papers and notes are not given in the following.

The results may be grouped into (i) sample studies of not closely related individuals in various primate species and (ii) pedigree studies of small groups of relatives.

Sizeable population samples in respect of PTC testing have been reported for 4 groups only, namely *Papio* sp. (75), *Cercopithecus* sp. (102), *Macaca* sp. (68) and *Pan* sp. (104). Of these only *Pan*, the chimpanzee, shows un- equivocal bimodal threshold distributions, though the antimodes between tasters and non-tasters of the several authors vary somewhat as do the population frequencies between the two phenotypes. The samples of the other species were too small for establishing—or excluding—a bimodal threshold distribution, though many of them show a wide span of individual taste thresholds. In a few species no tasters occurred, for instance in *Ateles* (sample of 10) and *Lagothrix* (12), while from 8 *Callithrix* and 8 *Leonto- cebus* all were "tasters". Most species sampled contained both "pheno- types". It must be pointed out, however, that in the absence of a clear cut bimodality a division into "tasters" and "non-tasters" is impossible and simple modes of inheritance very improbable.

Small groups of related animals have so far been investigated in only 13 species. In the chimpanzee offspring of two non-taster parents is reported to produce only non-tasters, which is in agreement with the situation in Man, where non-tasting is a recessive autosomal character. The numbers in the family groups of the other species as well as in hybridization experiments (Chiarelli, 1963b) are too small to allow any definite conclusions.

It thus remains reasonable to summarize all the evidence with some modest, if tentative, positive statements. Taster polymorphisms, as shown by PTC taste-testing may have existed and been maintained in the order of primates for many millions of years, though we cannot be certain whether it is or was always caused by homologous genes. Even assuming that we are dealing with frequencies at homologous "loci" in the various species it is still not possible to infer an unequivocal evolutionary tree from frequency considerations (Kalmus, 1970, this volume). On the other hand it is very probable that the tasting character in the primates may in some way be connected with the diet of the various primate species at different times and in particular with food plants (mostly *Brassicae*) containing NC = S substances. It has been demonstrated by Astwood *et al.* (1949) that these substances are goitrogenic, and that the taster status of humans is affecting certain thyroid diseases has been established by Harris *et al.* (1949) and later confirmed by others (Kalmus, 1961). Whether selection of a thyroid function through nutritional factors is the instrumental factor of maintaining and modifying the taster polymorphism in the various primate species is, however, far from certain.

REFERENCES

Astwood, E. B., Greer, M. A. and Ettlinger, M. G. (1949). L-S-Vinyl-2-thioxazolidone, an antithyroid substance from yellow turnips and from Brassica seeds. *J. Biol. Chem.* **181**, 121–127.

Chiarelli, B. (1959). Sensibilità alla Phenil-Thiocarbamide da parte delle Scimmie. *Arch. anthropol. etnol.* **89**, 1–9.

Chiarelli, B. (1963a). Sensitivity to PTC (Phenylthiacarbamide) in Primates. *Folia primatol.* **1**, 88–94.

Chiarelli, B. (1963b). Observations on PTC tasting and on hybridization in primates. *Symp. Zool. Soc. Lond.* **10**, 277–279.

Eaton, J. W. and Gavan, J. A. (1965). Sensitivity to PTC among primates. *Amer. J. Phys. Anthropol.* **23**, 381–388.

Fisher, R. A., Ford, E. B. and Huxley, T. (1939). Taste testing the anthropoid ape. *Nature* **144**, 750–51.

Fox, A. L. (1932). The relationship between chemical constitution and taste. *Proc. Nat. Acad. Sci.* **18**, 115–120.

Harris, H. and Kalmus, H. (1949). The measurement of taste sensitivity to phenylthiourea (PTC). *Ann. Eugen., Lond.* **15**, 32–45.

Harris, H., Kalmus, H. and Trotter, W. R. (1949). Taste sensitivity to phenyl-thiourea in goitre and diabetes. *Lancet* **2**, 1038–1041.

Kalmus, H. (1959). Genetical variation and sense perception. *CIBA Symp. on Biochemistry and Human Genetics* 60–72.

Kalmus, H. (1960). "The Chemical Senses in Health and Disease". C. C. Thomas, Springfield, Ill.

Kalmus, H. (1961). Genetical taste polymorphism and thyroid disease. *Proc. 2nd Intern. Congr. Hum. Genet., Rome* 1856–1862.

Kalmus, H. (1969). The sense of taste of chimpanzees and other primates. *In* "The Chimpanzee", Vol. 2, pp. 130–141. Karger, Basel, New York.

Vavilov, N. I. (1922). The law of homologous series in variation. *J. Genet.* **12**, 47–89.

Weiner, J. S. and Lowrie, J. A. (ed.) (1969). Human Biology. A Guide to Field Methods. Blackwell, Oxford, Edinburgh. IBP Handbook, No. 9, pp. 115–122.

Blood Groups of Non-human Primates and Their Relationship to the Blood Groups of Man*

ALEXANDER S. WIENER AND J. MOOR-JANKOWSKI

Department of Forensic Medicine, New York University of Medicine, New York, U.S.A.; Laboratory for Experimental Medicine and Surgery in Primates (LEMSIP), New York University School of Medicine, New York, U.S.A.

The heredity of the blood groups in Man has been investigated by family studies along the lines of classical Mendelian genetics, and also by studies on the distribution of the blood groups using the Hardy-Weinberg formula, i.e., by the method of population genetics (Wiener and Wexler, 1958; Wiener, 1962; Prokop and Uhlenbruck, 1968). Since blood groups closely resembling those of Man have been found in apes and monkeys, a third approach to the problem has become available, namely, the investigation of human-type blood factors in non-human primates (Moor-Jankowski *et al.*, 1964).

Conversely, in this paper, inferences will be drawn in the opposite direction concerning the heredity of the blood groups in non-human primates. This approach is particularly important because breeding of non-human primates in captivity with identified parents is only in its incipient stages, and production of families for genetic studies is still far away because of the relatively long sexual maturation time and the infrequency of multiple births in primates, in contrast to the situation with other common experimental animals.

All areas of primate genetics suffer from these same limitations. However, the advantage of blood groups as a tool for investigating primate genetics is that they are one of the longest known and the most thoroughly investigated genetic markers not only in Man but also in non-human primates.

THE A–B–O BLOOD GROUPS AND BASIC PRINCIPLES OF BLOOD GROUP GENETICS

Man

The A–B–O blood groups, discovered by Karl Landsteiner (1901), are determined by 2 antigens arbitrarily designated as A and B. Depending on

* Supported by USPHS, NIH grants GM09237 and 12074.

which of these agglutinogens are present (or absent) on the red cells, 4 blood groups exist, namely, O, A, B and AB, where the symbol O represents the absence of A and B. The A–B–O groups are unique among the human blood groups in the regular occurrence of isoagglutinins of specificities* anti-**A** and anti-**B** in the serum in conformity with Landsteiner's law, namely, those isoagglutinins and only those are present in the serum for which the corresponding agglutinogens are absent on the red cells. Thus, the complete formulas of the 4 blood groups are O$\alpha\beta$, Aβ, Bα and ABo, where small Greek letters are used to represent isoagglutinins and capital letters to represent agglutinogens.

The existence in Man of all 4 of the possible A–B–O blood groups led earlier workers to postulate 2 independent pairs of genes, *A–a* for agglutinogen A and its absence, and *B–b* for agglutinogen B and its absence, and family studies appeared at first to support this hypothesis. It remained for Bernstein (1925), using the methods of population genetics, to introduce the presently accepted multiple allele theory, with gene *A* determining agglutinogen A, gene *B* determining agglutinogen *B*, and gene *O* an amorph, determining neither A nor B. Thus, 6 genotypes were possible: group O, genotype *OO*; group A, genotypes *AA* and *AO*; group B, genotypes *BB* and *BO*; group AB, genotype *AB*.

According to the multiple allele theory, group AB parents could not have group O children, and group O parents could not have group AB children, but this could occur if the blood groups were inherited by independent pairs of genes. Analysis of the published family studies up to 1925 was misleading because of the large number of errors in blood grouping giving rise to numerous apparent exceptions to the Bernstein theory. Bernstein pointed out, however, that if agglutinogens A and B were inherited by separate gene pairs, they should be distributed in the population independently of one another and thus show no association.

A contingency table is drawn up as shown below, where *a*, *b*, *c* and *d* represent the number of persons of groups AB, B, A and O, respectively, in a population of size *N*.

	A+	A−	Totals
B+	*a*	*b*	*a + b*
B−	*c*	*d*	*c + d*
Totals	*a + c*	*b + d*	*N*

* To avoid ambiguity, symbols for serological specificities are printed in **bold** type, symbols for genes and genotypes in *italic* type, while symbols for agglutinogens and phenotypes are printed in regular type.

Then, absence of any association implies that $ad - bc = 0$, i.e., $O \times \overline{AB} = \overline{A} \times \overline{B}$, where $\overline{O}, \overline{A}, \overline{B}$ and $\overline{AB}$ are the frequencies of the 4 blood groups in the population. However, this relationship among the frequencies of the blood groups was found by Bernstein not to hold; instead $\overline{O} \times \overline{AB} < \overline{A} \times \overline{B}$.

In presenting his theory, Bernstein assumed that the distribution of the allelic genes in the population was the same in males and females and that mating occurred at random (panmixia). If germ cells bearing the genes A, B and O are assigned the frequencies p, q, and r, respectively, so that $p + q + r = 1$ or 100%, then random fertilizations produce 6 kinds of zygotes with the following frequencies: $OO = r^2$; $AA = p^2$; $AO = 2pr$; $BB = q^2$; $BO = 2qr$ and $AB = 2pq$, as shown by the table below.

		Sperm	
	$O = r$	$A = p$	$B = q$
$O = r$	r^2	pr	qr
$A = p$	pr	p^2	pq
$B = q$	qr	pq	q^2

Ova

This obviously makes use of the extended Hardy-Weinberg formula, $(p + q + r)^2 = 1$.

Thus, the distribution of the 4 blood groups in terms of the 3 gene frequencies must satisfy the following formulae:

$$\overline{O} = r^2 \tag{1}$$
$$\overline{A} = p^2 + 2pr \tag{2}$$
$$\overline{B} = q^2 + 2qr \tag{3}$$
$$\overline{AB} = 2pq \tag{4}$$

The Bernstein theory is tested by using the 4 equations above to estimate the 3 gene frequencies, thus leaving one degree of freedom. With the aid of these estimated gene frequencies, the expected distribution of the four blood groups is determined and then the x^2 value calculated using the standard formula, $x^2 = \sum \frac{(x - x_0)^2}{x_0}$, where x_0 is the expected number of persons of a given A–B–O group and x the observed number. Then P, the likelihood of a x^2 value of the observed magnitude is determined by consulting tables, taking d.f. (degrees of freedom) equal to 1 (Documenta Geigy). The estimated gene frequencies used must be those that give the smallest x^2 value, and a computer program for obtaining such "maximum likelihood" estimates of gene frequencies has been devised by Dr John Edwards.* Calculations using data on the distribution of A–B–O blood groups available for numerous human populations, support the validity of the Bernstein theory, and justify our application of similar calculations to non-human primates, as pointed out later.

The regular occurrence of isoagglutinins anti-**A** and anti-**B** is attributed to the ubiquitous presence in many animals, plants and micro-organisms of A-like and B-like substances, so that exposure to such antigens with the resulting production of antibodies is virtually inevitable (Wiener, 1951; Springer, 1970). These isoantibodies exist not merely as a single uniform species of molecules, but as an entire spectrum of antibody molecules as has been shown in fractionating them by absorption with appropriate red cells.† In this way von Dungern and Hirszfeld (1911) discovered the sub-groups of A, the most important of which are A_1 and A_2. These subgroups are detected in that anti-**A** serum (from group B persons) absorbed with A_2 cells retains a fraction of antibodies designated as anti-**A$_1$**. Group A red cells clumped by this anti-**A$_1$** reagent are subgroup A_1; those that fail to react are A_2. To account for the 6 resulting groups, O, A_1, A_2, B, A_1B and A_2B, 4 allelic genes must be invoked, O, A^1, A^2 and B. As will be shown later, the existence of homologous subgroups of A in some of the anthro-poid apes is one of the common characteristics which points up the paral-lelisms among these genetic markers in Man and non-human primates.

Bernstein's theory led to a search for an anti-**O** reagent which would be specific for the product of gene O and which could be useful for genotyping, e.g. for distinguishing between group B persons of genotypes BB and BO. Schiff's discovery (1927) that bovine sera contained an antibody specific for group O red cells was hailed as the predicted anti-**O**; similar reagents were obtained later from eel sera and extracts of seeds (lectins) like Ulex europeus. However, family studies failed to confirm the supposed anti-O specificity. Moreover, the reagents reacted regularly with red cells of sub-group A_2 and less regularly with blood of other blood groups (Wiener et al., 1966).

Presently such reagents are designated anti-**H** and it is considered that agglutinogen H is of almost universal distribution in Man due to homo-zygosity for a gene H independent of the A–B–O genes.

Due to the work of Kabat (1956) and of Watkins (1970), the difference

---

* Formulae derived by Bernstein which are suitable for the χ^2 test can be found in the book of Wiener (1962).

† This general phenomenon of fractionation of antibodies by absorption with red cells has been used by us to demonstrate differences as well as similarities between human agglutinogens and the corresponding agglutinogens in non-human primates, e.g. M, N, Rh, as will be shown later.

in serological specificity between agglutinogens A and B has been traced to different terminal monosaccharides in the agglutinogen molecule, namely, N-acetyl-D-galactosamine for agglutinogen A and D-galactose for agglutinogen B. The respective genes A and B are postulated to produce corresponding enzymes (transferases) which transfer the appropriate simple sugar from a donor molecule to the blood group precursor molecule. However, no evidence for a gene dose effect has been elicited whereby, for example, red cells from group B individuals of the 2 different genotypes BB and BO could be distinguished. Wiener and Karowe (1944) has suggested that the serological differences between A_1 and A_2 are due to the greater length of the subjacent heterosaccharide chain in A_1 blood, resulting in less steric interference and enabling agglutinogen A_1 to combine with a wider range of agglutinins from the anti-A spectrum of antibodies. It is significant that the same gene is not equally expressed in all blood groups, e.g. the reactivity of subgroup A_2B (genotype A^2B) red cells with anti-A reagents is considerably weaker than that of subgroup A_2 (genotype A^2O) even though both genotypes have the same number of A^2 genes. This can be explained by the postulation that whereas gene O is an amorph, genes A^2 and B compete for a common substrate; in addition there may also be steric interference between their respective determinant groups on the agglutinogen molecule (Wiener and Karowe, 1944).

The already mentioned reciprocal relationship of H to A–B, i.e. the usual absence of H reaction on red cells of group A_1B and the regular presence of H on red cells of group O, has been attributed by Watkins (1970) to the action of gene H to produce a blood group H precursor substance which is converted by genes A and B to their respective gene products; thus, since gene O is an amorph the H substance remains unconverted and is therefore fully expressed in group O blood (Watkins, 1970). An alternative hypothesis, proposed by Wiener et al. (1966) on the basis of our work on non-human primates, to be discussed later, is that gene H and genes $A–B$ compete for a common precursor substance, and that the reciprocity is due to steric interference between the corresponding determinant groups, as in the case of A_2B blood already discussed. For completeness it is mentioned that gene H (like A and B) is believed to produce an enzyme (transferase) which transfers a simple sugar (this time L-fucose) from a donor molecule to the acceptor precursor blood group substance.

The A–B and H blood group substances are not restricted to the red cells but occur also in water-soluble form in body fluids, and in high concentrations in secretions of mucus-producing glands. Not all individuals produce significant amounts of these water-soluble blood group substances —only the so-called secretors, who represent 75–80% of Caucasoids and 60–70% of Negroids.

While the terminal simple sugar responsible for blood group specificity appears to be the same for the blood group substance in secretions as well as on the red cells, the carrier molecules differ in that on red cells the blood group substances are lipopolysaccharides (glycolipids) while in secretions they are mucopolysaccharides (glycoproteins). All secretors have in their secretions the blood group substance H together with the appropriate A or B substance. The tests for secretor type are usually carried out on saliva using the inhibition technique. Family studies (Schiff and Sasaki, 1932) prove that the secretor type is transmitted by a pair of allelic genes *Se* and *se*, located on a pair of chromosomes different from that of the A–B–O genes, so that a person of genotype *sese* is a non-secretor. For completeness, it must be mentioned that rare individuals exist of the so-called "Bombay" type, who have no H, A or B on their red cells or in their secretions, due presumably to homozygosity for a rare "suppressor" gene, but have anti-**H** as well as anti-**A** and anti-**B** in their sera.

Chimpanzee

Among the anthropoid apes, chimpanzees have been studied most intensively by us—we have typed a total of about 250 animals. As was demonstrated clearly by Landsteiner and Miller (1925) more than 40 years ago, chimpanzee blood gives isoagglutination reactions indistinguishable from those of human blood. Their tests, as well as our extensive tests, however, have revealed the presence of only 2 blood groups, O and A. The red cells of almost all group A chimpanzees are clumped by anti-A_1 reagents, though not as strongly as human A_1 cells; of these about one-tenth react more weakly but not quite as weakly as human A_2 cells. Thus, chimpanzees have subgroups of A, and the reactivity of chimpanzee red cells compared to those of Man can be arranged as follows (Wiener, 1965):

$$A_1 \text{human} > A_1 \text{chimpanzee} > A_2 \text{chimpanzee} > A_2 \text{human}$$

As far as chimpanzee agglutinins are concerned, Landsteiner's rule holds, though a few chimpanzees of group A did not show the expected anti-**B**. In group O chimpanzees the isoagglutinin titers were almost regularly higher than in group A chimpanzees (Wiener and Gordon, 1960; Wiener et al., 1963). This behavior, encountered also in Man, has been attributed by Wiener (1953) and Wiener and Ward (1966) to the presence in the serum of group O of antibodies of the so-called anti-**C** specificity lacking from the serum of individuals of other groups.

No clear-cut serological differences have been demonstrated by reciprocal absorption tests between chimpanzee and human group A red cells. Certain differences were observed, however, in tests with anti-**H** reagents: the great majority of red cells of group O chimpanzees were clumped by anti-**H** reagents, but to distinctly lower titers than human O cells, and some chimpanzee O red cells failed to agglutinate at all. In this respect, therefore, chimpanzee group O cells resemble human group O cells from newborn babies. Red cells of group A chimpanzees, even those of sub-group A_2, in general were not clumped by anti-**H** reagents.

Saliva from about 100 randomly chosen chimpanzees has been tested by us; all the animals proved to be secretors, i.e., the group A chimpanzees had A and H in their saliva while group O chimpanzees had only H.

As in Man, there are "racial" differences in the distribution of the blood groups in chimpanzees (cf. Table I). The subdivision of this sample of 91 chimpanzees into 4 racial groups was carried out in agreement with morphological characteristics as established by W. C. Osman Hill (Moor-Jankowski, et al., 1966). Schmitt (1968) has tested 6 dwarf chimpanzees, *Pan paniscus*, of which one was group O and 5 were group A. Four dwarf chimpanzees tested by us all reacted as group A, and, unlike *Pan troglodytes*, gave reactions entirely comparable to human subgroup A_1.

TABLE I. Distribution of the human-type A-B-O blood groups among 4 sub-species of chimpanzees

Subspecies[a]	O		A		Totals	Gene frequencies	
	Number	%	Number	%		O	A
Pan troglodytes troglodytes	2	14·3	12	85·7	14	37·9	62·1
Pan troglodytes schweinfurthi	13	39·3	20	60·6	33	62·7	37·3
Pan troglodytes verus	4	9·5	38	90·5	42	30·8	69·2
Pan troglodytes koola-komba	1	—	1	—	2	—	—
All chimpanzees tested, including those above[b]	33	14·47	195	85·53	228	38·0	62·0

[a] Classified on the basis of morphological characteristics, as established by W. C. Osman Hill.
[b] The gene frequencies calculated for the total series are artificial because they deal with a racially non-homogeneous population.

Gibbon

We have found that isoagglutination occurs regularly in gibbons in conformity with Landsteiner's rule, giving rise to blood groups serologically indistinguishable from the A–B–O groups in Man. Among 52 gibbons (*Hylobates lar lar*) Wiener et al. (1968) found 10 group A, 20 group B and 22 group AB, so that group O either does not occur among these animals or is quite rare. If one assumes that group O is absent, one need postulate only 2 alleles, A and B, so that all the group A and group B gibbons would have to be homozygous. Therefore, the gene frequencies can readily be estimated by direct count as follows:

$$p(A) = \overline{A} + \tfrac{1}{2}\overline{AB} = 0 \cdot 1924 + \tfrac{1}{2}(0 \cdot 4230) = 0 \cdot 4039 \text{ or } 40 \cdot 4\%$$

Similarly,

$$q(B) = \overline{B} + \tfrac{1}{2}\overline{AB} = 0 \cdot 3846 + \tfrac{1}{2}(0 \cdot 4230) = 0 \cdot 5961 \text{ or } 59 \cdot 6\%$$

If the expected distribution of the blood groups among the 52 gibbons is now calculated using these gene frequencies, one finds that $X^2 = 0 \cdot 58$, and since d.f. $= 1$, P $= 0 \cdot 5$, approximately, which is in good support of the genetic theory.

Tests with anti-A_1 reagents demonstrated the existence in gibbons subgroups of A serologically indistinguishable from the A_1–A_2 subgroups in Man. Among the same 52 gibbons the distribution of the subgroups proved to be: $A_1 = 10$; $A_2 = 0$; $B = 20$; $A_1B = 20$ and $A_2B = 2$. The seeming paradox of the occurrence of subgroup A_2B despite the absence of subgroup A_2 among the 52 gibbons demonstrates the value of gene frequency analysis, without which this finding might have been attributed incorrectly to laboratory error. Proportion of A_2B among the AB's $=$

$$\frac{\overline{A_2B}}{\overline{A_1B} + \overline{A_2B}} = \frac{2p_2q}{2p_1q + 2p_2q} = \frac{p_2}{p_1 + p_2}$$

Similarly,
$$\frac{\overline{A_2}}{\overline{A_1} + \overline{A_2}} = \frac{p_2^2}{(p_1^2 + 2p_1p_2) + p_2^2} = \left(\frac{p_2}{p_1 + p_2}\right)^2$$

Thus, the proportion of A_2 among the A's is the square of the proportion of A_2B among the AB's.

Since one-eleventh of our group AB gibbons were subgroup A_2B, only one among $(11)^2$ or one among 121 group A gibbons could reasonably be expected to be subgroup A_2, but so far our series has included only 10 group A gibbons. Obviously, many more gibbons may have to be tested before a subgroup A_2 gibbon is found.

All gibbons from whom saliva was obtained proved to be secretors; H was regularly present in the saliva in high titer together with A or B or both, depending on the blood group. On gibbon red cells, H was regularly demonstrated in group B animals; in fact, group B gibbon red cells reacted strongly with anti-H in titers equal to those of human group O red cells. Gibbon red cells of groups A and AB reacted with anti-H less frequently and generally in lower titers (cf. Table II).

TABLE II. Relationship of blood factor **H** on the red cells to the A-B-O blood groups

Species	Blood group	H reactivity[a]			Totals
		Strong	Moderate	Weak or negative	
Man	O	168	0	0	168
	A₁	1	42	66	109
	A₂	30	1	0	31
	B	6	50	14	70
	A₁B	0	0	17	17
	A₂B	1	2	1	4
Chimpanzees	O	0	15	2	17
	A	0	2	48	50
Gibbons	A	0	0	3	3
	B	11	0	0	11
	AB	0	2	8	10
Orang utans	A	0	0	22	22
	B	0	0	1	1
	AB	0	0	3	3

[a] Numbers represent the number of individuals.
From Wiener et al. (1966).

Four gibbons of the subspecies *Hylobates lar pileatus* have been tested—all were group B and secretors. Since 20 of the 52 gibbons of subspecies *Hylobates lar lar* were group B, the chance of having 4 group B randomly selected gibbons in succession should be $(20/52)^4$ or about 1 in 100, which makes it reasonable to conclude that this was due to a racial difference in the distributions of the A–B–O blood groups. This calculation again demonstrates the value of population genetics studies, based on experience in Man and applied to findings in non-human primates. The division of gibbons into subspecies by means of population genetics was later confirmed by our cross-immunization experiments (Moor-Jankowski et al., 1965) between the 2 species, and by the difference in hemoglobin types observed by Hoffman and Gottlieb (1969). It was our findings on gibbons as well as the already described racial subdivision of chimpanzees that led to the establishment of the science of seroprimatology (Moor-Jankowski and Wiener, 1967).

Orang Utan (*Pongo pygmaeus*)

In orang utans, as in chimpanzees and gibbons, we have observed regular isoagglutination reactions in conformity with Landsteiner's rule. Among 26 animals we found 22 group A, 1 group B and 3 group AB. The group A and AB orang utan red cells reacted regularly with anti-A_1 reagents but generally somewhat more weakly than subgroup A_1 red cells of Man. Thus the subgroups of A and Man and apes may be arranged according to reactivity as follows:

A_1(Man and gibbon) > A_1(orang utan) > A_1(chimpanzee)
$\quad\quad\quad\quad\quad\quad > A_2$(chimpanzee) > A_2(Man and gibbon)

Since the determinant chemical group of group A appears to be the same irrespective of subgroup, namely, N-acetyl-D-galactosamine, the observations in primates support Wiener's hypothesis that the various subgroups are due to differences in the length of the genetically determined subjacent heterosaccharide chain.

In contrast to the situation in gibbons and Man, the red cells of all 26 orang utans failed to react with anti-**H** reagents, irrespective of their A–B–O blood groups, even though strong inhibition reactions for H were given by the saliva (Wiener et al., 1966). As shown in Table II, group B orang utan red cells failed to react with anti-**H**, group B gibbon red cells reacted regularly and as strongly as human red cells of group O, whereas group B human red cells gave all grades of reactions from negative to moderately strongly positive. This lack of any consistent relationship between B and H reactivity in anthropoids and in Man tends to refute the idea that H is the precursor of B (or A). Instead, it seems that the relative position of the determinant groups on the blood group substances could be the decisive factor in these serological reactions; this could readily occur as a result of biochemical evolution in the molecular structure of homologous blood group substances in different species. Our findings in orang utans and chimpanzees indicate that the precursor for the action of the *H* gene can be present in the secretions while absent on the red cells of one and the same animal, even when this does not hold for the *A–B* genes.

Gorilla

In contrast to the situation in other anthropoid apes, our tests failed to show any isoagglutination reactions when the red cells and sera of 11 lowland gorillas (*G. gorilla gorilla*) were cross-tested one against the other (Wiener et al., 1971). In addition, gorilla red cells fail to react with anti-**A** or anti-**B** sera, or with anti-**H** lectin, so that gorillas evidently have no A–B–H blood group substances on their red cells. Nevertheless, all the lowland gorillas tested by us had anti-**A** but not anti-**B** in their sera as was readily demonstrated in tests against human red cells of groups A and B.*

---

* Before such tests are performed the gorilla sera must be absorbed with human group O red cells in order to remove anti-human species heteroagglutinins.

Inhibition tests on the saliva of the same gorillas showed the regular presence B and H group substances and the absence of A. Thus, all lowland gorillas tested were of group B and Landsteiner's rule does hold, but the reciprocal relationship is between serum antibodies and blood group substances in secretions instead of on red cells. Accordingly, the A–B–O blood groups of gorillas differ from those of other anthropoid apes and of Man, and resemble those of monkeys, which will be described later.

Only 2 mountain gorillas (*G. gorilla berengei*) have been examined, both by Candela *et al.* (1940) to whom only urine samples were available. Inhibition tests showed both animals to be group A.

Old World Monkey (Catarrhina)

Numerous tests carried out on blood of many species of monkeys have failed to demonstrate the presence of any regularly occurring isoagglutination reactions comparable to those of Man, chimpanzee, gibbon and orang utan. Moreover, tests with anti-**A**, anti-**B** and anti-**H** reagents indicate that A–B–H blood group substances are absent on the red cells. Nevertheless, it was found by tests in various species that certain monkeys had only anti-**A** in their sera while others had only anti-**B**, while still others had both or neither. This led Wiener *et al.* (1942) to inquire whether the A–B–H blood group substances, not found on their red cells, might nevertheless be present in their body fluids and secretions. Indeed, he did find, by tests on saliva, the presence of the expected A, B and H substances. Thus, for Old World monkeys the Landsteiner rule does hold but in modified form, in that the reciprocal relationship holds for antibodies in the serum and blood group substances in secretions. As has already been shown the same situation was found by us to exist in gorillas.

Among the Old World monkeys some species such as baboons and crab-eating macaques have all 4 groups, O, A, B and AB, other species, notably rhesus monkeys have only group B, while in others, such as patas monkeys, only group A was found.

From the point of view of genetics, the most interesting results were obtained in baboons, of whom the largest number of animals have been tested. In the initial investigations (Moor-Jankowski *et al.*, 1964) only groups A, B and AB were found, leading us to the conclusion that group O was probably absent, so that, as in orang utans and gibbons, all the group A and group B baboons would have to be homozygous. It was puzzling therefore when in a few small families of baboons a group A mother was found with a group B baby, and a group B mother with a group A offspring (Moor-Jankowski *et al.*, 1964). Application of the methods of population genetics seemed at first to fit the hypothesis of heredity by a pair of allelic genes, A and B, except for a slight deficit of group AB animals in comparison with the expected number. Later among a total of about 680 baboons tested 3 were found (Wiener and Moor-Jankowski, 1969) which were clearly of group O.

Table III compares the distribution of the A–B–O blood groups among 4 different species of baboons. So far in only one of these species, *Papio papio*, have group O baboons been found. For this species, using the method of gene frequency analysis already described, the X^2 test shows an

TABLE III. A-B-O blood groups of baboons[a]

Species of baboons		O	A	B	AB	Totals	O	A	B
			Blood groups				Estimated gene frequencies		
Papio cynocephalus	No.	0	18	20	22	60			
	%	0	30·0	33·3	36·7	100·0	8·9	44·0	47·1
Papio anubis	No.	0	53	56	65	174			
	%	0	30·5	32·2	37·4	100·0	8·4	45·0	46·6
Papio ursinus (South Africa)	No.	0	4	59	26	89			
	%	0	4·5	66·3	29·2	100·0	1·8	18·8	79·3
Papio papio (Senegal)	No.	2	27	93	66	188			
	%	1·1	14·4	49·5	35·1		10·6	29·9	59·5
Papio (species undetermined)	No.	1	42	65	65	173			
	%	0·6	24·3	37·6	37·6		10·0	39·0	51·0

[a] Modified after Wiener and Moor-Jankowski (1969).

excellent fit between the observed distribution and the expected distribution under the theory of heredity by triple alleles O, A, and B, as in Man. With respect to the other 3 species there are two possibilities, namely, either that group O is absent, or else present in such low frequency that no such animals have yet been encountered. This would be comparable to the already discussed situation for group O and subgroup A_2 in gibbons. Maximum likelihood calculations on an IBM machine using a program devised by John Edwards gave an excellent fit under the hypothesis of triple alleles, and a far less satisfactory fit under the assumption that the gene O was missing. A simple algebraic method of making such estimations of gene frequencies when the frequency of group O is very low has been devised by Wiener as follows

Since $$\overline{A} = p^2 + 2pr$$

and $$\overline{B} = q^2 + 2qr$$

Then $\overline{A} - \overline{B} = p^2 - q^2 + 2r(p - q) = (p - q)(p + q + 2r)$

$$= (p - q)(1 + r)$$

Therefore, if r is quite small

$$p - q = \overline{A} - \overline{B}, \text{ approximately}$$

but $2pq = \overline{AB}$

Solving these 2 equations for p and q and then taking $r = 1 - (p + q)$, fairly good estimates of the gene frequencies are obtained which are close to the maximum likelihood values given by Edward's computer program. These calculations, as shown in Table III, leave no reasonable doubt that all 4 species of baboons have group O, but the frequency is so low that by chance in 3 of the species no group O animals have been found by us so far.

New World Monkey (Platyrrhina)

The principles that have been described for Old World monkeys appear to hold in New World monkeys, though with certain important exceptions. These exceptions as well as other serological reactions indicate that New World monkeys are less similar to Man, and since this has made them of less interest to us, we have tested smaller numbers of these monkeys of various species. An important difference between the *Platyrrhina* and the higher primates is that in *Platyrrhina* the Landsteiner's rule holds with less regularity. Moreover, as first shown by Landsteiner and Miller (1925), several species of New World monkeys have a B-like agglutinogen on their red cells quite apart from their human-type A–B–O blood groups as determined by tests on their serum and saliva. The small number of animals of each species tested precludes the application even of the methods of population genetics, but differences in genetics appear to account for the different serological findings.

What has been stated regarding Old World monkeys applies to an even greater degree for the relatively few prosimians tested.

LEWIS TYPES

These blood types are of interest and importance because of their close serological, genetical and biochemical relationship to the A–B–O blood groups. We have therefore carried out extensive studies on the Lewis types in apes and monkeys which have been reviewed in some of our recent publications (Wiener *et al.*, 1964; Wiener and Shapiro, 1965). Since no new insight has since been acquired, the reader is referred to our earlier papers for a discussion of this aspect of blood group genetics.

M–N TYPES

The M–N blood group system of Man, discovered by Landsteiner and Levine (1928) is now known to be quite complex, and comprises at the present stage of knowledge more than 20 distinct serological specificities (blood group factors). In our studies on apes and monkeys, we have re- stricted our investigation to the original 2 agglutinogens, M and N. As early as 1937, it was realized that the serological behavior even for this very limited range of reactions is exceedingly complex (Wiener, 1938). For example, anti-**M** reagents prepared from different rabbit immune sera which give parallel reactions in tests on human blood are not necessarily all identical in specificity when testing red cells of non-human primates. In these early studies it was already found that the red cells of chimpanzees all have M-like agglutinogens. However, not all anti-**M** reagents aggluti- nated chimpanzee red cells and in the case of anti-**M** reagents that did, at least 2 different fractions of anti-**M** could be separated by absorption. One fraction of anti-**M** was reactive for both human M cells and chimpanzee red cells while the other fraction was reactive only for human M cells. Similar fractions could be separated from anti-**N** sera by absorption with chimpanzee red cells (Wiener, 1938). When the experiments were extended to other apes and to monkeys, the situation was found to be considerably more complex (Landsteiner and Wiener, 1937).

A particularly useful reagent in our studies in the M–N types of non- human primates has been the lectin from *Vicia graminea* seed, discovered by Ottensooser and Silberschmidt (1953) to have the specificity anti-**N** in tests on human red cells; we prefer to designate this reagent as anti-**N**[V] lectin.

Ape

Perhaps the most instructive results for M–N types in apes have been obtained by us in gibbons. We have found that some anti-**M** sera agglutin- ate the red cells of all gibbons; other anti-**M** sera do not agglutinate red cells of any gibbon, while still other anti-**M** sera agglutinate red cells of some gibbons but not of others. Thus, all gibbons apparently have some kind of M-like antigen on their red cells, whereas some but not all gibbons have a different kind of agglutinogen M which is detected only by parti- cular anti-**M** sera. This latter agglutinogen M of gibbons may be considered homologous to agglutinogen M in Man, but it has only some of the numerous serological specificities of the human M agglutinogen. Anti-**N** sera can be classified into similar 3 categories with regard to their reactions with gibbon red cells. For showing blood type differences among gibbons most clearly anti-**N**[V] lectin, though not of high titer, proved to be the most suitable. Tests with anti-**M** reagents selected for their ability to show blood type differences among gibbons alongside with anti-**N**^{V} lectin disclose 3 blood types in gibbons, (M)^{Gi}, (N)^{Gi} and (MN)^{Gi}, homologous to the 3 types, M, N and MN in Man. However, it must be emphasized that whereas the A–B–O groups of gibbons are serologically indistinguishable from the A–B–O groups of Man, the 3 M–N types of gibbons are readily distinguishable serologically from the M–N types in Man, since only *particular* anti-**M** and anti-**N** reagents are usable for typing gibbon blood.

Among 52 gibbons (*Hylobates lar lar*), Wiener *et al.* (1968) found in this way 9 type M, 31 type N and 12 type MN. If, as in Man, heredity of the M–N types of gibbons is by a pair of alleles, the gene frequencies could be determined by direct count as follows

$$M = \overline{M} + \tfrac{1}{2}\overline{MN} = 0\cdot1730 + \tfrac{1}{2}(0\cdot2308) = 0\cdot2884 \text{ or } 28\cdot8\%$$

and $\quad N = \overline{N} + \tfrac{1}{2}\overline{MN} = 0\cdot5961 + \tfrac{1}{2}(0\cdot2308) = 0\cdot7119 \text{ or } 71\cdot2\%$

Using these gene frequencies to calculate the expected number of the 3 types according to the Hardy-Weinberg law, one finds $X^2 = 10\cdot5$ for only 1 degree of freedom so that $P = 0\cdot001$, approximately. Thus, the observations do not fit well with the genetic expectations, mainly due to a deficit of the type MN and a corresponding excess of type M. Four possible explanations may be considered: (i) because of the limited potency of our anti-**N**^{V} reagent some of type MN gibbons might have been erroneously classified as type M; (ii) our series was composed of a number of sub-series of animals and if all these sub-series did not have the same distribution of M–N types the conglomerate would show a shortage of type MN when the Hardy-Weinberg formula was used; (iii) since the M–N types of gibbons are serologically different from the M–N types in Man, the genetic mechanism could also be different; a new genetic theory for the M–N types which could explain the results also in gibbons will be presented at the end of this chapter; (iv) the Hardy-Weinberg formula is applicable only to populations undergoing panmixia (random mating), and we do not know the breeding pattern of the gibbons, which is not necessarily one of panmixia.

The situation in chimpanzees with respect to the M–N types is different. As has been pointed out, with most anti-**M** reagents the red cells of chimpanzee react strongly, like human blood having agglutinogen M. However, since anti-**M** reagents exist that do not agglutinate chimpanzee red cells, it is evident that all chimpanzees have on their red cells an agglutinogen similar to but not identical with the human agglutinogen M. With regard to tests for N, most of the anti-**N** reagents tested by us fail to agglutinate chimpanzee red cells. However, anti-**N**^{V} lectin and occasional anti-**N** rabbit immune sera agglutinate the red cells of about 40% of the chimpanzees tested. Thus, 2 types are defined by tests on chimpanzee red cells with anti-**M** and anti-**N** reagents, $(MN_1)^{Ch}$, positive for N^V factor, and $(MN_2)^{Ch}$, negative for N^V factor.

In orang utans (Wiener *et al.*, 1964), tests with anti-**M** rabbit immune sera suitably absorbed with orang utan red cells to remove species-specific heteroagglutinins, revealed polymorphism in that some orang utans were M positive while others are M negative. That the M agglutinogen of orang utans though very similar to, is not identical with the human agglutinogen M was demonstrated by tests with 3 kinds of anti-**M** reagents prepared from sera of rabbits immunized with human M red cells, chimpanzee red cells and baboon red cells, respectively. By absorption tests using the anti-chimpanzee serum it could be shown that orang utan M has serological specificities lacking from human M blood (Wiener *et al.*, 1964). In tests with anti-N^V lectin, on the other hand, red cells of all orang utans gave negative reactions. Thus tests with anti-**M** and anti-**N** reagents on orang utan red cells have disclosed only 2 blood types, $(M)^{Or}$ or M positive, and $(m)^{Or}$ or M negative.

In gorillas, the situation is quite different (Wiener *et al.*, 1971). Firstly, the red cells of all gorillas tested were strongly agglutinated by anti-N^V lectin. Moreover, the red cells of most but not all gorillas have an agglutinogen serologically very similar to but not identical with the human agglutinogen M. Thus, based on tests with anti-**M** and anti-**N** reagents, only 2 blood types can be distinguished in gorillas tentatively designated as MN^V and N^V, respectively.

To summarize, our investigations on apes with anti-**M** and anti-**N** reagents disclose both similarities and differences from the reactions of human red cells. Only in gibbons were 3 types found analogous to the 3 M–N types of Man, and then only by use of selected anti-**M** and anti-**N** reagents. In the other apes only 2 types could be defined: in chimpanzees all had M-like red cells but showed polymorphism for N^V, in orang utans none had N^V but the red cells showed polymorphism for M, whereas in lowland gorillas all had N^V but the red cells showed polymorphism for M. A general genetic theory for the M–N types which encompasses the situation in apes as well as in Man will be briefly described at the end of this section (Wiener, 1970).

Monkey

Wiener was led to investigate the presence of the agglutinogen M on monkey red cells as long ago as 1937 by conflicting reports, some of which asserted that M was present on the red cells of rhesus monkeys, while others asserted that it was absent. This paradox was resolved by the demonstration that red cells of Old World monkeys and some species of New World monkeys have agglutinogens similar to but not identical with the human agglutinogen M as shown by their ability to react with some but not all anti-**M** reagents (Landsteiner and Wiener, 1937; Wiener, 1938). Depending on the number of anti-**M** reagents used and the number of species of monkeys tested the existence of a multiplicity of anti-**M** specificities in anti-human M reagents prepared in rabbits could be demonstrated, even though all the reagents gave parallel reactions in tests with human red cells. By arranging their findings as shown in Table IV, Wiener (1938) and Landsteiner and Wiener (1937) were able to show that according to the number of serological specificities shared by the red cells of monkeys species with human type M red cells, Old World monkeys are more closely related to Man than are New World monkeys. Tests on red cells of all monkeys with anti-**N** reagents including anti-N^V lectin have up to now consistently given negative results.

TABLE IV. The reactions of red cells of apes and monkeys[a] with various anti-**M** reagents[b]

| Red cells | Rabbit anti-**M** sera | | | | | |
	No. 1	No. 2	No. 3	No. 4	No. 5	No. 6
Human M	+++	+++	++	++	++	++
Human N	—	—	—	—	—	—
Human MN	+++	+++	++	I I	++	++
Chimpanzee *Pan troglodytes*	+++	+++	++	++	++	—
Rhesus monkey *M. mulatta*	++	++	++	+	—	—
Green monkey *Cercopithecus pygerythrus*	+++	+++	—	—	—	—
Black spider monkey *Ateles ater*	+++	—	—	—	—	—
White spider monkey *Ateles*	+	—	—	—	—	—
Capuchin monkey *Cebus capucinus*	—	—	—	—	—	—

[a] The blood samples of these animals were obtained in 1937 from the Central Park Zoo, New York, and the exact species in the case of the white spider monkey could not be ascertained.

[b] Abridged and modified from Wiener (1938).

D

To account for the observations on the M–N blood factors of non-human primates, as well as of Man (cf. Table V), Wiener (1970) has proposed a new genetic theory. According to this theory, M and N instead of being determined by allelic genes are each determined by separate pairs of genes located on separate pairs of chromosomes. In Man, all individuals are assumed to be homozygous for *N*, i.e., genotype *NN*, but there is polymorphism for M producing three genotypes *MM*, *Mm* and *mm*, corresponding to the 3 phenotypes M, MN and NN respectively. Actually, all persons appear to have N on their red cells but the degree of agglutinability

TABLE V. The M and N blood factors in apes

| Species | M-like factors | | |
	Present	Absent	Totals
Chimpanzees			
(*Pan troglodytes*)	130	0	
Gibbons			
(*Hylobates lar lar*)	21	31	52
(*Hylobates lar pileatus*)	4	0	4
Orang utans			
(*Pongo pygmaeus*)	12	12	24
Gorillas			
(*Gorilla gorilla gorilla*)	9	3	12
	N^V factor (*Vicia graminea*)		
Chimpanzees			
(*Pan troglodytes*)	42	62	104
Gibbons			
(*Hylobates lar lar*)	43	9	52
(*Hylobates lar pileatus*)	0	4	4
Orang utans			
(*Pongo pygmaeus*)	0	24	24
Gorillas			
(*Gorilla gorilla gorilla*)	12	0	12

by anti-**N** reagents varies according to the degree of steric interference from the M combining group which may or may not be present and in varying degrees depending on zygosity. The relationship of N to M, therefore, is analogous to the relationship of H to A–B. In fact by using anti-**H** and anti-**C** (obtainable for certain catfish) 3 types of human blood can be defined comparable to the 3 M–N types. In orang utans only two types are defined, (M)^or and (m)^or, because all animals are homozygous for the absence of N. In chimpanzees, on the other hand, all animals appear to be homozygous for the presence of M but there is polymorphism for $N(N^v)$. Only in gibbons is the situation such that 3 types are defined as in Man. In Old World monkeys, again, all animals may be assumed to be homozygous for the absence of N but homozygous for presence of M, but the M genes and their corresponding products appear to be qualitatively different according to the species (Wiener et al., 1966, 1968).

Rh–Hr TYPES

The Rh–Hr types of primates including Man present a paradox in that whereas the first anti-**Rh** reagents were prepared by immunizing rabbits with red cells of rhesus monkeys, all anti-**Rh** reagents made in Man, irrespective of titer, avidity and specificity, fail to react with red cells of any monkey species, including those of rhesus monkeys. For this reason, as will be seen, our studies on the Rh–Hr types of non-human primates has been limited mainly to apes.

We have already pointed out how Wiener demonstrated clearly for the first time the presence of M-like antigens on the red cells of rhesus monkeys (Wiener, 1938). This led Landsteiner and Wiener (1937) to immunize rabbits with red cells of rhesus monkeys, and in this way to produce reagents having the specificity anti-**M** in tests on human red cells. Continuing this line of investigation, Landsteiner and Wiener (1940) found that by absorbing the anti-rhesus monkey rabbit sera with human M cells, antibodies could be fractionated that agglutinated the red cells of about 85% of human Caucasoid individuals, thus defining a blood type hitherto unknown in man (Landsteiner and Wiener, 1940). Due to the manner in which it had been discovered, the human agglutinogen detected by the anti-rhesus reagent was named Rh; persons having the agglutinogen are Rh positive while those lacking it are Rh negative (Wiener, 1954, 1969).

Subsequent study by Wiener and Peters (1940) of human sera from patients having hemolytic reactions following transfusions of blood of the correct A–B–O group, disclosed the presence of isoantibodies giving reactions with human red cells paralleling those of the animal anti-rhesus sera. Therefore, these antibodies have also been designated anti-**Rh** antibodies. Subsequent study on sera of mothers of erythroblastotic babies (Burnham, 1941; Levine et al., 1941) provided still another source of anti-**Rh** sera and since the human anti-**Rh** sera proved to be of higher titer and avidity than the animal anti-rhesus monkey sera, they are the preferred reagent at present. In fact, Wiener found that the titer and avidity of the sera from human patients could be raised by an injection of a small amount of Rh-positive blood, and he also showed that larger amounts of potent sera could be even more conveniently produced by deliberate immunization of previously non-sensitized volunteer male blood donors (Wiener, 1949; Wiener and Sonn-Gordon, 1947).

Further study of sera from human patients having transfusion reactions or erythroblastotic babies showed that not all gave identical reactions. Using newly developed methods of testing, isoantibodies of new specificities were discovered (Wiener, 1954). Most of these proved to be related to the original rhesus factor and were assigned appropriate symbols to indicate this. Thus, while the original most antigenic Rh factor (serological specificity) is now called Rh_o, other less antigenic related factors exist, namely, **rh′**, **rh″**, **rh**^{W}, **Rh**^{A}, **Rh**^{B}, etc., as well as reciprocally related factors **hr′**, **hr″**, **hr**, etc. It is noteworthy that while **rh′–hr′** and **rh″–hr″** make up reciprocally related pairs, no factor reciprocally related to the original, most antigenic factor Rh_o has been found, although factor **hr** closely approximates the specifications of the hypothetical factor Hr_o. Detailed tables presenting in logical order the serology, genetics and nomenclature of the Rh–Hr types in Man have been prepared by Wiener (1967; Wiener and Wexler, 1963), and are readily available in up to date and authoritative textbooks on the subject.

As many as 30 distinct Rh–Hr blood factors have been identified to date in tests on human red cells. In our studies on apes and monkeys, however, the tests have been limited mainly to human reagents of the specificities anti-Rh_o, anti-**rh′**, anti-**rh″**, anti-**hr′** and anti-**hr″**, as well as the original anti-rhesus monkey experimental serum (Wiener and Gordon, 1961). It should be mentioned, however, that in place of anti-rhesus monkey rabbit sera, it has been found that more satisfactory reagents can be prepared by immunization of guinea pigs (Landsteiner and Wiener, 1941). Moreover, we have found that the injection of guinea pigs with baboon red cells yield reagents of similar specificity to those obtained with injections of rhesus monkey blood (Wiener et al., 1969). Thus, whatever is said in this discussion for anti-rhesus monkey sera applies also to the anti-baboon guinea-pig sera.

As has already been pointed out none of the human reagents, anti-Rh_o, anti-**rh′**, anti-**rh″**, anti-**hr′** or anti-**hr″** agglutinates monkey red cells. That monkeys, at least baboons and rhesus monkeys, must nevertheless have some sort of Rh-like agglutinogen follows from the fact that usable anti-**Rh** reagents can be produced by immunizing rabbits or guinea pigs with red cells of baboons or rhesus monkeys. In contrast, red cells of all apes except gibbons react to a greater or lesser degree with anti-Rh_o human reagents (Wiener et al., 1964); red cells of all apes react with anti-**hr′** but not with anti-**rh′**, whereas red cells of all apes tested fail to react with either anti-**rh″** or anti-**hr″**. The absence of both the pair of factors **rh″** and **hr″** from the red cells of all apes, despite the reciprocal relation between the two factors in Man, has been indicated by us by placing a bar above the $\bar{R}$ (capital or lower case) of the symbol for the apes Rh type.

In contrast to the situation for the A–B–O groups, the red cells of gorillas most closely resemble those of Man in their reaction for the Rh–Hr blood factors (Wiener *et al.*, 1971). Red cells of all gorillas, as indicated above, are strongly agglutinated by anti-**hr'** reagents, to about the same titers as human **hr'**-positive red cells. On the other hand, while red cells of most gorillas react strongly with anti-**Rh$_0$** reagents, a minority react only weakly. Absorption of human anti-**Rh$_0$** reagents with the weakly-reacting gorilla red cells yields a reagent which reacts only with the more strongly reacting gorillas red cells, and to the same titer as with human **Rh$_0$**-positive cells. Thus, 2 types of gorilla red cells have been defined, namely, **Rh$_0$** positive and a second type which is either **Rh$_0$** negative or a weakly reacting **Rh$_0$** variant. Treatment of **Rh$_0$**-positive gorilla red cells with anti-**Rh$_0$** serum coats gorilla red cells to the same degree as human **Rh**-positive red cells, indicating that the number of Rh combining groups on the surface of gorilla red cells is of the same order of magnitude as on human red cells, and on chimpanzee red cells (Wiener and Gordon, 1953). Summarizing, therefore, tests with Rh reagents on gorilla red cells define 2 types which may be designated $\bar{R}h^{Go}$ and $\overline{\mathfrak{R}}h^{Go}$ (or, possibly, $\bar{r}h^{Go}$), respectively. That the gorilla Rh$_0$ agglutinogen while similar to the human Rh$_0$ agglutinogen is not identical with the human Rh$_0$ agglutinogen follows from the fact that the anti-rhesus monkey sera react only weakly with gorilla **Rh$_0$**-positive red cells.

Chimpanzee red cells, like gorilla red cells, react regularly with anti-**hr'** sera, and to almost the same titers as human **hr'**-positive red cells. Moreover, like gorilla red cells, they are not agglutinated by anti-**rh'**, anti-**rh''**, or anti-**hr''** sera. On the other hand, red cells of all chimpanzees are agglutinated by all human anti-**Rh$_0$** reagents. Thus all chimpanzees may be designated as type $\bar{R}h_0^{Ch}$. That the Rh$_0$ agglutinogen of chimpanzee differ from the human Rh$_0$ agglutinogen follows from (i) the failure of chimpanzee red cells to agglutinate at all with anti-rhesus monkey guinea-pig sera, (ii) that absorption of anti-**Rh$_0$** human sera with chimpanzee red cells leaves behind a fraction of **Rh$_0$** antibodies reactive for human **Rh$_0$**-positive red cells though not reactive for chimpanzee red cells. More recently we have found that when anti-**Rh$_0$** sera of human origin are absorbed with red cells of selected chimpanzees, a reagent results which determines blood type differences also in the chimpanzee (Wiener *et al.*, 1966). It should be noted, however, that in contrast to the situation in gorilla, the red cells of chimpanzee used for absorption do not necessarily show a lower titer in the direct tests with anti-**Rh$_0$** reagents. It is of interest that these specially absorbed anti-**Rh$_0$** reagents give parallel reactions in tests on chimpanzee red cells with a chimpanzee isoimmune reagent defining the simian-type blood factor c^c of the chimpanzee C–E–F system. Since the present review deals only with the human-type blood groups of non-human primates, the reader is referred to our previous publications for information regarding this C–E–F system.

In gibbons, the tests on red cells with Rh–Hr antisera have uniformly given negative reactions with anti-Rh_0, anti-rh', anti-rh'' and anti-hr'' sera, but positive reactions with anti-hr'. Therefore the symbol $\bar{r}h^{Gi}$ is used to designate the Rh–Hr of gibbon red cells. Red cells of orang utans, on the other hand, reacted positively with anti-Rh_0 and anti-hr' sera and negatively with the other reagents, and the orang utan Rh–Hr type has therefore been assigned the symbol $\bar{Rh}_0^{or}$.

Summarizing, among the apes, the red cells of gorillas give reactions with Rh–Hr anti-sera most closely resembling those of Man, and are the only species that exhibit clear polymorphism in direct tests with anti-Rh_0 reagents. Red cells of all apes lack *both* factors of the pair rh''–hr'' even though these two factors are antithetical in Man. Red cells of gibbons give reactions closely resembling that of human Rh-negative blood, as indicated by the designation $\bar{r}h^{Gi}$, in contrast to the symbol $\bar{Rh}_0$ used for blood from other apes. Actually, the symbol $\Re h_0$ is preferable to $\bar{Rh}_0$ since absorption tests indicate that Rh-positive ape red cells have many but not all of the serological specificities of the human Rh agglutinogen. Red cells of rhesus monkeys and baboon fail to react with *any* Rh–Hr reagents even though they must have some kind of Rh agglutinogen on them as was proved by the original production of anti-Rh reagents by the immunization of rabbits and guinea pigs with red cells of rhesus monkeys. Thus, there are differences as well as similarities among the Rh agglutinogens of Man and non-human primates, and this is almost surely reflected by corresponding differences in their genetics.

REFERENCES

Bernstein, F. (1925). Zusammenfassende Betrachtungen über die erblichen Blutstrukturen den Menschen. *Ztschr. indukt Abstamm. u. Vererbungs.* **37**, 223.

Burnham, L. (1941). The common etiology of erythroblastosis and transfusion reactions in pregnancy. *Amer. J. Obstet. Gynecol.* **42**, 389.

Candela, P. B., Wiener, A. S. and Goss, L. J. (1940). New observations on the blood group factors in Simidii and Cercopithecidae. *Zoologica* **25**, 513.

Documenta Geigy. Wissenschaftliche Tabellen. 7. Auflage (Konrad Diem and Cornelius Lentner, eds). J. R. Geigy A.G., Pharma, Basel.

Hoffman, H. A. and Gottlieb, A. J. (1969). Investigations of non-human primate hemoglobin: electrophoretic variation. *Ann. N.Y. Acad. Sci.* **162/1**, 205.

Kabat, E. A. (1956). "Blood Group Substances". Academic Press, New York.

Landsteiner, K. (1901). Ueber Agglutinationserscheinungen normalen menschlichen Blutes. *Klin. Wchnschr.* **14**, 1132.

Landsteiner, K. and Levine, P. (1928). On individual differences in human blood. *J. Exp. Med.* **47**, 757.

Landsteiner, K. and Miller, C. P., Jr. (1925a). Serological studies on the blood of primates. II. The blood groups in anthropoid apes. *J. Exp. Med.* **42**, 853.

Landsteiner, K. and Miller, C. P., Jr. (1925b). Serological studies on the blood of the primates. III. Distribution of serological factors related to human isoagglutinogens in the blood of lower monkeys. *J. Exp. Med.* **42**, 863.

Landsteiner, K. and Wiener, A. S. (1937). The presence of M agglutinogens in the blood of monkeys. *J. Immunol.* **33**, 19.

Landsteiner, K. and Wiener, A. S. (1940). An agglutinable factor of human blood recognizable by immune sera for rhesus blood. *Proc. Soc. Exp. Biol. Med.* **43**, 223.

Landsteiner, K. and Wiener, A. S. (1941). Studies on an agglutinogen (Rh) of human blood reacting with anti-rhesus sera and human agglutinins. *J. Exp. Med.* **74**, 309.

Levine, P., Burnham, L., Katzin, E. M. and Vogel, P. (1941). The role of iso-immunization in the pathogenesis of erythroblastosis fetalis. *Amer. J. Obstet. Gynecol.* **42**, 925.

Moor-Jankowski, J. and Wiener, A. S. (1967a). Sero-primatology: a new discipline. Neue Ergeb. Primatol. Progress in Primatol, pp. 373–381. First Congress Internat. Primatol. Soc., Frankfurt, 1966. Gustav Fischer, Stuttgart.

Moor-Jankowski, J. and Wiener, A. S. (1967b). Blood groups of apes and monkeys: human-type and simian-type. Neue Ergebnisse der Primatologie. "Progress in Primatology", pp. 382–410. First Congress of the International Primatological Society. Frankfurt, 1966. Gustav Fischer, Stuttgart.

Moor-Jankowski, J., Wiener, A. S. and Gordon, E. B. (1964a). Blood groups of apes and monkeys. I. The A-B-O blood groups in baboons. *Transfusion* **4**, 92.

Moor-Jankowski, J., Wiener, A. S. and Rogers, C. (1964b). Human blood group factors in non-human primates. *Nature, Lond.* **202**, 663.

Moor-Jankowski, J., Wiener, A. S. and Gordon, E. B. (1965). Simian blood groups. Two new blood factors of gibbon blood. A^g and B^g. *Transfusion* **5**, 235.

Moor-Jankowski, J., Wiener, A. S., Kratochvil, C. H. and Fineg, J. (1966). Chimpanzee blood groups; blood group distribution as racial characteristics. *Science, N.Y.* **152**, 219.

Ottensooser, F. and Silberschmidt, K. (1953). Hemagglutinin anti-N in plant seeds. *Nature, Lond.* **172**, 914.

Prokop, O. and Uhlenbruck, G. (1968). "Human Blood and Serum Groups". Wiley Interscience, New York.

Schiff, F. (1927). Ueber den serologischen Nachweis der Blutgruppeneigenschaft O. *Klin. Woch.* **6**, 303.

Schiff, F. and Sasaki, H. (1932). Die Ausscheidungstypus, ein auf serologischen Wege nachweisbares mendelndes Merkmal. *Klin. Wchnschr.* **11**, 34, 1426.

Schmitt, J. (1968). Immunbiologische Untersuchungen bei Primaten. Karger, Basel, New York.

Springer, G. F. (1970). Occurrence of A-B-H like substances in micro-organisms and plants. *In* "Advances in Blood Grouping III" (A. S. Wiener, ed.). Grune & Stratton Inc, New York.

von Dungern, E. and Hirszfeld, L. (1911). Ueber gruppenspezifischer Strukturen des Blutes. *Ztschr. f. Immunitätsforch.* **8**, 526.

Watkins, W. M. (1970). Blood-group substances. *In* "Advances in Blood Grouping III" (A. S. Wiener, ed.). Grune & Stratton, New York.

Wiener, A. S. (1938). The agglutinogens M and N in anthropoid apes. *J. Immunol.* **34**, 11.

Wiener, A. S. (1949). Further observations on isosensitization to the Rh factor. *Proc. Soc. Exp. Biol. Med.* **70**, 576.

Wiener, A. S. (1951). Origin of naturally occurring hemagglutinins and hemolysins: a review. *J. Immunol.* **66**, 287.

Wiener, A. S. (1953). The blood factor **C** of the A-B-O system, with special reference to the rare blood group **C**. *Ann. Eugen.* **18**, 1.

Wiener, A. S. (1954). "Rh–Hr Blood Types. Application in Clinical and Legal Medicine and Anthropology". Grune and Stratton, New York.

Wiener, A. S. (1962). "Blood Groups and Transfusion". 3rd edition, 1943. Reprinted by Hafner Publ. Co., New York.

Wiener, A. S. (1965). Blood groups of chimpanzees and other non-human primates: their implications for the human blood groups. *Trans. N.Y. Acad. Sci.* Series II, **27**, 488.

Wiener, A. S. (1967). Elements of blood group nomenclature, with special reference to the Rh–Hr blood types. *J. Amer. Med. Assoc.* **199**, 985.

Wiener, A. S. (1969). Karl Landsteiner, M.D. History of the Rh–Hr blood group system. *N.Y. State J. Med.* **69**, 2915.

Wiener, A. S. (1970). A new theory of heredity of the M-N types. Unpublished.

Wiener, A. S. and Gordon, E. B. (1953). Quantitative test for antibody globulin coating human red cells and its practical applications. *Amer. J. Clin. Path.* **23**, 429.

Wiener, A. S. and Gordon, E. B. (1960). The blood groups of chimpanzees: A–B–O groups and M–N groups. *Amer. J. Phys. Anthropol.* **18**, 301.

Wiener, A. S. and Gordon, E. B. (1961). The blood groups of chimpanzees: The Rh–Hr (CDE/cde) blood types. *Amer. J. Phys. Anthropol.* **19**, 35.

Wiener, A. S. and Karowe, H. (1944). The diagrammatic representation of the A–B–O blood group reactions. *J. Immunol.* **49**, 51.

Wiener, A. S. and Moor-Jankowski, J. (1969). The A–B–O blood groups of baboons. *Amer. J. Phys. Anthropol.* **30**, 117.

Wiener, A. S. and Peters, H. R. (1940). Hemolytic reactions following transfusions of blood of the homologous group, with three cases in which the same agglutinogen was responsible. *Ann. Int. Med.* **13**, 2306.

Wiener, A. S. and Shapiro, M. (1965). "Advances in Blood Grouping, II", pp. 156–177. Grune and Stratton, New York.

Wiener, A. S. and Sonn-Gordon, E. B. (1947). Simple method of preparing anti-Rh serum in normal male donors. *Amer. J. Clin. Path.* **17**, 67.

Wiener, A. S. and Ward, F. A. (1966). The serological specificity (blood factor) C of the A–B–O blood groups: theoretical implications and practical applications. *Amer. J. Clin. Path.* **46**, 27.

Wiener, A. S. and Wexler, I. B. (1958). "Heredity of the Blood Groups". Grune and Stratton, New York.

Wiener, A. S., Candela, P. B. and Goss, L. J. (1942). Blood group factors in the blood, organs and secretions of primates. *J. Immunol.* **45**, 225.

Wiener, A. S., Baldwin, M. and Gordon, E. B. (1963). Blood groups in chimpanzees. *Exp. Med. Surg.* **21**, 159.

Wiener, A. S., Gordon, E. B. and Moor-Jankowski, J. (1964a). The Lewis blood groups in Man. A review with supporting data on non-human primates. *J. Forensic Med., S. Africa* **11**, 67.

Wiener, A. S., Moor-Jankowski, J. and Gordon, E. B. (1964b). Blood group antigens and cross-reacting antibodies in primates, including Man. II. Studies on the M–N types of orang utans. *J. Immunol.* **93**, 101.

Wiener, A. S., Moor-Jankowski, J. and Gordon, E. B. (1964c). Blood groups of apes and monkeys. IV. The Rh–Hr blood types of anthropoid apes and monkeys. *Amer. J. Human Genet.* **16**, 246.

Wiener, A. S., Moor-Jankowski, J. and Gordon, E. B. (1966a). The relationship of the H substance to the A–B–O blood groups. *Int. Arch. Allergy* **29**, 82.

Wiener, A. S., Moor-Jankowski, J., Gordon, E. B. and Kratochvil, C. H. (1966b). Individual differences in chimpanzee blood demonstrable with absorbed human anti-Rh_0 sera. *Proc. Nat. Acad. Sci.* **56**, 458.

Wiener, A. S., Moor-Jankowski, J., Gordon, E. B. and Shell, N. F. (1966c). Human-type blood factors in gibbons, with special reference to the multiplicity of serological specifities of human type M blood. *Transfusion* **6**, 313.

Wiener, A. S., Moor-Jankowski, J., Cadigan, F. C., Jr. and Gordon, E. E. (1968). Comparison of the A–B–H blood group specificities and M–N blood types in Man, gibbons (*Hylobates*) and Siamangs (*Symphalangus*). *Transfusion* **8**, 235.

Wiener, A. S., Moor-Jankowski, J. and Brancato, G. J. (1969). LW factor. *Haematologia, Budapest* **3**, 385.

Wiener, A. S., Gordon, E. B. and Moor-Jankowski, J. (1971). Blood groups of gorillas. In press.

Leukocyte Groups of Non-human Primates; their Relation to Histocompatibility and to Human HL-A Antigens*

H. BALNER†, B. W. GABB‡, H. DERSJANT, W. v. VREESWIJK,
A. v. LEEUWEN§, AND J. J. v. ROOD§

Radiobiological Institute TNO, Rijswijk Z.H., The Netherlands

INTRODUCTION

Recent interest in the leukocyte antigens of man (Ceppellini *et al.*, 1967; Dausset *et al.*, 1967; Histocompatability Testing, 1970; Kissmeyer-Nielsen and Thorsby, 1970; Payne *et al.*, 1964; van Rood and van Leeuwen, 1963) has stimulated the study of tissue antigens of other primate species. It has been shown that chimpanzee leukocytes share numerous antigenic determinants with white blood cells of man and that chimpanzee iso-antisera can be used for human tissue typing (Balner *et al.*, 1967c; 1970c; Metzgar and Zmijewski, 1966; Shulman *et al.*, 1965). Further evidence for the sharing of antigens comes from the relatively long survival of chimpan-zee organs grafted to man (Reemtsma *et al.*, 1946). These observations make the chimpanzee a particularly attractive species to study the phylogeny of leukocyte antigens. Efforts have therefore been made to define the chim-panzee's own system of histocompatibility antigens and to compare this

* Part of the research was supported by a contract from the Dutch Government and contract nr. F 6 1052 67 C 0099 from the European Office of the Aerospace Research, Brussels. B. W. Gabb holds a C. J. Martin Travelling Fellowship from the N.H. and M.R.C. of Australia. Part of the work of the Department of Immuno-haematology was performed with the financial support of the National Institutes of Health (contract PH 43-65-992), the Dutch Organization for Pure Research (FUNGO), the Dutch Organization for Applied Research (TNO), the J. A. Cohen Institute for Radiology and Radiation Protection, the Whitehall Foundation and the World Health Organization. Euratom Publication No. 652.

† Member of the Biology Division of Euratom.

‡ J. A. Cohen Institute for Radiology and Radiation Protection, Leyden.

§ Department of Immunohaematology, University Hospital, Leyden, the Netherlands.

with the human HL–A system. Tissue antigens of lower monkeys, such as *M. Rhesus*, have also been studied because these animals are often used in transplantation research for which tissue typing is a prerequisite. Considerable interest in the histocompatibility antigens of lower monkeys exists also in connection with the phylogeny of leukocyte antigens. In fact, suggestive evidence has already been obtained showing that certain leukocyte specificities of Rhesus monkeys may have antigenic determinants in common with the two "precursor" specificities of human leukocytes, 4a and 4b (Balner *et al.*, 1967a). Thus, a comparison of the major histocompatibility antigens of Rhesus monkeys, chimpanzees and man should contribute to the knowledge of the comparative genetics of primates. In this chapter we shall describe the current knowledge about the histocompatibility antigens of Rhesus monkeys (RhL–A), then review the available data for leukocyte specificities of chimpanzees (ChL–A). This will be followed by a comparison of the reactivity patterns of chimpanzee typing sera with patterns of human sera of HL–A specificity and a brief discussion of the phylogeny of primate tissue antigens in general.

THE LEUKOCYTE ANTIGENS OF RHESUS MONKEYS

The study of leukocyte antigens of Rhesus monkeys started in 1964. At that time the relevance of leukocyte antigens to histocompatibility had not yet been proven in primates. Evidence for such relevance was first obtained in skin grafting experiments on immunized monkeys (Balner *et al.*, 1965) and shortly thereafter, with the same method, for leukocyte antigens in man (Dausset *et al.*, 1965). The results shown in Fig. 1 indicate that skin grafts from donors whose leukocytes were agglutinated or "killed" by the sera from pre-immunized recipients were rejected in an accelerated fashion. This provided evidence that the specificities recognized by the recipient's sera were transplantation antigens (Balner and Dersjant, 1965). Clearly, these reagents had to be thoroughly analysed before the leukocyte specificities in question could be identified.

Following these observations, isoantisera were produced by random and "planned" immunizations. Over the past 5 years the available sera were used to type between 80–150 monkeys annually. Each year, reactivity patterns were analysed by computer and the better reagents sorted into groups recognizing single or associated specificities; occasionally some absorptions were performed to improve the specificity of the reagents (Balner *et al.*, 1967a). By the end of 1969, 12 groups of sera, each having similar reactivity patterns, were available. The leukocyte specificities identified by these groups of sera were numbered 1–12. Figure 2 shows the computer

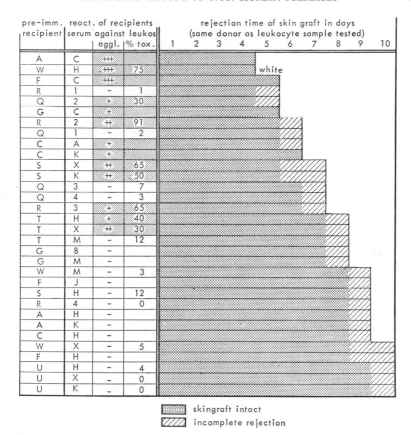

pre-imm. recipient	react. of recipients serum against leukos aggl.	% tox.	rejection time of skin graft in days (same donor as leukocyte sample tested)
A	C	+++	
W	H	+++	75
F	C	+++	
R	1	−	1
Q	2	+	30
G	C	+	
R	2	++	91
Q	1	−	2
C	A	+	
C	K	+	
S	X	++	65
S	K	++	50
Q	3	−	7
Q	4	−	3
R	3	+	65
T	H	+	40
T	X	++	30
T	M	−	12
G	B	−	
G	M	−	
W	M	−	3
F	J	−	
S	H	−	12
R	4	−	0
A	H	−	
A	K	−	
C	H	−	
W	X	−	5
F	H	−	
U	H	−	4
U	X	−	0
U	K	−	0

skingraft intact
incomplete rejection

FIG. 1. Comparison of serum reactivity and skin graft rejection times. Recipients A, C, S, T, U and W were pre-immunized with skin grafts only; F and G received skin grafts and blood intravenously; Q and R, skin grafts and leukocytes intradermally.

analysis of the reactivity patterns; the numerals indicate T values (square roots of x^2) for pairwise comparison of serum reactivities against a panel of 165 monkeys. The underlined numerals in the body of the table represent negative correlations between the reactivity of individual sera; crosshatched areas therefore indicate the possibility of allelic control of these specificities. Figure 3 represents the distribution of these 12 specificities in 118 monkeys.

Population Study

A statistical analysis of serological data for 198 unrelated monkeys by one of us (BWG) led to a number of interesting conclusions. The monkeys

CORRELATION BETWEEN REACTIVITY PATTERNS OF RHESUS ISO-ANTISERA

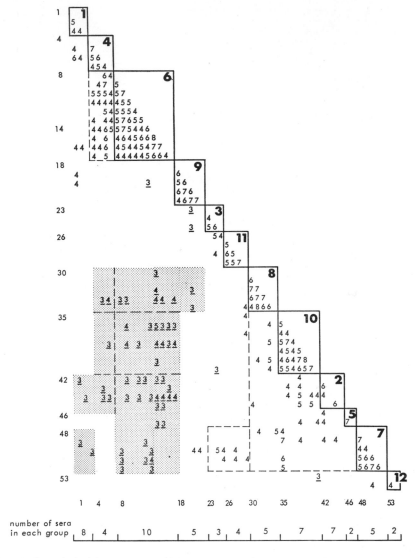

FIG. 2. Correlation between reactivity patterns of Rhesus isoantisera. Each serum was tested with 165 lymphocyte samples (microcytotoxicity). The numerals in the body of the table represent square roots of chi-square values for a two by two contingency test (between all individual sera) for positive and negative reactivity;

lymphocyte samples

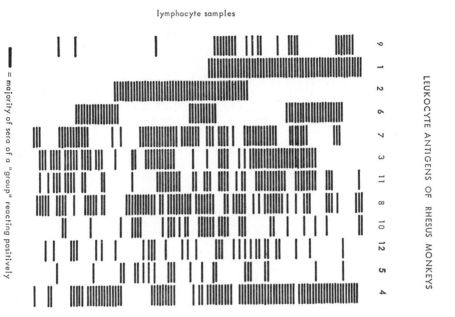

FIG. 3. Leukocyte antigens of Rhesus monkeys. Horizontal bars indicate that cell samples reacted positively with the majority of the sera defining a particular group. Numerals above columns indicate group or specificity; they correspond with those in bold type in Fig. 2. Cell samples are arranged for optimal visualisation of relationship between specificities 1, 2 and 6.

underlined numerals indicate that the relation between reactivity patterns of 2 sera was negative, for instance, antigen 6 with antigen 8, 10, 2 and 5 and 7 (cross-hatched).

Numerals at the left and below the figure represent single sera except for numbers 1 and 42, which stand respectively for combined data of 6 sera typing for group 1 and 4 for group 2; this explains the difference with the actual number of sera per group (bottom of table). Three of the sera defining group 9 were kindly provided by Dr G. N. Rogentine of N.I.H., Bethesda, U.S.A.

were typed with the 62 sera depicted in Fig. 2, and classified for the 12 specificities according to whether their lymphocytes reacted positively or negatively with the majority of the sera in each group. Most of these monkeys were received in the Institute in 8 shipments of between 7 and 33 animals each. However, the possibility of related animals within a shipment was not taken into account in this initial analysis.

Table I shows the significant x^2 values obtained from pairwise comparison of groups 1–12. Values of $x^2 > 3, 8$ indicate a positive association. It would seem that groups 9, 6, 1 and 4 are positively associated and groups 8, 2, 10, 12, 5, 7 and 3 are members of a second positively associated set.

TABLE I. Significant x^2 values from the pairwise comparison of 12 groups of sera defining the leukocyte specificities in 198 unrelated Rhesus monkeys (underlined figures indicate negative associations)

	9	6	1	4	8	2	10	12	5	7	3
6											
1	40·6										
4	53·5	15·9									
8	5·4	5·0		11·5							
2		16·8	5·3	6·9	12·7						
10	6·6	5·3		4·3	41·7	11·2					
12						7·4	5·5				
5						5·1		3·8			
7		5·0									
3							11·8		18·3		
11			4·7	5·3	10·6				12·3	37·2	

Negative associations between certain groups of the two sets (9, 6, 1, 4 versus 8, 2, 10) as well as between groups 10 and 12 of the second set are consistent with the hypothesis of an allelic relationship similar to the ABO system in man. Assuming random mating in the wild population, the hypothesis of independence and allelism were tested in a 2×2 comparison for several selected pairs, and the data were consistent with the assumption that the groups are controlled by a series of multiple alleles (Table II). Since this was so for the pairs (9, 10), (6, 10) and (6, 9), the numbers of animals showing various reactivity patterns with groups 9, 6 and 10 were investigated. It is clear that the observations shown in Table III are in agreement with the hypothesis that these specificities are controlled by a series of allelic genes ($x_3^2 = 4, 1$; $p > 0, 2$). The population data would thus support the theory that the majority of our sera react with specificities belonging to one complex system.

TABLE II. Analysis of the genetic relationships between selected pairs of groups of antisera defining leukocyte specificities of Rhesus monkeys

Specificities	X^2 Independence[a]	X^2 Allelism[b]	2 × 2 table data + +	+ −	− +	− −
9 + 8	5·4[c]	0·2	14	26	88	70
9 + 10	6·6[c]	0·8	4	36	47	110
6 + 8	5·0[c]	3·9[c]	36	49	66	47
6 + 2	16·8[d]	1·2	13	72	48	65
6 + 10	5·3[c]	0·3	15	70	36	76
6 + 9	0·6	3·4	15	70	25	88
1 + 2	5·3[c]	1·3	23	76	38	61
4 + 8	11·5[d]	4·7[c]	59	77	43	19
4 + 2	6·9[c]	2·4	34	102	27	35
4 + 10	4·3[c]	2·8	29	106	22	40

[a] Assuming the numbers of observations in each cell of the 2 × 2 table to be proportional to the marginal totals.

[b] Based on gene frequencies for each specificity calculated from

$$1 - \sqrt{\text{Frequencies of negatives.}}$$

[c] significant at 5 % level.

[d] significant at 1 % level.

Note: When the X^2 for independence is high (>3, 8) and the X^2 for allelism is low ($< 3·8$), the data are consistent with the hypothesis that the specificities are controlled by a series of multiple alleles.

TABLE III. The distribution of specificities 9, 6 and 10 in 197 random rhesus monkeys

Groups 9	6	10	Numbers observed	Numbers expected
+	+	+	—	—
+	+	−	15	10·4
+	−	+	4	6·1
−	+	+	15	13·2
+	−	−	21	24·5
−	+	−	55	59·6
−	−	+	32	32·0
−	−	−	55	51·2
			197	197·0

$$X_3^2 = 4,1; p > 0,2$$

The maximum likelihood estimate of gene frequencies for alleles 9, 6 and 10 are 0·11, 0·24 and 0·14, for the null allele 0·51. Yasuda and Kamura (1968).

Family Studies

The mode of inheritance of the described specificities was studied in 13 families with 3–5 offspring (4 different fathers, 13 mothers, 44 offspring). Groups 1–12 were determined in the same fashion as for the unrelated animals. The numbers of individuals in families showing segregation for each specificity were too small for significance tests without combining the data. Therefore, data from back-cross type matings (+ x —, where the positive parent was heterozygous) and intercross matings (+ x +, with both parents heterozygous) were analysed. After allowing for the bias introduced by the small size of the families (Maynard-Smith *et al.*, 1961), the frequency of negative offspring agreed with the Mendelian expectations of 0·5 and 0·25 respectively (Balner *et al.*, 1970a).

On the assumption that the 12 specificities belong to a single system, an attempt was made to assign the specificities to parental chromosomes (haplotyping). Figure 4 demonstrates how haplotypes could be "construed" from an informative family in which 3 of the offspring were of different genotype. Figure 5 tabulates the haplotypes that could be deduced from 13 families. Due to the small numbers of offspring per family, it was possible to allocate the determinants for all 12 specificities on only 16 haplotypes (from 10 parents). The combined results of analysing the population and family data would thus suggest that the Rhesus monkey has like mice (Amos *et al.*, 1955; Iványi, 1970; Snell *et al.*, 1953), rats (Palm, 1964; Stark *et al.*, 1966), Man (Ceppellini *et al.*, 1967; Dausset *et al.*, 1967; Payne *et al.*, 1964; Rood *et al.*, 1967; Terasaki *et al.*, 1966) and probably dogs (Ferreby *et al.*, 1970; Vriesendorp *et al.*, 1971), one genetic system controlling the leukocyte antigens; by analogy with HL–A this has been called the RhL–A system (Balner *et al.*, 1970a).

Relevance to Histocompatibility

An important issue in the study of the Rhesus monkey's leukocyte antigens was to establish whether genotypic identity of sibs (same parental haplotypes) stands for a high degree of histocompatibility. If the genetic information for all the important transplantation antigens is located on one chromosome then skin grafts exchanged between sibs with identical genotypes for that system should survive longer than grafts exchanged between non-identical sibs.

Haplotyping was done as described above, by assigning phenotypic specificities to parental haplotypes in agreement with the segregation observed in the offspring. In some cases, the antigenic composition of the parental haplotypes could not be determined because segregation of a particular specificity did not occur in any of the offspring, while at other

RhL-A HAPLOTYPING IN A RHESUS FAMILY

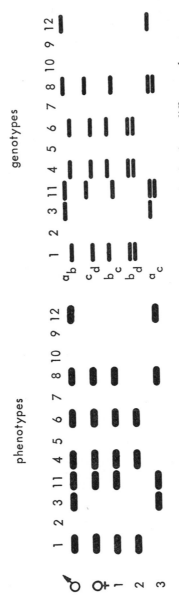

Fig. 4. "Haplotyping" of members of a Rhesus family in which the 3 offspring have different phenotypes (and genotypes). On the left, the phenotypes, black bars indicating the presence of the leukocyte specificities (see Figs. 2 and 3). On the right, deduction of RhL–A "alleles" or haplotypes (father a and b, mother c and d). The possibility to deduce haplotypes compatible with the distribution of the antigens on the cells of the offspring supports the assumption of one genetic system for the recognizable antigens.

times serological ambiguities interfered with a reliable assignment of antigens to a particular haplotype. In spite of these limitations regarding the exact configuration of the parents' haplotypes, it was usually possible to identify the offspring as ac, ad, bc or bd, if ab and cd are the fathers' and mothers' haplotypes, respectively.

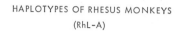

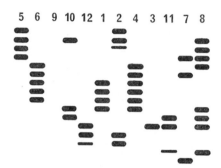

FIG. 5. Analysis of phenotypic data for the members of 13 families of Rhesus monkeys led to the assignment of the 12 leukocyte specificities to 16 parental haplotypes (from 10 parents). Thin horizontal bars indicate ambiguous serological evidence for the phenotypic presence of a particular specificity on the lymphocytes.

The data presented in Table IV demonstrate the influence of genotypic identity on allograft survival in inter-sib skin grafting; the mean survival time for 31 non-identical host/donor combinations was 8, 7 ± 0·2 days, that for 14 identical combinations 14, 3 ± 0, 2 days (t = 10, p < 0, 001). The significance of the differences in survival time can also be clearly seen in Fig. 6 where the data are presented as a histogram. These findings are in agreement with the assumption that the RhL–A system is a major histocompatibility locus in this species.

For couples 1 and 2 of family 599–581 prolongation of graft survival was only minimal although survival times of 12 days without immunosuppression have not been found in over 200 transplants between random monkeys. The marginal prolongation of survival time for skin grafts exchanged between some of the RhL–A identical sibs, might indicate that Rhesus monkeys also possess transplantation antigens that segregate independently of the main locus. Variations in the graft survival times found for identical sib combinations may therefore reflect the combined influence of antigens other than RhL–A.

TABLE IV. Skin grafts exchanged between sibs of rhesus families geno-typed for RhL-A

Family 381

sib	genotype	graft donors				
381♂	a b					
584♀	c d	1	2	3	4	5
1	b d		9	9	9	9
2	b c	9		14	14	
3	b c	9	18		18	
4	b c	8	16		16	
5	b c					

sib	genotype	graft donors			
494♀	e f	6	7	8	9
6	a	?		9	9
7	b ?	?	9		9
8	a	?			
9	*				

sib	genotype	graft donors		
432♀	g h	10	11	12
10	a g		8	8
11	b h	9		8
12	a h			

Family 600

sib	genotype	graft donors			
600♂	a b				
597♀	c d	1	2	3	4
1	a		?	13	8
2	a		? 15		9
3	b c				
4	*				

sib	genotype	graft donors		
669♀	e f	5	6	7
5	b f		13	13
6	b f	14		14
7	b f			

Family 598

sib	genotype	graft donors		
598♂	a b			
589♀	c d	1	2	3
1	b c		10	10
2	b d	10		9
3	a c	9	9	

sib	genotype	graft donors		
834♀	e f	4	5	6
4	a e		15	9
5	a e	13		9
6	b f			

Family 599

sib	genotype	graft donors				
599♂	a b					
581♀	c d	1	2	3	4	5
1	a c			12	8	8
2	a c		11		7	10
3	b ?	8	8	8		12
4	b ?	8	8	12		
5	*					

RhL–A haplotypes of various fathers always indicated with symbols a and b, of mothers with symbols c, d, e, f etc., although antigenic composition of haplotypes are probably different for all mothers; the question marks indicate uncertainty regarding the allocation of the maternal haplotype.
Graft survival period underlined when genotypic donor/recipient identity was uncertain.
Skin graft survival in days; open spaces indicate that grafts have not yet been exchanged.
Offspring that has not yet been typed indicated with asterisk.

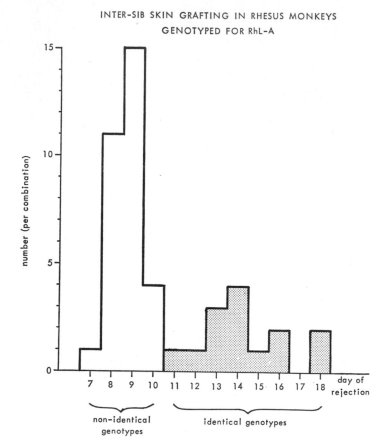

FIG. 6. Skin grafting survival times between sibs of Rhesus families genotyped for RhL–A. Shaded areas represent host/donor combinations with identical RhL–A genotypes (mean survival 14, 3 ± 0·6 days); empty areas those from combinations differing for one or both parental haplotypes (mean survival 8, 7 ± 0·2 days).

Relation between Leukocyte and Erythrocyte Antigens

The elaborate bone marrow transplantation programme in this Institute (van Bekkum and van der Waaij, 1970) led us to study also the red cell iso-antigens of Rhesus monkeys. In rodents, leukocyte antigens can usually be identified on red cells (Hildeman, 1970; Iványi, 1970). However, leuko-cyte antigens of primates are not as a rule serologically identifiable on mature erythrocytes although erythroblasts and reticulocytes have been shown to carry certain HL–A antigens in Man. (Harris and Zwerwas, 1969). In 1969 we described a Rhesus isoantiserum that probably recognizes a similar antigen on white and red cells. Reactivity patterns for red and white cells were similar and by separate absorptions with Rhesus erythrocytes and leukocytes it could be shown that the same or similar antigens were probably identified by this reagent (Hirose and Balner, 1969). Recently, similar data for Rhesus monkeys have been described by Bogden et al. (1968) and for Man by Morton et al. (1969) for Bg[a] and HL–A7.

Another point of interest was whether any of the red cell specificities of Rhesus monkeys might also be relevant for histocompatibility. Except for the antigens of the ABO system (Ceppellini et al., 1969; Rapaport et al., 1968), this does not seem to be the case for Man. Experiments of the type described for leukocyte antigens (comparing skin graft survival times on individuals presensitized against red cell antigens of the donors) have convinced us that the majority of the red cell antigens of Rhesus monkeys are not likely to be transplantation antigens (Balner et al., 1966; Hirose and Balner, 1969). Apparently, this species does not possess red cell specificities analogous to those of the human ABO system. Other investigators studying red cell antigens of Rhesus monkeys (Bogden et al., 1968; Duggleby, 1970; Moor-Jankowski et al., 1967; Owen and Anderson, 1962; Wiener et al., 1968) did not attempt to prove the relevance of these specificities to histocompatibility.

The occurrence of Rhesus Leukocyte Antigens in Other Primate Species

While the sharing of red cell antigens between Man, apes and monkeys has been the subject of intensive studies (Moor-Jankowski et al., 1964; Wiener, 1965), less attention has been paid to the sharing of leukocyte antigens between the primate species. As briefly mentioned in the Introduction, we reported in 1967 that the first recognizable Rhesus leukocyte groups (specificities 1 and 2) might have something in common with specificities 4a and 4b of Man. Results of absorptions of human anti-4a and 4b sera with Rhesus leukocytes carrying antigens 1 and 2 respectively, provided indirect evidence in favour of this hypothesis (Balner et al., 1967a). It was speculated, therefore, that 4a and 4b might represent "precursor" antigens from which other primate specificities could be derived. On the other hand, in simple cross-species typing of 12 chimpanzee lymphocyte samples with Rhesus sera of anti-1 and anti-2 specificity, chimpanzee cells always reacted positively with anti-2 sera but negatively with sera of anti-1 specificity. This would seem to refute the former speculation, because some of the chimps in question carried specificity 4a on their leukocytes, others 4a/4b or 4b only.

More recently, Rhesus antisera defining the 12 described specificities were tested with a panel of lymphocytes from other macaques (*M. Speciosa;* 40 individuals). Using the same microcytoxicity test, several antigens (1, 4, 6 and 12) were distinctly identifiable on some cells from Speciosa monkeys, others were less clearly represented (2 and 8), or not at all. To obtain a better picture of the degree of sharing of antigens between various primates, a programme has been started whereby a standard panel of individuals from 4 primate species (humans, chimpanzees, *M. Mulatta* and *M. Speciosa*) are typed with human, chimpanzee and Rhesus iso-antisera. Appropriate absorption studies to prove or disprove the presence of particular antigens are also planned.

Investigators at Duke University, Durham, N.C., have used different techniques to demonstrate the sharing of tissue antigens between primate species. They used a mixed agglutination technique on short-term cultures of epidermal cells and were also able to demonstrate the presence of several human HL–A antigens on cells of a variety of other primate species (Metzgar *et al.*, 1967). Combined data from several laboratories using different techniques should make it possible to clarify some of the phylo-genetic aspects of primate tissue antigens in the future.

THE LEUKOCYTE ANTIGENS OF CHIMPANZEES

The study of the leukocyte antigens of chimpanzees has not reached the stage of advanced immunogenetic analysis that it has for man and, to a lesser extent, for the Rhesus monkey. Before a thorough analysis of a species' leukocyte antigens can be carried out, it is obviously necessary to have reliable typing reagents and a reasonably large number of cell samples from unrelated individuals and, if possible, from pedigreed families. For chimpanzees, neither large populations nor families were available and typing reagents are not easily obtained. Therefore, rather than starting with the identification of the chimpanzee's own antigens, our first efforts were directed towards establishing similarities between chim-panzee and human lymphocyte antigens. This was done mainly by com-paring reactivity patterns of selected chimpanzee and human isoantisera with "informative" panels of chimpanzee and human lymphocytes and by absorption studies (Balner *et al.*, 1967c). In the course of 1970, however, a collaborative research project permitted the typing of several large colonies of chimpanzees in the U.S.A.*; this brought the total number

---

* 72 chimpanzees at the 6571st Aeromedical Research Laboratory, Holloman Air Force Base, New Mexico, U.S.A.

38 chimpanzees at Bionetics Research Labs., Bethesda, Md., U.S.A.

42 chimpanzees at the Laboratory for Experimental Medicine and surgery in Primates (LEMSIP), New York, U.S.A.

of typed chimpanzee cell samples to 198 and enabled us to start a preliminary analysis of the distribution of leukocyte antigens in a chimpanzee population.

Typing Chimpanzees with Chimpanzee Sera

Figure 7 depicts the reactivity pattern of 27 selected isoantisera available in 1968, against the lymphocytes of the chimpanzees maintained in this Institute. It can be seen that 5 groups of sera (obtained from 15 isoimmunized animals) showed distinct patterns in the microcytotoxicity test;

REACTIVITY OF CHIMPANZEE ISO-ANTISERA

(lymphocytotoxicity)

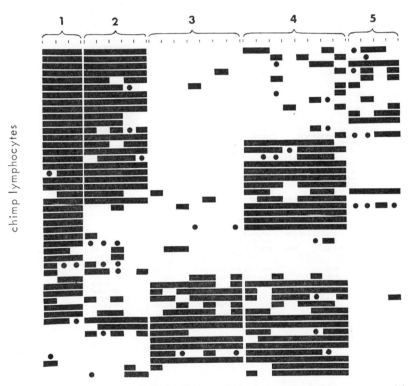

FIG. 7. Reactivity of 27 sera obtained from 15 isoimmunized chimpanzees with 47 chimp cell samples. Groups of sera with similar reactivity patterns (lymphocytotoxicity) are indicated by numerals, each group consisting of several individual sera (see sub-divisions). A solid black bar indicates a strongly positive reaction between serum and cell sample, a dot stands for a weak reaction; negatives are left open. Some of the sera belonging to the same or another "group", were different batches of serum raised in the same chimpanzee.

the specificities presumably identified were tentatively called groups 1–5 (Balner *et al.*, 1970b). Absorption studies revealed that the sera defining groups 1 and 4 were not monospecific, whereas the sera defining groups 2, 3 and 5 gave no evidence of polyspecificity; cytotoxic reactivity was absent after absorption of group 2 sera with 12 positive samples, of group 3 and group 5 sera after absorption with all positive samples of the panel. Technical details about production of sera, methods of typing and absorption, etc., have been published previously (Balner *et al.*, 1967b, c).

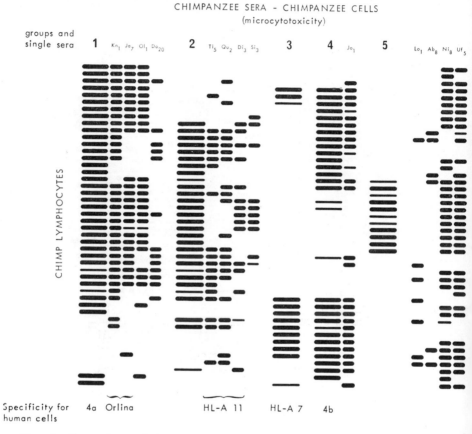

FIG. 8. Comparison of the reactivity patterns of 13 "new" chimp isoantisera (narrow bars) with the patterns of the sera constituting the 5 leukocyte groups depicted in Fig. 7 (a bold, broad bar indicates that the majority of sera of a group reacted strongly positive, a thin bar stands for weak or partly negative reactivity). Cell samples have been arranged for optimal visualization of related reactivity patterns. The symbols at the bottom of this figure indicate specificities of the same sera for human lymphocytes (compare Figs. 12 and 13).

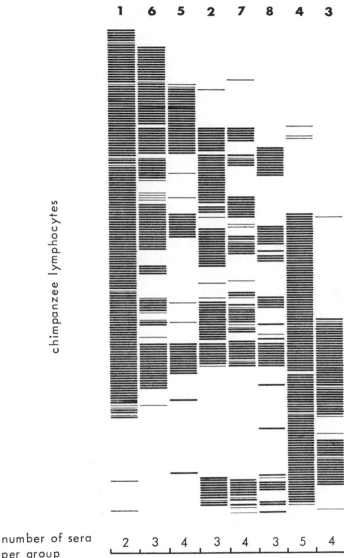

FIG. 9. Schematic representation of the cytotoxic reactivity patterns of 8 groups of antisera (28 sera from 20 chimpanzees) with lymphocyte samples from 198 chimpanzees. Bold bars indicate that the majority of sera of the group reacted strongly positive, thin bars correspond with weak or partly negative reactivity. Cell samples have been arranged for an optimal visualization of the relationships between some of the groups (see also Table V and text). Of the new groups 6, 7 and 8, specificity 6 is defined by sera Kn_1, Ja_7, Ol_1; specificity 7 by sera Qu_2, $Di_{3,4}$, Si_3, specificity 8 by Do_{20}, $Ti_{5.6}$ (compare Fig. 8 and text).

In the course of 1969, a further 13 discriminating chimpanzee sera were added to our panel of typing reagents. As can be seen in Fig. 8, the reactivity pattern of several new reagents was "included" in that of some of the sera defining certain groups (Kn_1, Ja_7, Ol_1 included in group 1; Do_{20} and Ti_5 in group 2; Qu_2, Di_3 and Si_3 probably also in group 2) while others had near-identical patterns with the sera of some of the groups, for instance Jo_1 with group 4. The validity of these similarities and inclusions (apparent from typing our own 46 animals) was corroborated when the results of typing the additional 152 cell samples from in majority unrelated* chimpanzees became available. Computer analysis of this raw data (x^2 values for associations between reactivity patterns) led to minor rearrangements in the grouping of sera defining specificities 1–5 and to the "establishment" of 3 separate specificities 6, 7 and 8. Figure 9 visualizes the distribution of the tentatively delineated chimpanzee specificities in 198 chimpanzees. It can be seen that group 5 is, as before, included in group 1 as well as in the new group 6; the new groups 7 and 8 are closely associated with group 2, while group 3 remained included in group 4. It was gratifying to observe that the similarities and inclusions predicted on the basis of typing the 46 chimpanzees available at our Institute (Fig. 8) were confirmed by the results of typing the 152 additional chimpanzees. The reactivity patterns of 11 additional sera showed certain complex associations with some of the groups but have not been sufficiently analysed.

Population Study

Statistical analysis of the data for the 193 unrelated chimpanzees (leaving out the 5 offspring) indicates that groups 1, 6, 5, 4 and 3 are probably controlled by the same genetic system. On the basis of the present population data, it is not possible to be certain at this stage whether the antisera of groups 2, 7 and 8 also belong to the same system; however, this possibility cannot be excluded. Table V shows the significant x^2 values obtained from the pairwise comparison of the reactivity patterns of groups 1–8. It can be seen from Table VI that reactivity patterns of the positively associated groups 1, 6 and 5 versus groups 4 and 3, are usually consistent with the hypothesis that these specificities are controlled by multiple alleles of a system. The genetic locus controlling this system has previously been designated as ChL–A (Balner *et al.*, 1971).

---

* There were 9 related individuals, namely 4 fathers and 5 offspring from different mothers.

TABLE V. Significant X^2 values from the pairwise comparison
of 8 groups of sera defining leukocyte specificities in
193 unrelated chimpanzees (underlined figures
indicate negative associations)

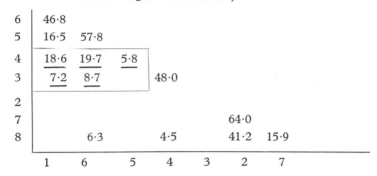

Evidence for the Sharing of Antigens between Chimpanzee and Man

Two approaches were applied for this study.

1. Typing chimpanzees with human sera, including absorption of the sera with chimpanzee cells, to indicate the presence of specificities similar to HL–A antigens on chimpanzee cells.

2. Typing human cells with chimpanzee sera. If sera that identify certain chimpanzee specificities show HL–A specificity when used for human tissue typing, this can be regarded as indirect evidence for a similar configuration of the chimpanzee and human antigens.

1. *Typing chimpanzee leukocytes with human isoantisera*

Figure 10 demonstrates results of typing chimpanzees with human sera using agglutination and agglutination inhibition. These data were obtained before 1967 (Balner *et al.*, 1967c) which explains the use of the original nomenclature for specificities that have since been given an official HL–A designation (Rood *et al.*, 1965, 1967, 1970). These data provided evidence that chimp leukocytes show polymorphism for several defined human antigens, a finding which introduced the chimpanzee as an eligible producer of tissue typing reagents for man. Some of the human specificities seemed absent from chimpanzee cells (for instance 8a or HL–A2); others were likely to be present (for instance 7c or HL–A7, see below).

Figure 11 shows the cytotoxicity pattern of human sera of HL–A specificity (usually 2–4 monospecific sera per group) when reacting with the cells of the same chimpanzee panel. For comparison, the reactivity of the chimpanzee sera defining the 5 initial chimpanzee groups is shown in the

HUMAN LEUKOCYTE ANTIGENS ON CHIMPANZEE LEUKOCYTES

	4a	4b	6a	6b	7a	7b	7c	7d	8a	9a	5a	5b
Henk	+ +											+ +
Debbie	+ +											+ +
Fred	+ +											+ +
Caroline	+ +											+ +
Renee	+ +											+ +
Abe	+ +						+ +					+ +
Diana	+ +						+ +					+ +
Engelbert	+ +					+ +						+ +
Pascale	+ +					+ +						+ +
Odette	+ +					+ +						+ +
Gina	+ +					+ +						+ +
Frits	+ +					+ +	+ +					+ +
Wilma	+ +					+ +	+ +					+ +
Helga	+ +		+ +			+ +	+ +					+ +
Tuttie	+ +		+ +			+ +	+ +					+ +
Liesbeth	+ +		+ +			+ +	+ +					+ +
Regina	+ +		+ +			+ +						+ +
Mario	+ +		+ +			+ +						+ +
Veronica	+ +		+ +			+ +						+ +
Yvonne	+ +		+ +			+ +						+ +
Ufford	+ +		+ +		+ +	+ +						+ +
Belinda	+ +		+ +		+ +		+ +					+ +
Anita	+ +		+ +	+ +								+ +
Tineke	+ +			+ +	+ +		+ +					+ +
Louisa	+ +	+ +										+ +
Ursula	+ +	+ +										+ +
Isac	+ +	+ +				+ +						+ +
Nico	+ +	+ +				+ +						+ +
Wanda	+ +	+ +				+ +						+ +
Arnold	+ +	+ +				+ +	+ +					+ +
Elvis	+ +	+ +	+ +									+ +
Jolanda	+ +	+ +	+ +			+ +						+ +
Katie	+ +	+ +	+ +			+ +						+ +
Quarles	+ +	+ +	+ +			+ +						+ +
Nina	+ +	+ +	+ +		+ +	+ +						+ +
Simon	+ +	+ +	+ +		+ +	+ +						+ +
Marco	+ +	+ +	+ +		+ +	+ +						+ +
Olivier	+ +	+ +	+ +		+ +	+ +	+ +					+ +
Indira	+ +	+ +	+ +		+ +	+ +	+ +					+ +
Zelma	+ +	+ +	+ +	+ +	+ +	+ +						+ +
Susie	+ +	+ +	+ +	+ +			+ +					+ +
Claudia		+ +	+ +	+ +	+ +	+ +	+ +					+ +
Dorus		+ +	+ +		+ +	+ +	+ +					+ +
Karin		+ +	+ +		+ +	+ +	+ +					+ +
Jan		+ +	+ +		+ +	+ +						+ +
Piet		+ +	+ +			+ +						+ +
Gerrit		+ +	+ +				+ +					+ +
Victoria		+ +				+ +	+ +					+ +
Brigitte		+ +				+ +	+ +					+ +

FIG. 10. Distribution of a number of defined human leukocyte specificities on leukocyte samples from 49 chimpanzees. Results are based on repeated agglutination reactions with human sera, as well as on the inhibition of the agglutinating activity of human sera with human cells by absorption with chimpanzee leukocytes. Note that 4a and 4b show, as in Man, an alternative distribution. For use of nomenclature, see text.

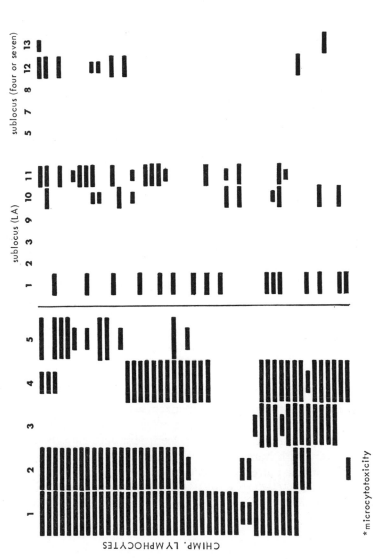

* microcytotoxicity

Fig. 11. Comparing reactivity patterns of chimpanzee and human isoantisera with chimpanzee lymphocytes. The chimpanzee groups 1–5 are defined by the sera of Fig. 7 (cell samples are arranged in the same order; broad bars: strong reactions with sera of group, narrow bars: weak reactions; 3–8 sera per group). The human specificities were defined by 2 or 3 sera which were monospecific for HL–A specificities in human tissue typing (as indicated).

same figure. Absorptions of the cytotoxic human sera with chimpanzee cells have not yet been carried out; therefore we cannot be sure that the HL–A antigens present according to Fig. 11 are actually carried by these cells.* Similarly, negative reactions with 3 HL–A7 sera, also depicted in Fig. 11, do not prove the absence of this specificity and might be attributable to the CYNAP phenomenon as described for numerous species (cytotoxicity – negative – absorption – positive; (Ceppellini et al., 1967; Rood et al., 1965)). In fact, a specificity defined with anti–7a sera (which cross-react with HL–A7, FJH and AA (Rood et al., 1970)) was found on chimpanzee cells by agglutination and absorption studies (Fig. 10). Furthermore, certain chimpanzee isoantisera have HL–A7 specificity for human cells (see below), which also suggests polymorphism for an antigen identical with or very similar to the human HL–A7 on chimpanzee cells.

2. *Typing of human cells with chimpanzee antisera*

Whether chimpanzee sera are useful for human tissue typing can be investigated by screening these sera against human cells, preferably from a panel which includes a sufficient number of individuals positive for all currently recognizable HL–A antigens. The reactivity patterns (if any) of the chimpanzee sera are then compared by computer analysis with those of the human HL–A reagents and possible similarities of reactivity pattern can be established. Using this approach, we have been able to find positive correlations which can be regarded as circumstantial evidence that chimpanzees are polymorphic for the same or similar tissue antigens.

Specificities 4a and 4b

Figure 12 demonstrates that chimpanzee sera that define chimpanzee groups 1–5 were clearly discriminating when reacting with a panel of human lymphocyte samples. As can be seen, sera defining groups 1 and 4 reacted positively with human cells of 4a and 4b specificity, respectively. These correlations had already been established with a somewhat smaller panel of human cells (Balner et al., 1967c) and were recently confirmed by testing another 100 human cell samples from unrelated individuals (Balner et al.,1971).

---

* The appropriate absorption studies have since been carried out; a detailed report was presented by J. J. van Rood at the Symposium on Transplantation Genetics of Primates, Rijswijk, Sept. 71. To be published in the March 1972 issue of "Transplantation Proceedings," Grune and Stratton Inc., Med Publishers, 757 Third Avenue, New York, N.Y. 10017, U.S.A.

Absorption studies of these chimpanzee sera were also performed. One of the sera defining chimp group 1 was absorbed with 30 selected samples of human leukocytes; reactivity was removed by human cells homozygous and heterozygous for 4a, but not by cells homozygous for 4b. Similarly, one of the chimp sera defining chimp group 4 (anti-4b in chimps and man) was absorbed with 30 human leukocyte samples; again, homozygotes and heterozygotes for 4b absorbed out all activity while 4a/4a cells removed virtually none of the antibodies. There seems to be little doubt

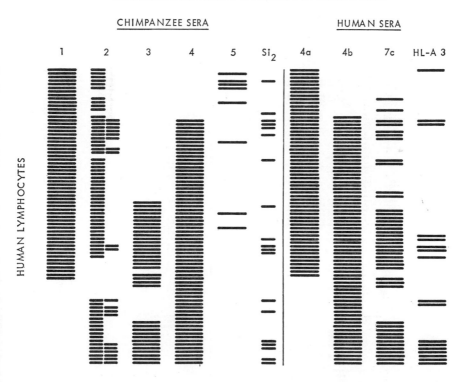

FIG. 12. Comparing reactivity of chimpanzees and human isoantisera with human lymphocyte samples. On the left are sera selected from those defining chimp groups 1–5 of Fig. 7. The sera of chimp group 2 showed two distinct patterns: reactivity of sera from a particular chimp (right column) were "included" in the reactivity of sera from two other chimps (left column). The results of HL–A typing of the same cell samples are depicted on the right (the results for specificities 4a, 4b and 7c are based on agglutination as well as cytotoxicity data).

E

therefore that specificities 4a and 4b, whatever their exact nature may be, are of similar configuration and rather similar distribution in man and chimpanzees (compare Fig. 12 with Fig. 9).

Specificity 7c or HL–A7 and Specificity HL–A3

Figure 12 also shows that chimpanzee sera of group 3 have anti-7c specificity for human cells. In subsequent studies on 100 human cell samples, a highly significant positive correlation (x^2 values of around 50 in a 2×2 contingency test for positive and negative cytotoxic reactions) was

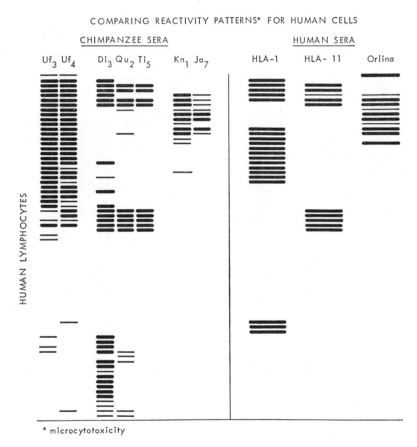

COMPARING REACTIVITY PATTERNS* FOR HUMAN CELLS

CHIMPANZEE SERA   HUMAN SERA

Uf_3 Uf_4 Di_3 Qu_2 Ti_5 Kn_1 Ja_7 HLA-1   HLA-11   Orlina

HUMAN LYMPHOCYTES

* microcytotoxicity

FIG. 13. Comparing reactivity patterns of chimpanzee and human isoantisera with 71 human lymphocyte samples; bold type bars indicating strong, thin bars weak cytotoxic reactivity. Wide bars on the right depict results of HL–A typing as routinely performed at Leiden University (two or more monospecific sera per antigen).

found for the reactivity of group 3 sera and several human HL–A7 sera.

Whether the reactivity pattern of serum Si_2 correlates significantly with human HL–A3 sera remains to be seen (Fig. 12). It is noteworthy that Dorf and Metzgar (1970) have also found chimpanzee sera of HL–A3 related specificity but further typing and absorption studies will be required to ascertain the occurrence of an HL–A3 like antigen in chimpanzees.

Specificity HL–A11

Figure 13 demonstrates the positive correlation between the reactivity pattern of sera Di_3 and Qu_2 (chimp group 7) and Ti_5 (chimp group 8) and those of human HL–A11 sera (71 cell samples from unrelated human individuals); the reactivity of serum Ti_5 was virtually identical with that of mono-specific human HL–A11 sera, also in the hands of Kissmeyer-Nielsen (personal communication).

Other Defined Human Antigens

Also in Fig. 13, a positive correlation can be observed between the reactivity of sera Uf and human HL–A1 sera; HL–A1 positive cells were missed in only 3 instances and this might be due to the CYNAP phenomen.

Sera Kn_1 and Ja_7 reacted similarly to human sera defining antigen "Orlina". The latter has as yet not been given an official HL–A designation but recent human family data suggest that this is also an antigen controlled by the HL–A locus (Allen et al., 1970).

THE PHYLOGENY OF PRIMATE TISSUE ANTIGENS

As has been stated in the preceding sections there is a considerable degree of antigenic similarity between human HL–A antigens and some of the recognizable chimpanzee specificities. By analogy with HL–A, we have therefore assumed the existence of one genetic locus "ChL–A" governing the expression of important tissue antigens of chimpanzees (Balner et al., 1970c).

This brings us to one of the interesting aspects of tissue typing in primates, namely the phylogeny of leukocyte antigens, particularly with regard to HL–A. It may be worthwhile to recall some of the data already available. For instance, specificities 4a and 4b were the first system of human leukocyte antigens recognized (van Rood and van Leeuwen, 1963). The original interpretation of a simple allelic system of two mutually exclusive antigens had to be abandoned and although several new explanations have been put forward (Kissmeyer-Nielsen et al., 1968; Rood et al.,

1970; Zmijewski *et al.*, 1970), none of them seems to be entirely satisfactory. Chimpanzees show such alternative distribution of 4a and 4b as pronounced as human beings and, with a few exceptions, cytotoxic anti-4a and anti-4b sera (for Man and chimp) have come from isoimmunized chimpanzees (Rood *et al.*, 1970). As we have said before (see section on Rhesus monkeys), antigens 4a and 4b can probably be traced further down the evolutionary ladder: by appropriate absorptions of human sera, we obtained suggestive evidence that two of the important transplantation antigens of the Rhesus monkey are possibly of similar antigenic configuration as 4a and 4b (Balner *et al.*, 1967a). These and other data led to the speculation that 4a and 4b may well be the "basic substance" from which the important tissue antigens of primates, including the two well known series of allelic transplantation antigens of Man, evolved (Balner *et al.*, 1970c).

Now that the chimpanzee antigens have been somewhat further analysed one might be tempted to speculate also about the origin of the two allelic series of HL–A antigens. There is evidence that 4a and 4b are closely related to and possibly cross-reacting with antigens of only one of the sub-loci, the "seven" or "four" series (Kissmeyer-Nielsen *et al.*, 1968; Kissmeyer-Nielsen and Thorsby, 1970). The current analysis of the chimpanzee

TABLE VI. Analysis of genetic relationships between pairs of groups of antisera from box in Table V, defining leukocyte specificities of chimpanzees

Specificities	X^2 Independence[a]	X^2 Allelism[b]	2 × 2 table data + +	+ −	− +	− −
4 + 3	48·[d]	116·1[d]	55	58	2	78
4 + 1	18·6[d]	4·6[c]	79	34	76	4
3 + 1	7·2[c]	3·0	39	18	116	20
4 + 6	19·7[d]	0·1	41	72	55	25
3 + 6	8·7[d]	0·1	19	38	77	59
4 + 5	5·9[c]	0·9	22	91	28	52
3 + 5	0·4	3·3	13	44	37	99

[a] Assuming the number of observations in each cell of the 2 × 2 table to be proportional to the marginal totals.

[b] Based on gene frequencies for each specificity calculated from
$1 - \sqrt{\text{Frequencies of negatives}}$.

[c] Significant at 5 % level.

[d] Significant at 1 % level.

Note: When the X^2 for independence is high (>3,8) and the X^2 for allelism is low (< 3·8), the data are consistent with the hypothesis that the specificities are controlled by a series of multiple alleles.

leukocyte specificities revealed that groups 1 and 4 (the equivalent of 4a and 4b) are associated with groups 3, 5 and 6 and are likely to represent an allelic series of chimpanzee specificities (Table VI). It is tempting, there-fore, to regard this series as the possible equivalent of the "seven" or "four" series of HL–A. Such reasoning is supported by the fact that sera from chimpanzee group 3 show HL–A7 like reactivity for chimpanzee and human cells (HL–A7 being an antigen of the "seven" series). We would expect that groups 5 and 6 should correspond to other antigens of the seven series but so far no evidence to test this hypothesis is available.

The interesting observation that chimpanzee groups 2, 7 and 8 did not fit into the described series of allelic genes remains to be explained. This finding and the observation that sera of group 8 show distinct HL–A11 reactivity in human tissue typing (HL–A11 belonging to the LA–series), would be consistent with the hypothesis that the principal leukocyte anti-gens of chimpanzees are also controlled by two series of closely linked allelic genes. A critical test of this hypothesis depends upon the production of new chimpanzee reagents that will have to be tested in sufficiently large popula-tions of chimpanzees and human beings. In theory, we would thus be dealing with chimpanzee specificities of two allelic series that might repre-sent the equivalents or precursors of the allelic antigens belonging to the known sub-loci of HL–A.

SUMMARY

The serological identification of leukocyte specificities of Rhesus monkeys is described. As in various species, the genetic information for these specificities seems to be located on a single chromosome. By analogy with the human HL–A system, it is suggested that the Rhesus monkey's main histocompatibility locus be called RhL–A. Statistical analysis of population data as well as preliminary results of genotyping a number of families, is consistent with the hypothesis that certain specificities are con-trolled by a series of allelic genes. The relevance of the RhL–A system for histocompatibility was proven by inter-sib skin grafting.

A number of leukocyte specificities of chimpanzees is also described; suggestive evidence was obtained that some of the isoantisera recognize specificities of a hypothetical genetic system of chimpanzee antigens, which, again in analogy with the human HL–A system, is tentatively called ChL–A. The similarity of several ChL–A specificities with human antigens of the HL–A system was demonstrated by typing chimpanzee cells with monospecific human sera and human cells with chimpanzee isoantisera. Several chimpanzee sera contained antibodies of HL–A specificity when used for human tissue typing (HL–A1, HL–A7, HL–A11); antigens 4a and

4b appear to be distributed in an alternative fashion also on lymphocytes of chimpanzees.

The phylogeny of primate leukocyte antigens is discussed in the light of the current knowledge regarding the sharing of leukocyte specificities between lower and higher primate species.

Acknowledgements

The authors wish to thank the following persons and/or Institutions for help and assistance.

Dr D. E. Kayhoe of NIH, Bethesda and Dr F. T. Rapaport of New York University, for financial and other support which made possible the typing of 152 chimpanzees at the following three Primate Centers in the U.S.A.

Holloman Air Force Base, 6571st Aeromedical Research Laboratory, Alamagordo, New Mexico.

Bionetics Research Laboratories, Inc., 7300 Pearl Street, Bethesda, Md.

Laboratory for Experimental Medicine and Surgery in Primates (LEMSIP) New York.

Mrs A. Raisbeck (New York University), Dr R. F. Ziegler and Dr Th. M. Butler (Holloman Base), Dr C. C. Darrow and Mrs A. Vaal (Bionetics Laboratory) as well as Dr J. Moor-Jankowski and Mr J. Davis (LEMSIP) are gratefully acknowledged for their kind cooperation in this programme.

Professor D. W. van Bekkum, Director of the Radiobiological Institute TNO.

Mr E. J. Beijersbergen, H. Brinkhof, J. J. Schonk, C. Schoonhoven, H. D. Wiersema and several other members of the maintenance staff of the Primate Center TNO.

Mr J. de Kler and Mr A. A. Glaudemans (Art Department, Radiobiological Institute TNO).

Mr R. Zweerus of the Rekencentrum, Leiden (Computer Service).

We also thank Dr G. N. Rogentine of NIH, Bethesda and Dr J. Moor-Jankowski of the New York University, School of Medicine for kindly providing several typing reagents.

REFERENCES

Allen, F. *et al.* (1970). Joint Report of Fourth International Histocompatibility Workshop. *In* "Histocompatibility Testing 1970", pp. 17–47. Munksgaard, Copenhagen.

Amos, D. B., Gorer, P. A. and Mikulska, Z. B. (1955). An analysis of an antigenic system in the mouse (the H–2 system). *Proc. roy. Soc. B* **144**, 369.

Balner, H. and Dersjant, H. (1965). Iso-antibodies against leukocytes as a tool to study histocompatibility in monkeys. *In* "Histocompatibility Testing 1965", pp. 103–112. Munksgaard, Copenhagen.

Balner, H., Dersjant, H. and Rood, J. J. van (1965). A method to relate leukocyte antigens and transplantation antigens in monkeys. *Transplantation* 3, no. 2, 230–234.

Balner, H., Dersjant, H. and Rood, J. J. van (1966). Leukocyte antigens and histocompatibility in monkeys. *Ann. N.Y. Acad. Sci.* 129, 541–557.

Balner, H., Dersjant, H., Leeuwen, A. van and Rood, J. J. van (1967a). Identification of 2 major leukocyte antigens of Rhesus monkeys and their relation to histocompatibility. *In* "Histocompatibility Testing 1967", pp. 267–276. Munksgaard, Copenhagen.

Balner, H., Leeuwen, A. van, Dersjant, H. and Rood, J. J. van (1967b). Chimpanzee iso-antisera in relation to human leukocyte antigens. *In* "Histocompatibility Testing 1967", pp. 257–265. Munksgaard, Copenhagen.

Balner, H., Leeuwen, A. van, Vreeswijk, W. van, Dersjant, H. and Rood, J. J. van (1967c). Defined leukocyte antigens of chimpanzees: Use of chimpanzee isoantisera for leukocyte typing in man. *Transplantation* 5, 624–642.

Balner, H., Gabb, B. W., Dersjant, H., Vreeswijk, W. van and Rood, J. J. van (1970a). The major histocompatibility locus of Rhesus monkeys (RhL–A). *Nature, Lond.* 230 (1971), 177–180.

Balner, H., Leeuwen, A. van, Dersjant, H., Vreeswijk, W. van and Rood, J. J. van (1970b). Leukocyte antigens of Rhesus monkeys and chimpanzees. *Proc. 2nd Conf. Exp. Med. Surg. Primates.* In press. Karger, New York.

Balner, H., Leeuwen, A. van, Vreeswijk, W. van, Dersjant, H. and Rood, J. J. van (1970c). Leukocyte antigens of chimpanzees and their relation to human HL–A antigens. *Transplantation Proceedings* 2, 454–462.

Balner, H., Vreeswijk, W. van, Leeuwen, A. van and Rood, J. J. van (1971). Identification of chimpanzee leukocyte antigens (ChL–A) and their relation to HL–A. *Transplantation* 11, 309–317.

Bekkum, D. W. van and Waaij, D. van der (1970). Total body irradiation and bone marrow transplantation in monkeys. *In* "Infections and Immunosuppression in Sub-human Primates", pp. 225–229. Munksgaard, Copenhagen.

Bogden, A. E., Gray, J. H. and Brule, M. E. (1968). Relationship of hemagglutinatiogens and leukotoxic antibodies in the serum of Rhesus monkeys immunized with Rhesus leukocytes. *In* "U.S. Air Force Sch. Aerospace Med. Tech. Rep. No. SAM-TR-68-35", Brooks Air Force Base, Texas.

Ceppellini, R., Curtoni, E. S., Mattiuz, P. L., Miggiano, V., Scudeller, G. and Serra, A. (1967). Genetics of leukocyte antigens: A family study of segregation and linkage. *In* "Histocompatibility Testing 1967", pp. 149–187. Munksgaard, Copenhagen.

Ceppellini, R., Bigliani, S., Curtoni, E. S. and Leigheb, G. (1969). Experimental Allotransplantation in Man: II. The Role of A_1, A_2 and B Antigens; III. Enhancement by Circulating Antibody. *Transplantation Proceedings* 1, no. 1, 390–394.

Dausset, M., Iványi, P., Colombani, J. and Feingold, N. (1967). The Hu-1 system. In "Histocompatibility Testing 1967", pp. 189–202. Munksgaard, Copenhagen.

Dausset, J., Iványi, P. and Iványi, D. (1965). Tissue alloantigens in humans: Identification of a complex system (Hu-1). In "Histocompatibility Testing 1965", pp. 51–62. Munksgaard, Copenhagen.

Dorf, M. E. and Metzgar, R. S. (1970). Serological relationships of human, chimpanzee and gorilla lymphocyte isoantigens. In "Histocompatibility Testing 1970", pp. 287–297. Munksgaard, Copenhagen.

Duggleby, Ch. R. (1970). Immunogenetic and Immunoreproductive Studies of Rhesus Monkeys (Macaca Mulatta). Thesis, University Wisconsin.

Ferreby, J. W., Cannon, F. D., Mollen, N. and St. John, D. (1970). Beagles for studies of histocompatibility and organ transplantation. Transplantation 9, no. 1, 68–69.

Harris, R. and Zwerwas, J. D. (1969). Reticulocyte HL–A antigens. Nature, Lond. 221, 1062–1063.

Hildeman, W. H. (1970). In "Immunogenetics". Holden-Day Inc., San Francisco.

Hirose, Y. and Balner, H. (1969). Red cell iso-antigens of Rhesus monkeys. Blood 34, no. 5, 661–681.

Histocompatibility Testing 1970. Report of an International Workshop and Fourth International Histocompatibility Conference, Los Angeles, U.S.A. Munksgaard, Copenhagen.

Iványi, P. (1970). The major histocompatibility antigens in various species. In "Current Topics in Microbiology and Immunology", pp. 1–90, Springer-Verlag, Berlin.

Kissmeyer-Nielsen, F. and Thorsby, E. (1970). Human Transplantation Antigens. In "Transplantation Reviews 4", pp. 125. Munksgaard, Copenhagen.

Kissmeyer-Nielsen, F., Svejgaard, A. and Hauge, M. (1968). Genetics of the human HL–A transplantation system. Nature, Lond. 219, 1116–1119.

Maynard-Smith, S., Pennrose, L. F. and Smith, C. A. B. (1961). In "Mathematical Tables for Research Workers in Human Genetics". Churchill, London.

Metzgar, R. S. and Zmijewski, C. M. (1966). Species distribution of human tissue isoantigens. 1. Detection of human tissue isoantigens in chimpanzees. Transplantation 4, 84–93.

Metzgar, R. S., Seigler, H. F. and Zmijewski, C. M. (1967). The cross-reactions of primate tissue antigens with human leukocyte isoantibodies. In "Cross-reacting antigens and neoantigens", pp. 98–104, The Williams and Wilkins Company, Baltimore, Md., U.S.A.

Moor-Jankowski, J., Wiener, A. S. and Rogers, Ch. M. (1964). Human blood factors in non-human primates. Nature, Lond. 202, 663–665.

Moor-Jankowski, J., Wiener, A. S., Gordon, E. B. and Davis, J. H. (1967). Blood groups of monkeys: demonstrated with isoimmune Rhesus monkey sera. Int. Arch. Allerg. 32, 373–377.

Morton, J. A., Pickless, M. M. and Sutton, L. (1969). The correlation of the

Bg^a blood group with the HL–A7 leukocyte group: Demonstration of antigenic sites on red cells and leukocytes. *Vox. Sang.* **17**, 536–547.

Owen, R. D. and Anderson, D. R. (1962). Blood groups in Rhesus monkeys. *Ann. N.Y. Acad. Sci.* **97**, 1, 4–8.

Palm, J. (1964). Serological detection of histocompatibility antigens in two strains of rats. *Transplantation* **2**, 603–612.

Payne, R., Tripp, M., Weigle, J., Bodmer, W. and Bodmer, J. (1964). A new leukocyte isoantigen system in man. *In* "Cold Spring Harbor Symposium on Quantitative Biology 29", p. 285.

Rapaport, F. T., Dausset, J., Barge, A., Horse, H. and Converse, J. M. (1968). ABO erythrocytes in transplantation-sensitization to skin allografts with soluble "A" substance. *In* "Advance in Transplantation", pp. 311–315. Munksgaard, Copenhagen.

Reemtsma, K., McCracken, B., Schliegel, J., Pearl, M., Pearl, C., De-Witt, C., Smith, P., Hewitt, R., Flinner, R. and Creech, O. (1964). Renal hetero-transplantation in man. *Ann. Surg.* **160**, 384–410.

Rood, J. J. van and Leeuwen, A. van (1963). Leukocyte Grouping. A method and its application. *J. Clin. Invest.* **42**, 1382–1390.

Rood, J. J. van, Leeuwen, A. van, Schippers, A. M. J., Vooys, W. H., Frederiks, E., Balner, H. and Eernisse, J. G. (1965). Leukocyte groups, the normal Flymphocyte transfer test and homograft sensitivity. *In* "Histocompatibility Testing 1965", pp. 37–50. Munksgaard, Copenhagen.

Rood, J. J. van, Leeuwen, A., Schippers, A. M. J., Pearce, R., Blankenstein, M. and Volkers, W. (1967). Immunogenetics of the group four, five and nine systems. *In* "Histocompatibility Testing 1967", pp. 203–219. Munksgaard, Copenhagen.

Rood, J. J. van, Leeuwen, A. van and Zweerus, R. (1970). The 4a and 4b antigens. Do they or don't they? *In* "Histocompatibility Testing 1970", pp. 93–104. Munksgaard, Copenhagen.

Shulman, N. R., Moor-Jankowski, J. and Hiller, M. C. (1965). Platelet and leukocyte isoantigens common to man and other animals. *In* "Histocompatibility Testing 1965", pp. 113–123. Munksgaard, Copenhagen.

Snell, G. D., Smith, P. and Gabrielson, F. (1953). Analysis of the histocompatibility-2 locus in the mouse. *J. Nat. Cancer Inst.* **14**, 457.

Stark, O., Kren and Frenzl, B. (1966). Histocompatibility locus in the rat. *In* "Polymorphism bioch. animaux". *Prox. Xth Europ. Conf. Animal Blood Groups, Paris*, pp. 501–506.

Terasaki, P. I., Vredevoe, D. L. and Goyette, D. R. (1966). Serotyping for homotransplantation. IV. Grouping and evaluation of lymphotoxic sera. *Vox Sang. (Basel)* **11**, 350–376.

Vriesendorp, H. M., Rothengatter, E., Bos, E., Westbroek, D. L. and Rood, J. J. van (1971). The production and evaluation of dog allolymphocytotoxins for donor selection in transplantation experiments. *Transplantation* **11**, no. 5, 440–445.

Wiener, A. S. (1965). Blood groups of chimpanzees and other non-human primates. *Trans. N.Y. Acad. Sci.* **27**, no. 5, 488–504.

E*

Wiener, A. S., Moor-Jankowski, J., Balner, H. and Gordon, E. B. (1968). Blood groups in monkeys, demonstrated with antisera produced by combined skin transplantation and isoimmunization with red cells. *Int. Arch. Allerg.* **34,** 386–391.

Yasuda, N. and Kamura, M. (1968). A gene counting method of maximum likelihood for certainty gene frequencies in ABO and ABO-like systems. *Ann. Hum. Genet.* **31,** 409–420.

Zmijewski, C. M., Fletcher, J. L. and Cannady, W. G. (1970). The serological relationship of 4a and 4b to the HL–A system. *In* "Histocompatibility Testing 1970", pp. 277–286. Munksgaard, Copenhagen.

Genetic Structure and Systematics of some Macaques and Men

MARK L. WEISS AND MORRIS GOODMAN

Department of Anthropology, Wayne State University, Detroit, Michigan, U.S.A.;
Department of Anatomy, Wayne State University, Medical School,
Detroit, Michigan, U.S.A.

INTRODUCTION

During the last decade a new sub-field of anthropology has fully emerged with the realization that comparative biochemical and immunological data provides a most promising angle of attack on many aspects of primate evolution. In 1963, Zuckerkandl utilized the term "molecular anthropology" in reference to amino acid sequence analysis as a mode of deciphering primate relationships, but it seems reasonable to expand the umbrella of this name to include studies utilizing other techniques directed at the informational macromolecules (proteins and nucleic acids). Now, in addition to the study of the external form, proportions or anatomy of the primates, anthropologists can survey qualitative and quantitative variation of traits much closer to the genetic material.

While there were several isolated attempts at this type of analysis prior to 1960 (Nuttall, 1904; Florkin, 1944), it was not until recently that large scale, systematic investigations were carried out by many researchers employing a host of techniques. In the last 10 years, however, comparative analysis plates (Goodman, 1967, 1968), micro-complement fixation (Sarich and Wilson, 1967), DNA hybridization (Hoyer and Roberts, 1967; Kohne et al., 1970), electrophoresis (many authors, e.g. Barnicot et al., 1966; Tashian et al., 1968) and the primary structures of protein (Buettner-Janusch and Hill, 1965; Fitch and Margoliash, 1967) have all been utilized as part of a many-pronged attack on questions of primate phylogeny. The proliferation of techniques results from the fact that each has its maximum utility for deciphering processes and relationships at specific taxonomic levels.

For example, the trefoil plate comparisons are most powerful above the species level since genetic variation below this taxon is unlikely to yield antigenic dissimilarities. Electrophoresis, on the other hand, is most useful at the lower (infrageneric) taxonomic levels. Polymorphism or allelism of

129

proteins within populations can usually be attributed to single point mutations. Similarly, in higher primates, interspecific differences in proteins
among closely related, congeneric sibling species, like intersubspecific differences, are also typically due to single point mutations. This provides
the rationale for using allelic frequency data derived from electrophoretic
population surveys to measure genetic divergence between conspecific
populations and congeneric species.

In this paper, we will attempt to demonstrate two uses of electrophoretic
surveys. First, we would like to estimate the percentage of the genome
which is polymorphic in higher primates. In other words, at how many
loci can an individual be expected to be heterozygous? This is a central
problem of population genetics, for the amount of variation in an evolving
population limits the rate at which evolution can proceed by natural
selection.

Secondly, phylogenetic relationships may become much clearer when
based on substances close to the genetic level rather than on a few, heavily
weighted morphological criteria, for the genetic data allows for quantitative
as well as qualitative estimates of relationship.

We have attempted to clarify these problems on the basis of our studies
of the genus *Macaca*. This group is of particular interest to anthropologists
since the macaques are thought to approximate the spatial distribution and
systematic position of early Man, thus possibly providing a model for the
interpretation of the racial differentiation of Man.

ESTIMATION OF GENETIC VARIABILITY

Theoretical Considerations

In 1966, R. C. Lewontin and J. L. Hubby published two articles dealing
with a new approach to the assessment of the amount of genetic variability
in natural populations of *Drosophila pseudoobscura*. Starch gel electrophoresis was utilized to screen the individuals for variant forms of 18
proteins chosen at random. In this way the authors felt that they could
take an unbiased sample of gene products, and therefore structural genes,
in an attempt to determine the number of loci which can be expected to
show polymorphisms. H. Harris, also in 1966, employed the same method
in a study of human red-cell enzyme polymorphisms. Here, three populations of *Macaca fascicularis* (the crab-eating or cynomolgous monkey)
from known geographic areas are utilized to estimate this parameter in
another higher primate.

Theoretically starch gel electrophoresis will discriminate between allelic
products if they differ in net charge and/or molecular size. Due to the
electric field applied to the gel, forms with different net charges will mi-

grate differentially. Additionally, the sieve-like effect of the gel itself will serve the same purpose when the proteins differ in molecular size. When dealing with allelic products within a species, the probability is low, however, that the molecular alternatives will differ sufficiently in size as to be detectable by this procedure, due to the fact that a single mutational event is not likely to reduce the size of the protein to a great extent without disrupting its functional capability. Thus, the variations which survive selection pressures will most often reflect differences in charge, not size.

As Lewontin and Hubby and Harris rightly point out, starch gels will most likely yield an underestimate of the underlying variability. Assuming that a mutational event produces the deletion, addition or substitution of at least one amino acid, the possibility exists that this change will effect the net charge of the protein so that it may be detected upon electrophoresis.

Dealing with the case in which one amino acid is substituted for another due to a single base change, it has been estimated that roughly 27% of the substitutions will produce a protein with an altered net charge (MacCleur in Shaw, 1965). MacCleur's approximation is based on the fact that of the 20 common amino acids, 15 are neutral, 3 positively charged and 2 negatively. Thus, typing on starch gels will give a minimum estimate of amino acid substitutions and may underestimate the full extent of genetic variability.

A further underestimation could be introduced if known polymorphic loci are ruled out at the beginning of the study. This would disproportionately increase the number of monomorphic loci in the population from which the sample loci are chosen.

Notwithstanding these limitations, electrophoresis provides an efficient means of assessing genetic variability within populations. In the above mentioned studies, as well as this one, the systems were chosen at random, not because they were expected to show a polymorphic condition. Thus, an unbiased estimate of the variability in, at least, the structural genes is obtained, allelic variation in operator or regulator genes presently being impossible to ascertain in mammals. The quantitative variation in the product of the associated structural gene is not necessarily proof of a mutant regulator. Qualitatively different alleles at the structural locus may produce this result (Harris, 1966) as well as environmental factors.

Electrophoretic screening also satisfies two additional criteria mentioned by Hubby and Lewontin (1966: 578).

1. Phenotypic differences cased by allelic substitutions at single loci must be detectable in single individuals.

2. Allelic substitutions at one locus must be distinguishable from each other.

In the present study, 2 serum proteins (albumin, prealbumin) and 5 red-cell enzyme systems (lactate dehydrogenase, malate dehydrogenase, 6-phosphogluconate dehydrogenase, esterase, carbonic anhydrase) were selected at random for screening, the choices being made without reference to an expectation of a polymorphic condition. It was felt that the choices, 3 dehydrogenases, 2 hydrolases and 2 serum proteins provide a reasonably wide range of physiological functions. The inclusion of serum proteins was made to reduce any bias introduced by a survey of red-cell enzymes. Although it would be desirable to sample proteins from other tissues, it was impossible in this case since only blood samples were available. It is heartening that many erythrocyte enzymes such as LDH and MDH can be demonstrated to exist in the same form in other tissues (Harris, 1966).

Methodology

The whole blood utilized in this study was obtained in early 1969 by Dr William Prychodko of Wayne State University while on a collecting trip to South-east Asia. While the animals sampled within a population cannot be considered members of the same troops, they do represent well-defined, restricted geographic areas. In this sense they can be considered members of 3 "local races" (Garn, 1965). These samples, called Thailand (N = 119), Negri Sembilan (N = 76) and Philippines (N = 58) for brevity, represent populations from the area south-west and west of Bangkok, Thailand, the area south-east of Kuala Lumpur, Malaysia, within the state of Negri Sembilan and the island of Mindanao in the Philippines. Actually it is advantageous to have samples from several troops within an area. A species such as *M. fascicularis*, with a troop size of 6–40 (Fuyura, 1961; Napier and Napier, 1967), would introduce a second underestimate of heterozygosity due to isolation and inbreeding. And, certainly, it is not the troop which evolves but the population, as Lindberg's (1969) field study indicates.

Electrophoretic methods

All screenings were carried out by starch gel electrophoresis, except for prealbumin which was found to be more amenable to immunoelectrophoresis.

Lactate dehydrogenase: The gel buffer, pH 7·0, was 8 mM tris (hydroxymethyl) amino methane (Tris), 2·7 mM citric acid and 3% sucrose. The electrode buffer, pH 6·0, was 0·19 M Tris, 0·068 M citric acid and 3% sucrose. Vertical electrophoresis was performed at 5·3 V/cm for 18 hours at 4°C. After electrophoresis, one-half of the gel was incubated for approximately 1 hour in 150 ml of a solution containing 45 mg β-diphosphopyridine nucleotide (DPN), 38 mg nitro blue tetrazolium (NBT), 3 mg phenazine methosulfate (PMS), 5 ml 0·5 M Tris-HCl pH 7·0 and 7·5 ml of a neutralized 1·0 M Na-lactate solution.

Malate dehydrogenase: The second half of the LDH gel was stained in an incubation solution identical to that used for LDH except 7·5 ml of a neutralized 1·0 M Na-L-malate solution was substituted for the Na-lactate.

6-Phosphogluconate dehydrogenase: The pH 8·0 gel buffer was 0·02 M Tris, 3 mM citric acid, 0·8 mM ethylenediaminetetraacetate-Na_2 (EDTA) and 5% sucrose. The electrode buffer was 0·3 M Tris, 0·56 M citric acid, 8 mM EDTA, 5% sucrose at pH 8·0. Gels were run horizontally at 1·5 V/cm for 18 hours in the cold. Variant forms were then run vertically (4·8 V/cm, 4°C, 18 hours) to assure proper typing. The gels were incubated at 37°C in 150 ml of a solution of 15 mg NBT, 25 mg TPN, 3 mg PMS, 15 ml 0·05 M Tris-HCl pH 8·0, 2·5 ml 0·3 M $MgCl_2$ and 4·5 ml 0·05 M 6-phosphogluconate.

Esterase and carbonic anhydrase (Shaw and Koen, 1968): The esterases and carbonic anhydrase were stained on the same gel. The gel buffer was 0·03 M boric acid, 0·0125 M NaOH. The pH was adjusted to 8·5 with 0·1 M HCl. The electrode wells contained a pH 8·0, 0·3 M boric acid, 0·5 M NaOH buffer. The gels were run vertically at 5·0 V/cm for 18 hours in the cold.

The 150 ml incubation solution contained 150 mg Fast Blue RR, 15 ml 0·5 M Tris-HCl pH 7·0, 4·5 ml 1% α-napthyl acetate, 4·5 ml β-napthyl acetate. Just prior to staining, the two substrates were added to the solution, mixed and the incubation solution was filtered.

Albumin (Barrett et al., 1962): The electrode buffer, pH 8·0, was 0·1 M LiOH and 0·38 M boric acid. A Tris-citrate buffer pH 8·0 was made 0·016 M Tris, 3·3 mM citric acid. The gel buffer was made by adding 1 part electrode buffer to 9 parts Tris-citrate buffer. The gels were run at 3·2 V/cm for 18 hours at room temperature.

Prealbumin: This protein was found to be more amenable to study by immunoelectrophoresis. This is due to the low concentration of the protein, yielding a weak stain on starch gels.

The electrode buffer, pH 8·2, was 0·065 M Na-acetate and 0·065 M Na-barbital. 100 ml of the electrode buffer, 200 ml H_2O, 3 g Ionagar No. 2 (Oxo Limited, London) and 0·3 ml 10% merthiolate were used to make the gel. The antisera was Behringwerke (Certified Blood Donor Service, Inc., Woodbury, New York) anti-human prealbumin made in rabbits. The stain was made with 1 g amido black dye powder, 54 ml glacial acetic acid, 12·5 g Na-acetate, 100 ml glycerine brought to one liter with water. The slides were run at 6·25 V/cm at room temperature for 1 hour. They were then fixed in a 2 ethanol: 2 water: 1 acetic acid solution.

Genetic and ecological considerations

Loci were considered to be polymorphic in a population if there was more than one allele present and the variant had a frequency of at least 1%. This is in accord with Ford's (1940) classic definition of a polymorphism as "the occurrence together in the same habitat of two or more discontinuous forms of a species in such proportions that the rarest of them cannot be maintained merely by recurrent mutation". The 1% cutoff was chosen somewhat arbitrarily but was consistently used throughout.

The gene frequencies are presented in Table I. All loci were found to be in Hardy-Weinberg equilibrium but it should be remembered that the chi-square test is relatively insensitive to even moderate departures from expectations (Workman, 1969).

TABLE I. Gene frequencies for loci tested in Thailand, Malayan and Philippine *M. fascicularis*[a]

Locus	Allele	Thailand N = 119 Gene frequency	Negri Sembilan N = 76 Gene frequency	Philippines N = 58 Gene frequency
Prealbumin	F	·996	·868	·941
	S	·004	·132	·059
Albumin		1·0	1·0	1·0
LDH–M	F	·008	0	0
	S	·992	1·0	1·0
MDH		1·0	1·0	1·0
6-PGD	A	·966	·855	1·0
	C	·025	·138	0
	D	·008	·007	0
Esterase A1	A	·996	1·0	·931
	B	·004	0	·069
Carbonic	A	·979	·967	1·0
Anhydrase 1	C	·004	·033	0
	G	·017	0	0

[a] All loci are in Hardy-Weinberg equilibrium.

The calculated gene frequencies provide data which can be applied to several theoretical problems. Two interrelated questions will be considered here. First, are these results comparable to those obtained on widely differing organisms? These comparisons are especially interesting since they test the generality of Lewontin's findings by comparing the frequency of polymorphisms in insects and mammals. A comparison with Man also gives information about animals relatively close phylogenetically. The second question is how much intra-specific, inter-population variability exists and to what might differences be due ? Essentially the data are being applied to the same question but on two levels, macro- and micro-evolutionary.

Lewontin and Hubby found that 39% (7/18) of the loci tested were variable in at least one of their groups. In the combined macaque samples 4 (6-PGD, prealbumin, esterase Al, CA l) of the 7 loci tested (57%) were polymorphic. Within any one population, however, the maximum proportion of polymorphic loci was 43% (Negri Sembilan). Harris' data showed 3 of the 10 loci tested to be variable in a sample of humans from England. In light of the small number of loci viewed, all of these results are in remarkably good agreement.

An interesting difference is noted if one compares the proportion of the genome heterozygous for one individual. This is done by use of the Hardy-Weinberg expectation for heterozygotes and then averaging over the number of loci observed. For instance, in Thailand at the 6-PGD locus the proportion of heterozygotes expected is calculated as:

$$2(\cdot966)(\cdot025) + 2(\cdot966)(\cdot008) + 2(\cdot025)(\cdot008) = \cdot064$$

This is then averaged with comparable figures for the other 6 loci. Obviously the monomorphic loci contribute no variability.

This figure is found to be equal to 8% in Negri Sembilan, 2% in Thailand and 3% in the Philippines, with a weighted overall average of 4%. Lewontin and Hubby, however, found that between 8 and 15% ($\bar{x} = 11\cdot5\%$) of the loci in an individual can be expected to be heterozygous. Harris does not give enough data to precisely calculate this statistic, but it is approximately 9%.

The disparity of the findings in *Drosophila* and macaques reflects the lack of loci with a high degree of polymorphism in the latter. In *Drosophila* 3 loci were found to have ubiquitous polymorphism; they were essentially without a wild type. One locus had 6 alleles with as many as 5 of these present in any single strain; accordingly, over 50% of the population was expected to be heterozygous for this one locus. In the macaque series the highest degree of heterozygosity for any one locus was 25% for 6-PGD in Malaya. Prealbumin was the only other protein to approach this degree of variability. Thus ubiquitous polymorphisms seem to be rarer than in *Drosophila*.

The most obvious explanation for the discrepancy could be sampling error. Due to the relatively small number of loci involved in both studies it is very possible that this degree of discordance could be due to chance factors.

A much more interesting and speculative explanation of the difference can be based on the hypothesis that evolution is decelerating in higher organisms. One of us (Goodman, 1961) originally set forward this idea in an attempt to show that the higher primates are evolving at a slower rate than the lower primates. This supposition is based in part on the fact that the hemochorial placenta provides an intimate apposition of maternal and fetal blood. Consequently, transplacental leakage, a phenomenon known to occur, allows fetal antigens to enter the maternal blood stream. The result may be the production of maternal antibodies which then may attack the fetus by crossing the placenta. The best understood example of iso-antigenic incompatibility is erythroblastosis fetalis or Rh incompatibility in Man.

The evolutionary result of such a process may be a reduction in the number of immunologically tolerated mutations. That is, a mutation which changes the antigenic structure of a fetal molecule might produce a maternal reaction which acts as an immunological selective agent. The Rh incompatibility certainly acts as a selective force and will tend to fix either the Rh + or Rh− allele in the population, thus increasing homozygosity.

The long gestation period in the higher primates is also a factor. It is known, again from Rh incompatibility, that antibody production is not immediately effective. Not until the second or third pregnancy, after the mother has been sensitized, is the anti-Rh+ present at a high enough titer to damage the fetus. A shorter gestation period would serve to increase the number of incompatible pregnancies required for an immunologically competent response. Since the average number of births per female macaque per lifetime is approximately 5 or 6 (Napier and Napier, 1967), this could be an important selective agent.

While Goodman applied this concept to the higher versus lower primates, it is also applicable to a comparison of a mammal versus an insect. In the latter, immunological rejection is not possible while it is a definite possibility in macaques, especially considering their long gestation period. This hypothesis may explain some of the variation observed between macaque and *Drosophila*, for it allows for the maintenance of more alleleomorphs in the fly.

All of this is not to say that multiple alleles cannot occur at high frequencies in macaques. At the transferrin locus, for instance, 11 alleles are found in Thailand. There also seems to be a gene duplication and allelic variation for hemoglobin. In Man many polymorphic loci are also known. However, it is possible that proportionately fewer loci are highly variable in placentals than in lower animals. Confirmation of this theory would depend on a much larger screening program of randomly chosen proteins.

One might also expect that an insect, with a generation time measured in weeks, might evolve more rapidly in a unit of time than a mammal whose generation is measured in years. This is based on the assumption that the mutation rate is better measured as a function of generation time rather than absolute time. However, conflicting opinions (Crow, 1963; Herskowitz, 1962) exist on this matter. If the mutation rates of insects and mammals are equivalent when measured on the basis of generations, then more mutations would be available for incorporation into the gene pool of the insect population during a given unit of time. This could result in a greater number of polymorphisms in the insect population, especially if neutral mutations do exist (King and Jukes, 1969). Recently, Carson (1970) has demonstrated the evolution of reproductively isolated species of *Drosophila* in the relatively short span of 750,000 years.

The second aspect of the data to be considered is the marked difference in polymorphism of Malaya and the peripheral populations of Thailand and the Philippines. The calculated degree of heterozygosity for Malaya is 8%, while Thailand is 2% and the Philippines is 3%. Another interesting point is that Thailand, while having only two polymorphic loci (6-PGD, CA 1), has variants for 3 additional proteins (esterase A1, LDH, prealbumin). Neither of the other populations exhibits variants at monomorphic loci.

Another measure of the difference between populations is attained by use of the Shannon-Weaver (1949) information coefficient. Basically, this measures the amount of information that can be transmitted at a given locus. A locus with one allele provides no information, there is no variation. A 2-allele locus with both forms at 50% would yield a coefficient of one. The sum of information coefficients for Malaya averaged over 7 loci is 0·096, for Thailand 0·038 and for the Philippines 0·047. These figures are on the same order as those for the degree of heterozygosity. It is not possible to compare these results with Lewontin's for his study was based on the variance between inbred lines.

The classic explanation given for the low genetic variability in island populations is the founder's effect, a form of genetic drift. Essentially, this states that an island population is inhabited by a small number of individuals, possessing a gene pool unrepresentative of the main population from which the emigrants were drawn. The probability is that an allele present at a low frequency in the main population will not be found among the migrants. Due to the extremely low rate of subsequent migration to the island, chances are that the allele will never be introduced. While this is a plausible explanation for the distribution of alleles in crab-eating macques, it is not especially satisfying. For one thing, it is a statement which can never be proven after the fact for it is essentially stating the negative—"no selection has occurred".

A more satisfying explanation of low genetic variability in island populations was put forward by Mayr (1954) and later confirmed by Dobzhansky and Pavlovsky (1957) and Carson (1970) in the laboratory and field. It is a well-known fact that evolutionary change is a function of both the external and internal environments. Included in the latter is the genetic make-up of the individual. That is, a gene's fitness is not solely a property of its own functioning in relation to the external conditions. Fitness is also dependent upon the interaction of a gene with genes at other loci. To use Dobzhansky's term, selection will produce a coadapted gene pool.

Mayr points out that a small population not only contains an uneven representation of allelic variation for one locus but for many loci. This sampling error of the original gene pool may well disrupt the coadapted character of the gene complex, consequently altering selection pressures. An allele which was selected because of its ability to function well on a wide range of genetic backgrounds may be disadvantageous in the genetically limited island situation. This sudden modification of the genetic environment constitutes a massive shock to the population since it affects all loci at once, setting off a chain reaction for a change at one locus affects the selective values at other loci. The initial reduction in genetic variability produced by drift will be followed by a further decline due to selection. Eventually a new coadapted gene pool will evolve, followed by an increase in variability.

Unfortunately, little is known about the entry of *M. fascicularis* into the Philippines. Wallace (1890) and Darlington (1957) feel that the monkey was brought to the islands as a pet, presumably in the last one to two thousand years. This is no more than an opinion based on the fact that Asians keep monkeys as pets.

On the other hand, there is no reason to discredit the idea that entry was gained during one of the Pleistocene glaciations, at which time most of the present-day Malaya archipelago was above sea level.

At the present time, it is impossible to make any statement about the possible difference in the ecology and comparability of selective agents in the Philippines and Malaya. To date, *M. fascicularis* has not been extensively studied in the wild. Only several brief notes (Bernstein, 1968; Fuyura, 1961–62) are available. Wernstedt and Spencer (1967) do note, however, that the flora and fauna of the Philippines is not quite as rich as that of the tropical areas of mainland Asia. Areas inhabitable by *M. fascicularis* seem to be limited and somewhat marginal. Darlington states that, "(The Philippines) are fringing islands with a fringing progressively depauperate . . . fauna" (1957: 506). This situation would tend to alter the selective pressures from the predominantly density-dependent ones (disease, competition for food, space, etc.) to density-independent agents, mainly climatic (Haldane, 1951). While it is presently impossible to relate the physiological functioning of these proteins to given density-dependent or density-independent selective agents, it is not unlikely that such relationships could exist (Fuller and Thompson, 1960).

The situation in Thailand is more difficult to explain. Here 5 of the 7 loci have variants but only 2 of these are polymorphic. Ecological surveys may help to provide a definite explanation of this situation. Thailand is the northern limit of the range of the crab-eating macaque (Napier and Napier, 1967) and is likely to be marginal. Prychodko (personal communication) states that *M. fascicularis* in Thailand has been forced into a marginal area due to Man's activities. Here, to a much greater extent than in Malaya, the crab-eating macaque is being forced into marginal mountainous areas. Darlington (1957) does note that this is roughly the area of transition between the tropical Oriental fauna and Paleartic faunas. As in the Philippines, therefore, the selective agents may be density-independent rather than density-dependent. This explanation is not entirely satisfactory for in contrast to the Philippines there is probably a fairly constant flow of genes from the central area into this outlying population. Thus, an equilibrium is produced between selection against certain alleles and gene flow into the population. The *S* allele of prealbumin, for instance, might be involved in a balanced polymorphism in Malaya but not in Thailand. While selection might tend to fix its frequency at zero in Thailand, gene flow from the south would prevent fixation.

Four species of macaque are sympatric in Thailand (*M. mulatta, M. nemestrina, M. speciosa, M. fascicularis*) while only 2 inhabit Malaya (*M. nemestrina, M. fascicularis*). As a result, selection pressures in the northern area of the crab-eating macaque's range may be different for two additional reasons. First, it is probable that interspecific competition is much more severe in this area especially in light of Man's activities. In fact, this may be the factor limiting the northward spread of the crab-eating macaques rather than climatic conditions. Also, introgression of alleles from the other 3 species may prevent the Thailand crab-eating group from establishing an equilibrated, well-adapted gene pool. Cross-breeding between species could produce a sizable proportion of selectively disadvantaged animals. We will return to the question of introgression later.

In the past few pages, selection pressures have been utilized to help explain the results. This is not necessarily a contradiction of the finding that all the variants are in Hardy-Weinberg equilibrium. While such an equilibrium implies that selection is not occurring and deviations from Hardy-Weinberg indicates selection, it is often very difficult to demonstrate the latter. Boyd (1955) has shown that extremely large sample sizes, of the order of 10,000, are necessary to show a significant departure from Hardy-Weinberg by the use of chi-square when the selection pressure is of a low magnitude. The allelic variations observed here may well be due to balanced polymorphisms.

The possibility also exists that the polymorphisms represent the effects of drift on neutral mutations. This question of neutrality of mutants has recently been reviewed by King and Jukes (1969). In the present study it is impossible to discriminate between the neutral mutants and balanced polymorphism since all loci are in Hardy-Weinberg equilibrium. This fact might suggest neutrality but as mentioned before much larger samples would be necessary to detect slight departures from Hardy-Weinberg due to small selection pressures. Since it is probable that mutations can both change the primary structure of a protein and yet be selectively neutral, it is possible that at least part of the variation observed is due to the random incorporation of neutral mutants into the gene pool.

MOLECULAR APPROACH TO SYSTEMATICS AT THE INFRAGENERIC AND POPULATION LEVELS

The Agglomerative Approach

The electrophoretically derived data can also be used to generate clado-grams of closely related lineages. Basically the approach is that of Fitch and Margoliash (1967) wherein the allelic frequencies are used to measure genetic divergence between populations. Two populations which show no common alleles at a gene locus have an index of dissimilarity (a mutation distance) of one at this locus; whereas two populations which share alleles at identical frequencies have a zero index of dissimilarity (I.D.) at the locus. If the two populations share the alleles at different frequencies intermediate I.D. values from 0 to 1 are found, depending on how different the frequencies are for each allele.

The clad computer program (developed by our colleague W. Moore) can process such I.D. values to generate a branching arrangement of the populations. If these I.D. values are based on several highly polymorphic loci, the dendrogram produced by the agglomerative procedure of the clad program can be considered a cladogram, i.e. a graph depicting the sequence of ancestral branching. The agglomerative procedure used is copied after the unweighted pair group method described by Sokal and Sneath (1963). It builds the tree from the smallest to the largest branches in a series of pairwise clustering cycles. The first cycle joins together the two singleton groups in the total collection with the smallest mutation distance between them; then these two joined populations are treated as one group in the next agglomerative cycle which again unites the two member groups separated by the smallest I.D.; this process being repeated until a complete tree is generated.

Macaque Systematics

The data now presented represents a fraction of our results on the systematics of the macaques, here dealing only with *M. mulatta*, *M. cyclopis* and *M. fascicularis*. During the past few years debates have arisen as to the use of the term "species" in relation to many macaque groups. Fooden (1964), for instance, has found naturally occurring hybrids between rhesus and crab-eating monkeys, Bernstein (1968) has noted monkeys which are evidently the result of hybridization of *M. nemestrina* and *M. fascicularis* and data on transferrin (Goodman, 1965, 1967, 1968) also supports the contention that introgression occurs between macaque species. As Bigelow (1965, p. 458) notes, however, "reproductive isolation should be considered in terms of gene flow and not in terms of interbreeding . . .". That is, species barriers are more a function of the actual passage of genes, rather than the ability to hybridize upon occasion, making the quantity of gene flow the important parameter. This accords with Washburn's (personal communication) contention that most of the "naturally" occurring hybrids are found under rather peculiar conditions due to Man's disruption of isolating barriers, thus being comparable to hybrids in zoos.

We felt that in order to ameliorate this problem we could generate a cladogram based on several polymorphic loci in an attempt to quantify the distance between the rhesus, crab-eating and Formosan macaques. The 5 loci picked were transferrin, carbonic anhydrase I, 6-phosphogluconate dehydrogenase, thyroxine-binding prealbumin and hemoglobin. The first cladogram or phylogeny generated encompassing populations of *M. mulatta* from Thailand, East Pakistan and India, *M. cyclopis* from Formosa and *M. fascicularis* from Thailand, Malaya and the Philippines, is depicted in Fig. 1. It should be noted that we assume that electrophoretic equivalence of gene products implies allelic equivalence. Since parallel evolution at the molecular level is thought to be rare, this assumption is not considered to be an obstacle to this approach, especially at the intrageneric level.

Perhaps the most striking observation is that *M. cyclopis* appears as a subgroup of *M. mulatta*. Indeed it is depicted as genetically closer to the central rhesus populations than is the rhesus found in Thailand. *M. mulatta* whose present range extends from West Afghanistan eastward to the South China, East China and Yellow Seas probably also inhabited Formosa in the not too distant past. Darlington (1957, p. 492) states, "It (Formosa) has evidently been connected to the mainland recently, for its fauna is mainly that of southern China . . .". As with the Philippine crab-eating monkey, Formosan macaques were probably isolated after one of the Pleistocene inter-glacials. Behaviorally the rhesus and Formosan macaques are also quite similar, being the most terrestrial members of the genus (Napier and Napier, 1967).

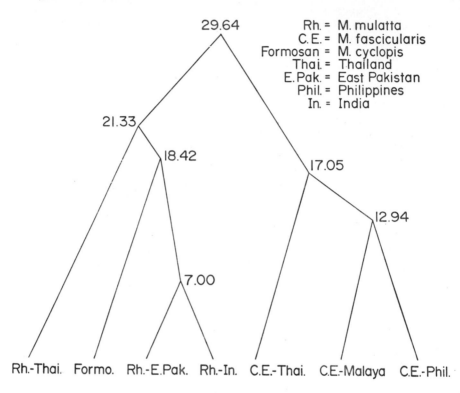

FIG. 1. Cladogram of macaque populations generated from the data on 5 gene loci. The nodal numbers are the average I.D. values.

Another important factor which is not immediately apparent from the cladogram is that the Thailand *M. mulatta*, while most closely tied to other rhesus groups, is significantly closer to the *M. fascicularis* of Thailand than to other cynomolgous populations. Unfortunately, a two-dimensional graph cannot express this relationship adequately, but Table II bears this out. As we mentioned previously, Thailand does seem to be a marginal area for macaques, especially of late, and the ecological condition is thought to account, in part, for the genetic divergence of both indigenous Thai populations from their conspecific central populations.

TABLE II. Matrix of macaque indices of dissimilarity

		a	b	c	d	e	f
a	*M. fascicularis* (Thailand)						
b	*M. fascicularis* (Malaya)	14·60					
c	*M. fascicularis* (Philippine)	19·50	12·94				
d	*M. mulatta* (Thailand)	27·40	35·36	38·62			
e	*M. mulatta* (E. Pakistan)	19·32	26·52	31·80	19·54		
f	*M. mulatta* (India)	20·50	29·18	32·24	23·96	7·00	
g	*M. cyclopis* (Formosa)	26·70	33·80	34·80	20·48	19·46	17·38

In fact, a pronounced cline of I.D. values emerges showing increasing genetic divergence as one travels eastward from India through the rhesus range and then south-eastward through the range of *M. fascicularis*. The only break in this cline occurs in Thailand, for the cynomolgous population here, not the rhesus, is closer to the Indian rhesus group, again supporting the marginality of Thailand and possibly gene flow between these species.

Another cline is apparent, notably the arcuate one within the crab-eating monkeys. Here the Thai group is closest to the Malayan and furthest from the Philippine. It would be most interesting if we could test animals from Borneo, Java and other islands of the area, the one major gap in our data, but as yet we have been unable to acquire the necessary samples.

Comparative Human Data

What then, one might ask, is the taxonomic and systematic significance of this cladogram. Of course, it is necessary to realize that the indices of dissimilarity are based on a very small proportion of the genome and that as the number of loci utilized increases so does the probability that the results reflect the reality of the situation. But, on the other hand, there is no *a priori* reason to feel that the loci used here provide a skewed picture of macaque relationships. In order to develop a feeling for the relationships shown in the macaque cladogram allelic data for 6 loci was gathered for 17 populations of humans from New Guinea, Australia and India (Table III). By using the divergence between populations of Man as a yardstick, we can better appreciate the levels of divergence between groups of macaques. It is pleasing to note that the cladogram shown in Fig. 2 generally agrees in topology with the classically accepted view that Australian Aborigines

TABLE III. Human populations and sources of data
utilized in the clad computer program

Area	Reference
India	
a Irulas	Kirk *et al.*, 1962b
b Oraons	Kirk *et al.*, 1962a
Australia	
c Edward River	Kirk, 1965
d Mitchell River	Kirk, 1965
e Elcho Island	Kirk *et al.*, 1969
	Simmons and Cooke, 1969
f Forest River	Kirk, 1965
g Port Hedland	Kirk, 1965
h Haast Bluff	Simmons *et al.*, 1957
	Nicholls *et al.*, 1965
i Western Desert	Kirk, 1965
New Guinea (Language Family)[a]	
j Fore (NAN)	Simmons *et al.*, 1961
	Curtain *et al.*, 1965
k Enga (NAN)	Simmons *et al.*, 1961
	Curtain *et al.*, 1965
l South Highlands (NAN)	Vines and Booth, 1964–65
	Curtain *et al.*, 1965
m Onga-Naruboin (MN)	Giles *et al.*, 1966
	Baumgarten *et al.*, 1968
n Bampa-Antir-Siats (MN)	Giles *et al.*, 1966
	Baumgarten *et al.*, 1968
o Kusing (NAN)	Giles *et al.*, 1966
	Baumgarten *et al.*, 1968
p Narumonke (NAN)	Giles *et al.*, 1966
	Baumgarten *et al.*, 1968
q Tumbuna (NAN)	Giles *et al.*, 1966
	Baumgarten *et al.*, 1968

[a] NAN, Non-Austronesian; MN, Melanesian.

are genetically close to the Dravidian peoples of India. Most importantly we see that the distance between New Guineans and the combined Indian-Australian group is just slightly less than that separating *M. mulatta–M. cyclopis* and *M. fascicularis*. In fact, *M. cyclopis* is closer to Indian and Pakistani rhesus than Australian Aborigines are to Indians. While we do not feel these trees are necessarily the best ones (see Fitch and Margoliash, 1967) they do indicate that the gap between *M. mulatta* and *M. cyclopis* is certainly not very deep nor is that between the rhesus–Formosan group on the one hand and the crab-eating monkeys on the other. On the basis of this type of data we cannot directly quantify gene flow between these groups during the recent past, but we certainly can indirectly approximate this parameter.

The data on the human populations have been gathered from the literature. Choice of loci, therefore, was primarily a function of the availability of the data and is mainly blood group allelic frequencies (ABO, MN, Rh, P) but also includes haptoglobin and transferrin. Efforts were made to use information derived from samples of at least 200 individuals, but it was not always possible to hold to this rule.

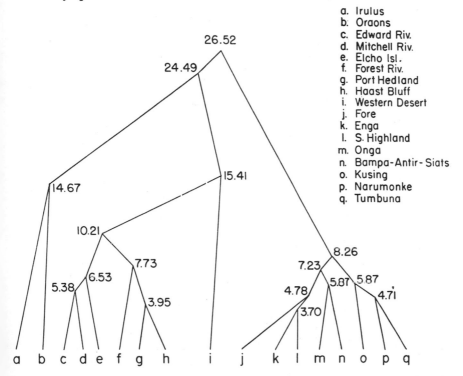

a. Irulus
b. Oraons
c. Edward Riv.
d. Mitchell Riv.
e. Elcho Isl.
f. Forest Riv.
g. Port Hedland
h. Haast Bluff
i. Western Desert
j. Fore
k. Enga
l. S. Highland
m. Onga
n. Bampa-Antir-Siats
o. Kusing
p. Narumonke
q. Tumbuna

FIG. 2. Ancestral branching sequence of 17 human populations from Oceania and India.

Interpretation of the human material (Table IV) is muddied by the fact that we really do not have a very good idea of the relationships uniting the Dravidians, New Guineans and Australian Aborigines. In general outline there is common agreement that indigenous Australians are derived from a South Indian ancestor, at least in part, and that New Guineans are related to Australians to a greater or lesser degree, with "some Mongoloid

TABLE IV. Matrix of human indices of dissimilarity[a]

	a	b	c	d	e	f	g	h	i	j	k	l	m	n	o	p
a																
b	14·67															
c	26·18	19·00														
d	26·47	17·91	5·38													
e	29·83	19·87	7·52	5·55												
f	28·25	20·22	8·10	6·98	9·67											
g	27·30	20·50	11·88	10·43	11·45	7·05										
h	28·13	22·45	12·67	9·72	11·00	8·40	3·95									
i	31·47	25·72	19·25	17·03	14·62	18·68	11·63	11·25								
j	37·80	25·73	24·03	20·48	20·97	19·63	19·47	19·72	18·98							
k	38·82	26·97	26·47	22·92	21·83	22·07	22·13	22·75	21·73	5·52						
l	38·37	26·18	25·90	22·35	21·90	21·50	21·48	22·10	20·42	4·05	3·70					
m	41·17	29·02	30·53	26·98	24·77	26·13	26·72	27·33	23·22	9·42	4·58	5·93				
n	37·78	26·52	31·03	27·48	25·27	26·63	27·65	28·27	24·08	9·33	6·48	7·63	5·87			
o	39·28	27·02	28·62	25·07	25·92	25·48	26·77	27·38	26·93	8·35	7·23	8·42	7·25	7·88		
p	40·02	28·72	26·33	22·78	21·30	23·27	27·02	27·30	27·10	11·45	6·80	10·10	7·43	7·38	6·70	
q	38·07	26·62	28·23	24·68	25·27	24·65	27·33	27·48	27·83	9·78	7·97	9·43	8·80	5·65	5·03	4·72

[a] The letters refer to populations listed in Table III.

and Negritic elements" (Birdsell, 1948; Graydon *et al.*, 1958; Howells, 1967), these considerations being primarily based on morphological comparisons. Linguistically, the situation also is unclear. Some have theorized that the Dravidian language family is related to the Melanesian languages spoken by some New Guineans and to the Australian family but this theory is not widely held today. Alternative relationships have not been hypothesized so the topology of our cladogram has hardly been discredited.

Most disagreement with the cladogram would probably center on the relation of the Indians to the other two major geographic groups. Some would probably feel that the Australian Aborigines and New Guineans should stand closer to each other than either is to the Dravidians. Due to the confusion surrounding the linguistic and morphological data, the relationships shown in Fig. 2 are by no means unlikely. In fact, as we have already mentioned, the technique employed here may provide the closest approach to reality since it is completely objective and is based on substances little removed from the genetic level.

One oddity noted is the relatively high degree of divergence of the Western Desert group from the other Australian Aborigines. The primary reason for this results from the peculiar frequencies at the MN and Rh loci in the Western Desert. Since the allelic data was gathered under far from ideal circumstances, it is probably safe to discount this deviation.

Within New Guinea the linguistic data agrees reasonably well with the geneaology. Kusing, Narumonke and Tumbuna are all non-Austronesian speakers and they all appear genetically closer to each other than they are to Melanesian speakers from Onga and Bampa-Antir-Siats both of which are combined in one of the early agglomerative cycles. The Fore, Enga and South Highland samples present some difficulty since they appear to be genetically closer to the Melanesian speakers but are in fact Non-Austronesian speakers. The fact that the Non-Austronesian groups (Fore, Enga, South Highland and Kusing, Narumonke, Tumbuna) are not as close to each other as the former group is to the Melanesian speakers is really not too surprising for Non-Austronesian is not a language family but rather an amalgam of residual languages (Grace, 1968). Furthermore, the Melanesian speakers are in close geographic proximity to the Fore.

Because of the likelihood of gene flow between some of these groups, a low I.D. value may be due, not to recent descent from a common ancestor, but rather to admixture. Thus, at low levels of comparison, such as within New Guinea, the relations pictured are best viewed as dendrograms or branching sequences based on genetic relatedness rather than cladograms. In considering major divisions, such as Australia versus India, this stipulation does not hold.

Additionally, the genetic dispersion of human populations should be thought of as more complex than that of macaques. It is not hard to imagine that cultural factors (marriage rules, settlement patterns, etc.) operate which can either retard or advance the rage of divergence. Obviously the human data must be more carefully analyzed in terms of these factors. For instance, the Edward and Mitchell River samples, which show a very low I.D., are probably drawn from the same breeding population (Pilling, personal communication). However, in broad outline the human data does provide a frame of reference for interpretation of the macaque data.

Alternatively, in many respects the modern day macaques provide a model of an early stage of hominid evolution. As did early Man, they have radiated throughout the warmer regions of the Old World and yet, on the basis of genetic data, have not undergone speciation in the classic sense. The two most successful groups of macaques, *M. mulatta* and *M. fascicularis*, have remained quite similar, seemingly no more different from each other than are races of men.

Thus it seems that in some of the higher primates factors are operating which retard speciation. In part this is due to the increasing intimacy between fetal and maternal blood systems, as well as the high degree of mobility and adaptability of these animals.

SUMMARY

The use of electrophoretically derived allelic frequency data can be of great aid in answering some persistent questions of interest to physical anthropologists. We have attempted to show how such information can aid in estimating both intra-specific and inter-specific variation.

In the former case we have found the central population of *M. fascicularis* to be the most heterozygous, while the ecologically marginal areas have greatly reduced variability. Our data on *M. mulatta*, while not completely analyzed, also show this tendency. Thus, it would seem that the greater degree of heterozygosity in the central range of a species would provide the variability for most adaptive changes.

At a higher level, the data are used to generate the ancestral branching sequences of *M. mulatta*, *M. fascicularis* and *M. cyclopis* populations, as well as for several human populations. Of major interest is our finding that species of macaques appear to be no more different from each other than are some races of men.

Assuredly, the increasing growth of molecular approaches to physical anthropology will provide for a more objective view of primate evolution and systematics.

REFERENCES

Barnicot, N. A., Huehns, E. R. and Jolly, C. J. (1966). *Proc. Roy. Soc. Lond. B* **165**, 224–244.

Barrett, R. J., Freisen, H. and Astwood, E. B. (1962). *J. Biol. Chem.* **237**, 432–439.

Baumgarten, A., Giles, E. and Curtain, C. C. (1968). *Am. J. Phys. Anthropol.* **29**, 29–37.

Bernstein, I. S. (1968). *Folia Primat.* **8**, 121–131.

Bigelow, R. S. (1965). *Evol.* **19**, 449–458.

Birdsell, J. B. (1948). *Rec. Queen Vic. Mus. Launceton II* **3**, 105–122.

Boyd, W. (1955). *Am. J. Phys. Anthropol.* **13**, 37–52.

Buettner-Janusch, J. and Hill, R. L. (1965). *Science, N.Y.* **147**, 836–842.

Carson, H. L. (1970). *Science, N.Y.* **168**, 1414–1418.

Crow, J. F. (1963). "Genetics Notes", 5th Edition. Burgess, Minneapolis.

Curtain, C. C., Gajdusek, D. C., Kidson, C., Gorman, J. G., Champness, L. and Rodrigue, R. (1965). *Am. J. Phys. Anthropol.* **23**, 363–379.

Darlington, P. J. (1957). "Zoogeography: The Geographical Distribution of Animals". John Wiley, New York.

Dobzhansky, T. and Pavlovsky, O. (1957). *Evol.* **11**, 311–319.

Fitch, W. M. and Margoliash, E. (1967). *Science, N.Y.* **155**, 279–284.

Florkin, M. (1944). "L'evolution Biochimique", 1st Edition. Mason et Cie, Paris.

Fooden, J. (1964). *Science, N.Y.* **143**, 363–365.

Ford, E. B. (1940). *In* "The New Systematics" (J. Huxley, ed.), pp. 493–513. Clarendon Press, Oxford.

Fuller, J. L. and Thompson, W. R. (1960). "Behavior Genetics". John Wiley, New York.

Fuyura, Y. (1961–62). *Primates* **3**, 75–76.

Garn, S. M. (1965). "Human Races", 2nd Edition. C. C. Thomas, Springfield.

Giles, E., Walsh, R. J. and Bradley, M. A. (1966). *Ann. N.Y. Acad. Science* **134**, 655–665.

Goodman, M. (1961). *Human Biol.* **33**, 131–162.

Goodman, M. (1967). *Am. J. Phys. Anthropol.* **26**, 255–275.

Goodman, M. (1968). *Primates Med.* **1**, 10–26.

Goodman, M., Kulkarni, A., Poulik, E. and Reklys, E. (1965). *Science, N.Y.* **147**, 884–886.

Grace, G. W. (1968). *In* "Peoples and Cultures of the Pacific" (A. P. Vayda, ed.), pp. 63–79. Natural History Press, Garden City, New York.

Graydon, J. J., Semple, N. M., Simmons, R. T. and Franken, S. (1958). *Am. J. Phys. Anthropol.* **16**, 149–171.

Haldane, J. B. S. (1951). *New Biol.* **15**, 9–24.

Harris, H. (1966). *Brit. Med. Bull.* **25**, 5–13.

Herskowitz, I. H. (1962). "Genetics". Little, Brown, Boston.

Howells, W. (1967). "Mankind in the Making", 2nd Edition. Doubleday, Garden City, New York.

Hoyer, B. H. and Roberts, R. B. (1967). *In* "Molecular Genetics" (J. H. Taylor, ed.), Vol. 2, pp. 425–479. Academic Press, New York.

Hubby, J. L. and Lewontin, R. C. (1966). *Genetics* **54**, 577–594.

King, J. L. and Jukes, T. H. (1969). *Science, N.Y.* **164**, 788–798.

Kirk, R. L. (1965). "The Distribution of Genetic Markers in Australian Aborigines", *Occasional Papers in Aboriginal Studies No. 4*. Australian Institute of Aboriginal Studies, Canberra.

Kirk, R. L., Lai, L. Y. C., Vos, G. H. and Vidyarthi, L. P. (1962a). *Am. J. Phys. Anthropol.* **20**, 375—385.

Kirk, R. L., Lai, L. Y. C., Vos, G. H., Wickremasinghe, R. L. and Perera, D. J. B. (1962b). *Am. J. Phys. Anthropol.* **20**, 485–497.

Kirk, R. L., Blake, N. M., Lai, L. Y. C. and Cooke, D. R. (1969). *Archaeol. Phys. Anthropol. Oceania* **4**, 238–251.

Kohne, D. E., Chiscon, J. A. and Hoyer, B. H. (1970). *Science, N.Y.* In press.

Lewontin, R. C. and Hubby, J. L. (1966). *Genetics* **54**, 595–609.

Lindberg, D. G. (1969). *Science, N.Y.* **166**, 1176–1178.

Mayr, E. (1954). *In* "Evolution as a Process" (J. Huxley, A. C. Hardy and E. B. Ford, eds), pp. 188–213. Collier, New York.

Napier, J. R. and Napier, P. H. (1967). "A Handbook of Living Primates". Academic Press, London and New York.

Nicholls, E. M., Lewis, H. B. M., Cooper, D. W. and Bennett, J. H. (1965). *Am. J. Human Genet.* **17**, 293–307.

Nuttall, G. H. F. (1904). "Blood Immunity and Blood Relationship". Cambridge University Press, London.

Sarich, V. M. and Wilson, A. C. (1967). *Proc. Natl. Acad. Sci.* **58**, 142–148.

Shannon, C. E. and Weaver, W. (1949). "The Mathematical Theory of Communication". University of Illinois Press, Urbana.

Shaw, C. R. (1965). *Science, N.Y.* **149**, 936–943.

Shaw, C. R. and Koen, A. L. (1968). *In* "Chromatographic and Electrophoretic Techniques. Vol. II: Zone Electrophoresis" (I. Smith, ed.), pp. 325–364. John Wiley, New York.

Simmons, R. T. and Cooke, D. R. (1969). *Archaeol. Phys. Anthropol. in Oceania* **4**, 252–259.

Simmons, R. T., Semple, N. M., Cleland, J. B. and Casley-Smith, J. R. (1957). *Am. J. Phys. Anthropol.* **15**, 547–553.

Simmons, R. T., Graydon, J. J., Zigas, V., Baker, L. L. and Gajdusek, D. C. (1961). *Am. J. Trop. Med.* **10**, 639–664.

Sokal, R. R. and Sneath, P. H. A. (1963). "Principles of Numerical Taxonomy". W. H. Freeman, San Francisco.

Tashian, R. E., Shreffler, D. C. and Shows, T. B. (1968). *Ann. N.Y. Acad. Sci.* **151**, 64–77.

Vines, A. P. and Booth, P. B. (1964–65). *Oceania* **35**, 208–217.

Wallace, A. R. (1890). "The Malay Archipelago", 10th Edition. Macmillan, New York.

Wernstedt, F. and Spencer, J. E. (1967). "The Philippine Island World: A Political, Cultural and Regional Geography". University of California Press, Berkeley.

Workman, P. L. (1969). *Human Biol.* **41**, 97–114.

Zuckerkandl, E. (1963). *In* "Classification and Human Evolution" (S. L. Washburn, ed.), pp. 243–272. Aldine Press, Chicago.

Acknowledgements

This study was made possible by National Science Foundation Grants GF–253, GB–7426 and GB–15060 of the U.S.–Japan Cooperative Science and Systematic Biology Programs, National Institutes of Health (National Institute of General Medical Sciences) Training Grant GM–1224 and Wayne State University Grant-in-Aid FR 07051–04.

We also wish to thank Drs W. Prychodko and G. Lasker for their aid and discussion. Mrs. K. Crooks provided invaluable assistance, as did Mrs. E. Poulik.

F

Evolving Primate Genes and Proteins*

MORRIS GOODMAN,†‡ ANN L. KOEN,§ JOHN BARNABAS†‡
AND G. WILLIAM MOORE§§

†*Department of Anatomy, Wayne State University, Medical School,
Detroit, Michigan, U.S.A.*

‡*Plymouth State Home and Training School, Northville, Michigan, U.S.A.*

§*Hawthorn-Plymouth Research Laboratory, Northville, Michigan, U.S.A.*

§§*Biomathematics Program, Institute of Statistics,
North Carolina State University at Rayleigh, U.S.A.*

INTRODUCTION

Much of the study of primate evolution has been carried out by paleonto-logists and physical anthropologists; it is only in the last few years that molecular biologists have entered the field. These new converts to prima-tology are attempting to view the evolutionary process and its long-range results in the Primates at the basic level of genes and proteins. Genes are segments of the winding double-stranded chains of deoxyribonucleic acid (DNA) located in the chromosomes. Each link of such a segment of a DNA chain is called a nucleotide base pair, and the sequences of these nucleotides carry the blueprint of inheritance from one generation to the next. Proteins are primary gene products; they can reflect, within certain limits, gene mutations of even a single nucleotide. Individually, the vast majority of such point mutations cause no phenotypic differences, yet, in sum, they delineate the process of species divergence.

DNA exists in the cell nucleus as a double-stranded helix. The sequence of its bases determines the primary structure of the proteins formed from it. There are 4 kinds of nucleotides in DNA: guanine (G), cytosine (C), adenine (A) and thymine (T). In the double-stranded state, adenine is paired with thymine and guanine with cytosine. It is known (Lengyel *et al.*, 1961, 1962; Speyer *et al.*, 1962a, b) that each amino acid of a protein is coded for by a sequence of 3 nucleotide bases (a codon). There are 64 codons (the possible permutations in a sequence of 3 bases of 4 kinds) but only 20 different amino acids. Thus, an amino acid is typically specified by

* Supported in part by NSF grants GB-7426 and GB-15060, with additional support from the Werner-Gren Foundation for Anthropological Research.

more than one codon. The nature of the genetic code is such that a point mutation substituting one nucleotide base for another in either the first or second position of a codon almost always produces a codon for a different amino acid, whereas a base substitution in the third position of the codon often produces a synonymous codon for the same amino acid.

During replication of the cell, the double strands of DNA separate, and new complementary strands are formed on each so that the 2 new double-helical DNA molecules each contain a daughter strand and the complementary parent strand producing, thereby, a replica of the original parent helix. In the conception of a new individual, assuming diploid inheritance, each parent contributes 1 of the 2 DNA strands. If the DNA contribution of both parents for a particular genetic locus is the same, then the progeny is said to be homozygous at the locus. If the parental contributions are different, the progeny is heterozygous at that locus, and two different (allelic) proteins will be produced from that gene locus.

The formation of a protein is initiated by the genic process of transcription. In this process, the instructions in the DNA base sequence of the gene specifying the primary structure of the protein are transcribed by an enzyme-mediated reaction into messenger ribonucleic acid (mRNA) whose nucleotide bases are complementary to the DNA bases, except that uracil replaces thymine. The mRNA migrates from the nucleus to the cytoplasmic particles, called ribosomes (rRNA), to serve there as a template for protein formation. At this time, another type of RNA called transfer RNA (tRNA) transports, in another enzyme-mediated reaction, the amino acids to the messenger RNA template on the ribosomes where these amino acids are then positioned according to the transcribed codon sequence in the mRNA and joined together to form a protein, or, rather, a polypeptide chain of a protein. In the jargon of molecular biologists, this process is called translation, since the genetic blueprint for the protein written in codon sequences is now rewritten in amino acid sequences. The mechanism of translation is complex; for example, for each of the 20 sets of synonymous codons, there is a unique set of matching transfer RNAs, also, there is a specific amino acyl activating enzyme which prepares the amino acid energetically for polymerizing with other amino acids. However, in this review, all that needs to be emphasized is that, with certain qualifications, the amino acid sequence of a protein is an accurate facsimile of the nucleotide sequence of a gene.

It has been recognized for some time that the amount of DNA present in mammalian cells is greatly in excess of that required to contain all the genetic information needed to specify the various proteins of a mammal, and a large proportion of the genome (the haploid complement) in higher cells has been shown to be inactive (Littau *et al.*, 1964; Bonner *et al.*, 1968) in the synthesis of mRNA. Nei (1969) believes that most of the excess DNA is the result of gene duplication and that higher organisms, including man, carry many non-functional genes in their genomes.

Following the demonstration by Schildkraut *et al.* (1961) that separated complementary strands of purified DNA recognize each other and can be reassociated, Bolton, McCarthy and Hoyer developed a DNA-agar technique for the hybrid reassociation of DNAs from different species. Since the technique measures the degree of nucleotide complementarity between such DNAs, it provides an effective means of determining the extent of genetic relatedness of the species. The method was first worked out on microbial DNAs (Bolton and McCarthy, 1962; McCarthy and Bolton, 1963) and then applied to mammalian DNAs (Hoyer *et al.*, 1964), especially to primates (Hoyer *et al.*, 1965; Hoyer and Roberts, 1967).

A surprising discovery was that the hybrid reassociation progressed as rapidly between DNAs of mammalian sources as between DNAs of micro-organisms, even though a mammalian genome contains about 1000 times more nucleotide bases than a bacterial genome. If each average-sized sequence of several hundred nucleotides in mammalian DNA had occurred only once in the haploid genome, as is the case in micro-organisms, the individual nucleotide sequences should have been so diluted by the large quantity of total DNA in the mammalian cell that their hybrid reassociation reaction in the DNA-agar method should have progressed much slower than that of bacterial DNA. The discrepancy between what was observed and expected led to the discovery that a considerable portion, about 40–60%, of mammalian DNA exists as families of repeated nucleotide sequences (Waring and Britten, 1966; Britten and Kohne, 1968) and the remaining portion as non-repeated sequences, each sequence of which occurs only once per haploid cell. It was only the families of repeated nucleotide sequences that were involved in the comparison of mammalian DNA's by the DNA-agar technique (Hoyer *et al.*, 1964, 1965). Much longer incubation times and different techniques involving hydroxyapatite columns, which can selectively separate reassociated from unreassociated DNA, are used to measure the correspondence between mammalian genomes of the non-repeated DNA fraction (Laird *et al.*, 1969; Kohne *et al.*, 1971).

Repeated nucleotide sequence families are thought to result from saltatory replications of previously-existing single genes. From a hundred to a million copies of single sequences are produced by such a saltatory event, with a typically-sized family containing about 100,000 members of a repeated sequence. There are certain families of repeated DNA in which all the member sequences within a family are almost identical copies of each other, such as in mouse satellite DNA, and there are other families in which the repeated sequence members are only similar to each other, showing various degrees of heterogeneity due to base-pair mismatching (Britten and Kohne, 1968). It was postulated that the former families are quite young in evolutionary time, while the latter are relatively old. A young family would not only have to have been recently produced in an organism by a saltatory replication (a type of mutational event), but would then have to have been incorporated rapidly into genomes throughout the evolving species by selection or genetic drift or the combined workings of both these mechanisms for fixing mutations in a population. In contrast, an old family would have a much longer evolutionary history, during which time a gradual accumulation of point mutations (i.e. single nucleotide substitutions) would have caused the member sequences to diverge from one another.

Competitive hybrid reassociation experiments on the repeated DNA sequences have been carried out by Martin and Hoyer (1967) with several primate and rodent DNAs which had first been separated into adenine + thymine-rich (AT-rich) and guanine + cytosine-rich (GC-rich) fractions. These experiments indicated that the AT-rich fragments possess greater species specificity than do the GC-rich fragments. This was interpreted to mean that the GC-rich fragments may be the more conservative parts of the genome, while the AT-rich fragments represent the regions where more nucleotide mutations have occurred during the course of evolution.

Recent DNA investigations (Laird *et al.*, 1969; Kohne *et al.*, 1971) have dealt with the non-repeated DNA sequences. Although these constitute from 40 to 60% of the DNA of mammalian cells, they actually represent most of its potential genetic information in that each sequence in the non-repeated fraction is different from all the others, whereas each sequence in the repeated fraction is repeated many-fold and, thereby, redundant with respect to the amount of potential information. Considering the great diversity possible in DNA sequences, the probability is very high that identical non-repeated sequences in two different species must be derived from the same common ancestor. This type of DNA is, therefore, highly suitable for investigation of the extent of nucleotide change since the divergence of two species. On the other hand, investigations of the repeated sequences can yield data on the incorporation of new DNA into the genomes of a phyletic line at various times during evolution.

An initial survey of primate relationships by the hybrid reassociation techniques revealed a striking parallel between the results obtained on the non-repeated DNA sequences and on the repeated sequences (Kohne, 1970). Degrees of genetic relatedness were effectively portrayed at the intermediate (generic through subordinal) taxonomic levels. Apparently, the overall number of nucleotide base pair mismatches is so great between the genomes of organisms separated at the higher (ordinal and above) taxonomic levels, that the DNA hybrid reassociation experiments, on the whole, fail to reveal significant degrees of genetic relationship at these higher taxonomic levels. However, the possibility exists that by working with enriched fractions of the mRNA which are produced by fetal organisms and represent evolutionarily conservative portions of the genome, phylogenetically-distant relationships can be revealed by the hybridization techniques. As currently practised, these techniques can provide a clear picture of evolutionary relationships among the genera of a mammalian sub-order but not relationships at the higher taxonomic levels, nor the fine details of species, sub-species and individual differences. Immunologic measurements of the degree of antigenic site matching between proteins in different organisms can also decipher evolutionary relationships at the intermediate taxonomic levels, while certain types of enzyme studies demonstrate some of the fine details of species and individual differences.

Amino acid sequence determinations, depending upon the proteins investigated, can be used to help decipher relationships among populations at all taxonomic levels, from individual differences to those of the animal and plant kingdoms. For example, the sequences of a slowly-evolving protein, cytochrome c, depict surprisingly well affinities between phylogenetic branches which separated over a billion years ago in evolution (Fitch and Margoliash, 1967a). Also, the most common event in the ongoing process of evolution, the point mutation, often results in an amino acid sequence difference between allelic proteins in a population, such as the substitution of valine for glutamic acid at position 6 of the β-polypeptide chain of hemoglobin which converts the normal chain into the sickle cell chain (Ingram, 1958).

It must be pointed out that although DNA and immunological studies cover the same general taxonomic levels, the observations made are not from the same vantage point, and, therefore, the conclusions drawn need not always agree. Nevertheless, as will be described later, they do agree remarkably well with regard to the evolutionary relationships depicted among primate genera. Except for the recent work of Laird et al. (1969) and Kohne et al. (1971), most DNA experiments have been performed on the repeating nucleotide sequences. Certain of these sequences, as has been hypothesized by Britten and Davidson (1969), may represent regulatory genes. Immunological comparisons (also amino acid sequence determinations), since they are performed on proteins, reveal relationships among products of certain structural genes only. Enzyme studies, although, of course, using the products of certain structural genes, have shown that enzyme patterns reveal the influence of regulatory mechanisms, especially in the distribution and concentration of isozymes.

Molecular biologists believe that evolution at the level of nucleotide and amino acid sequences is predominantly divergent with elapsed time being the most important parameter determining the accumulation of nucleotide substitutions in genomic DNA and amino acid substitutions and antigenic changes in proteins. The reasons for this belief are developed below, also the corollary to it (Goodman, 1961, 1963a, 1965) which holds that such molecular evolution decelerated in higher organisms, especially so in the line leading to Man. After these concepts are formulated, we present the results of amino acid sequence, immunological and DNA studies, which provide information on the phyletic relationships among contemporary primates and on the relative rates of molecular evolution in the primate lineages. Finally, we will review enzyme studies which add another dimension to our picture of the course of primate evolution.

DIVERGENT CODON EVOLUTION

Protein Structure and Neutral Mutations

A consideration of protein structure–function relationships will introduce the arguments for believing that evolution, at the level of codon sequences, is predominantly divergent. Proteins are large molecules made up of one or more polypeptide chains, each of which typically contains 150 or more amino acids of the 20 different kinds. For a polypeptide chain of 150 amino acids, the corresponding structural gene consists of a chain of 150 codons or 450 nucleotide residues. When the polypeptide chain is completely released from its mRNA after the process of translation is completed, it folds into a characteristic shape that can be mapped by X-ray diffraction studies. It has been found that many amino acid substitutions can accumulate in a protein's primary structure (its linear sequence of amino acids) without altering its tertiary (3-dimensional) structure (Margoliash *et al.*, 1968; Noland and Margoliash, 1968; Perutz *et al.*, 1965). Indeed, in the case of myoglobin compared to hemoglobin, 80% of the homologous positions in the polypeptide chains differ in amino acid composition, yet the tertiary structures of those globins are remarkably similar. The evidence shows that it is among the stereochemical sites which are exposed by the folded polypeptide chain at the protein's exterior surface and which appear to execute no physiological functions, that the largest number of amino acid changes have occurred during evolution. However, there are among the many surface sites a few which are known to be biologically active, such as the catalytic sites of enzymes, and these have not changed in composition through long stretches of evolutionary time, presumably because changes in such sites would interfere with physiological functions and be selected against. Thus it has been deduced that many of the amino acid substitutions in proteins which became fixed in different species were caused by selectively neutral mutations (Goodman, 1961; Kimura, 1968; King and Jukes, 1969) occurring at a fairly high rate.

The concept of selectively neutral amino acid substitutions needs to be qualified, for at any one amino acid position in a polypeptide chain (or corresponding codon position in the structural gene) the mutations assumed to be selectively neutral would only be so relative to one another, as cogently pointed out by Corbin and Uzzell (1970). For example, let us say at a particular position in the polypeptide chain any one of 5 different amino acids, all showing similar physical chemical properties, can serve equally well with respect to the functional properties of the protein, whereas each of the other amino acids, if substituted in this position, would alter in a detrimental way the functional properties of the protein. Thus, alleles for the 5 equivalent amino acids would be selectively neutral compared to each other but also selectively advantageous with respect to alleles for any of the remaining 15 amino acids. It would be misleading to describe the fixation in a population of one of these 5 equivalent amino acids as due to non-Darwinian evolution *a la* King and Jukes, even if the fixation resulted randomly from genetic drift, since Darwinian or adaptive evolution would constantly act to eliminate any mutant alleles which occurred for any of the 15 less satisfactory amino acids. On the other hand, there might be an occasional position in a polypeptide chain where any amino acid substitution of the 20 possible ones would be selectively neutral. Furthermore, at the DNA level there would be a number of selectively neutral mutations relative to one another, not producing any amino acid substitutions. These would result from those base substitutions which can produce at every third position of a codon a synonymous codon. (One-fourth of cell mutations in a structural gene would be expected to produce such codons.) Also, as discussed by Kohne *et al.* (1971), many DNA sequences may have no function at all, in which case all base substitutions in these sequences would be selectively neutral ones. The accumulation in evolutionary time of base substitutions in such DNA sequences can be described appropriately as an example of non-Darwinian evolution.

Paradoxically, the random fixation of a selectively neutral amino acid substitution in a population may largely result from the process of Darwinian evolution itself. Suppose, for example, a particular amino acid substitution due to a point mutation imparts a new beneficial property to the protein in which it occurs. Then, not only would this beneficial mutation spread through the population, but so would the selectively neutral codons in the same gene or recombination region and in closely-linked genes at those sequence positions in which isoallelic codons were found in other

F*

members of the population. (A newly-arisen family of repeated DNA sequences, depending on its chromosomal location, could also be incorporated, even if it had initially no useful function, in the genomes of a descendent population by such a mechanism.)

In view of the fact that there are 20 amino acids, a stereochemical site formed by only a few clustered amino acid groups at a protein's surface can be transformed during evolutionary time by point mutations within codons into a vast number of different configurations. Given the millions of possible choices for such codon changes at any time in even a single gene, the random order in which mutations occur, and the overall random way in which they are fixed in different populations, evolution at the level of codons and amino acid sequences must almost certainly be divergent between descending lineages. Conversely, the degrees of retained similarity among homologous proteins must reflect ancestral homologies and, as Fitch and Margoliash (1967a, b) have shown for cytochrome *c* by rigorous statistical procedures, cannot be due to convergent evolution, i.e. to an adaptive evolution in long-separated lineages producing similar structures to execute the same functions.

Antigenic sites on proteins are especially likely to reflect divergent codon evolution. This is because the stereochemical sites not playing any functional role for a protein, i.e., the evolutionarily labile sites produced by random selectively neutral substitutions of amino acids are more apt to show antigenic properties (induce an antibody response in a foreign host) than the evolutionarily conservative sites playing key functional roles for the protein; these latter conservative sites are apt to be shared by donor and host species and thus be non-antigenic (lack immunogenicity) in the host species due to the phenomenon of immunological tolerance (Cinader, 1963). In an evolving lineage, an occasional antigenic site may arise which matches another site in a widely-separated lineage purely on a random basis, i.e., by a fortuitous amino acid substitution. Such rare random matches should not distort the general pattern in which degrees of retained antigenic similarity among homologous proteins reflect ancestral relationships, provided the immunological comparisons sample enough antigenic sites. Such sampling can be accomplished in immunosystematic studies by using serum proteins as immunizing antigens, for these proteins (even single ones, such as serum albumin) expose many antigenic sites at their surfaces.

Deceleration of Molecular Evolution in Higher Organisms

Although elapsed time must be the most important parameter determining the accumulation of amino acid substitutions (Zuckerkandl and Pauling, 1965; Margoliash *et al.*, 1968) and antigenic changes (Goodman,

1962a; Sarich and Wilson, 1967) in proteins, the rate of evolution is definitely not the same for all proteins (see Table I of King and Jukes, 1969) and, for a particular protein, need not be the same in all lines of evolution nor constant in any one descending line throughout its history. Even if the accumulation of amino acid substitutions is primarily determined by the rate of selectively-neutral mutations, the proportion of harmless to harmful mutations, i.e. the actual rate of neutral mutations, need not be constant at all gene loci or in all evolutionary lines or at all stages of descent. The hypothesis has been proposed (Goodman, 1961, 1963a, 1965) that molecular evolution at a number of gene loci has decelerated in higher organisms, especially in the higher primates. This hypothesis, which preceded that of Kimura (1968) and King and Jukes (1969) in attributing divergent protein evolution to an extensive accumulation of relatively neutral mutations, argues that the proportion of the total mutation rate due to neutral mutations was greater in primordial organisms than in the higher organisms of today (Goodman, 1961).

Consider the evolution of globin chains in the vertebrate proteins, myoglobin and hemoglobin. During the past half billion years of vertebrate evolution, there was a series of duplications and translocations of the primordial globin gene (Ingram, 1961; Zuckerkandl and Pauling, 1965) and a marked progression of amino acid substitutions in the descending lines of globin which gradually encompassed most of the 150 or so amino acid sequence positions of the primordial globin chain. Although the amino acid sequence changes may have largely resulted from the random fixation of selectively neutral mutations, natural selection also influenced these changes, for during the phylogenetic development of the vertebrates, the biochemical complexity of globin systems increased producing multi-chained hemoglobins with a quarternary structure. The single-chained hemoglobin of the lamprey represents a primitive vertebrate globin. (The term teriary structure denotes the 3-dimensional structure of such single-chained proteins, whereas quarternary structure denotes the final 3-dimensional structure of multi-chained or polymeric proteins.) The hemoglobins of higher vertebrates are tetrameric proteins, each consisting of 4 globin chains of 2 types. Thus, the higher vertebrate globins contain not only heme binding sites, but other stereochemical sites for sub-unit combination. The conformation of the tetrameric hemoglobin also exposes sites on the β-type chains which react specifically with diphosphoglycerate in a way which helps regulate the oxygen binding capacity of these globin molecules (Benesch and Benesch, 1970; Tyuma and Shimizu, 1970). Furthermore, mammalian hemoglobins, when mixed with plasma, specifically combine with the plasma protein haptoglobin; consequently, there must be additional stereochemical sites for the specific binding of hemoglobin to haptoglobin. No doubt certain of the mutations in the globin genes which gave rise to these surface globin sites imparted an adaptive physiological advantage of their bearers; Darwinian evolution must have then fixed the responsible alleles in the lineages in which they occurred. It was the very success of such adaptive selection which caused protein evolution to decelerate, for certain surface sites on proteins, which at an earlier evolutionary stage were selectively neutral, ceased to be neutral and became fixed in certain lineages by acquiring useful functions.

Suppose a series of alleles for protein A and another series for protein B exist in a population. No combinations of these alleles produces proteins A and B which can specifically interact. However, a new point mutation at locus B does produce a site in protein B which will allow it to combine with one of the allelic forms of A. Until a mating in the population brings these particular A and B alleles together in the same offspring, they are still selectively neutral, but, once brought together, they produce the interacting A and B protein system which has an adaptive value for the individuals in which it occurs, resulting in the selective spread and fixation in the population of the beneficial interacting alleles at the A and B loci. In a related species in which polymorphisms due to selectively neutral alleles also exist for the A and B proteins, the same phenomenon might occur, i.e. a particular combination of alleles at the A and B loci will produce A and B proteins which specifically interact. However, by analogy to antigen–antibody reactions, a different set of alleles and complementary stereochemical sites could be responsible for the A–B interaction in the second species as compared to the first. Thus, codon divergencies at both the A and B loci will occur between the 2 species due to Darwinian evolution fixing alleles which initially had been neutral to natural selection. This generalized example of protein evolution illustrates that at the grosser levels of macromolecular structure, such as the quarternary structure of proteins, Darwinian evolution can produce parallel or even convergent similarities between phyletic lines while fixing divergent changes at the level of amino acid sequences and antigenic specificities.

The extensive divergencies which exist between lineages in amino acid sequences and the close relationship between observed amino acid frequencies in proteins and the expected frequencies predicted by the model of non-Darwinian evolution from random permutations of nucleic acid bases (King and Jukes, 1969) is clearly evidence for a high rate of selectively neutral mutations in earlier stages of evolution, but not necessarily at the present stage. In the phyletic lines in which Darwinian evolution produced more complexly organized macromolecular systems, the rate of neutral mutations to non-synonymous codons in structural genes could still be high but (according to the model of decelerating protein evolution) not so high as in the past. One can summarize this view by stating that *anagenesis* limits the possibilities for genetic divergence during *cladogenesis*.

Anagenesis is defined, after Rensch (1959) and Huxley (1964), as that process of phyletic change which increases the complexity of internal organization in the living systems of certain lineages. It gives these systems a greater independence of the outside world and a greater capacity to cope with changes in the external environment. *Cladogenesis*, or phylogenetic branching, again after Rensch (1959) and Huxley (1964), refers to the multiplication of phyletic lines (i.e. the production of new clads) and to their diverging adaptive specializations. Anagenesis has led to new deployments of life, i.e. to recurrent bursts of cladogenesis, because at each higher anagenetic stage the organisms with superior metabolic efficiency and more versatile behavior displaced some of the species of less advanced organisms from their habitats, colonized widely-separated geographic areas and entered diverse ecological zones. But, according to the theory being discussed here (Goodman, 1961, 1963a, 1965), the rate of genetic change and the degree of molecular divergence between clads in the adaptive radiations initiated by lineages in the main stream of anagenetic evolution decreased after each anagenetic advance. Finally the stage is reached where the rate of genetic change slows so much that allopatric speciation ceases, as exemplified by the radiation of the genus *Homo*.

During the span of primate evolution from the Cretaceous to the present, the progenitors of Man can be viewed as always being in the most anagenetically advanced lineage. This lineage must have had a large gene pool with numerous alleles upon which natural selection could act to bring about organisms with more complex internal organizations. In their advance towards Man these organisms did not specialize for one or another narrow ecological zone but, instead, inhabited a diversified environmental range which broadened rather than contracted as evolution progressed. With each increase in internal molecular organization the probability would be greater that any mutation affecting a functional macromolecule would cease to be neutral to natural selection. Then the conditions of the internal environment requiring a state of complex molecular coadaptions in these organisms would select for homozygosity of the particular gene set which established this state. At the same time, the broadened external environment would select for balanced polymorphism at certain gene loci. The net effect in the anagenetically-advancing lineage would be for homozygosity to increase among genes at a growing number of loci, while heterozygosity was still maintained at certain other loci. However, even with respect to the alleles at the polymorphic loci, increased resistance to change would characterize their further evolution, since new mutations in these genes would be less apt to be neutral to natural selection than before and more likely to disrupt an adaptively useful polymorphism rather than perfect it.

Clearly, the most important of the gene sets selected for in the lineage advancing towards Man were those involved in the progressive elaboration of the cerebral cortex, for organisms with better brains were able to cope more adequately with the numerous diverse challenges of their external environment. Furthermore, the great expansion of the cerebrum in higher primates may have depended on the evolution of the hemochorial placenta, for the intimate apposition of fetal and maternal blood streams in this placenta allows the developing fetal brain to obtain a rich supply of oxygen and nutrients. But such a placenta also allows the maternal immunological system to respond to fetal isoantigens and further restricts genetic change in evolving primate populations.

This postulated immunological factor (Goodman, 1961, 1962a, b, 1963a, 1965) would barely be active in mammals with placentas of the epithelio-chorial type, which separate the maternal and fetal vascular systems by several layers of tissue and thus constitute a firm barrier to the transplacental passage of proteins between mother and fetus. Such placentas are found in the lemuroid and lorisoid prosimian primates (LeGros Clark, 1959), also in ungulates, and may have been the first type to evolve in the primitive eutherian mammals (Hamilton *et al.*, 1952). However, maternal immunization to fetal isoantigens, known to occur in the case of the Rh and ABO incompatibilities in Man, could be an important factor in the higher primates, since these animals have not only hemochorial placentas but long gestation periods. While the hemochorial placenta places the fetal blood stream in intimate apposition to the maternal blood stream permitting a transplacental passage of proteins to occur between mother and fetus, the long gestation period provides ample time for maternal antibodies to be produced to fetal isoantigens and then attack the fetus by crossing the placenta. Due to this immunological selective factor, the possibility would be greater that any mutation affecting a stereochemical site on a circulating protein in a higher primate fetus would be detrimental. This would decrease the neutral mutation rate and thus decelerate the evolution of the pre-natally circulating proteins in the higher primates.

Evidence that protein evolution decelerated in higher primates is presented in the following sections which review protein data on primate phylogeny from amino acid sequence and immunological studies. Furthermore, the data reviewed in a later section on the evolution of primate DNA sequences also support a decelerating rate model of molecular evolution in the higher primates. However, non-Darwinian evolution appears to play a much greater role in nucleotide base sequence evolution than in amino acid sequence evolution. If the basic mutational mechanism depends in some way on the DNA replication cycle, i.e. if the mutation rate is generation time-dependent, then the rate of fixation of selectively neutral nucleotide substitutions might simply be a function of the generation times of the species involved, the longer the generation times the slower the rate of nucleotide change. As evolution progressed during the Tertiary Epoch in the primate lineage leading to Man, there was a marked lengthening of generation time and, apparently, a marked slowing of DNA sequence evolution (Kohne, 1970; Kohne *et al.*, 1971).

It would be a mistake to rule out selection as an important factor in the slowing down of nucleotide sequence evolution in higher primates, even within the families of repeated DNA sequences which are thought to contain many non-functional copies of genes. The production of these families is not a unique attribute of the mammals. Repeated DNA is found in all metazoans studied so far, and certain lower vertebrates, such as lungfish and salamanders, have much more DNA per cell than Man and other mammals. We might conjecture that the proportion of non-functional sequences is greater in lower vertebrates than in higher. If so, nucleotide sequence changes due to non-Darwinian evolution would progress much more rapidly in these lower organisms. The theory of Britten and Davidson (1969), which attributes a genetic regulatory role to repeated DNA sequences, lends itself to this speculation. This theory proposes that there are sensor genes which respond to signals carried in from the cytoplasm (by hormones, for example) so as to cause adjacent sets of integrator genes to produce complementary activator RNA molecules. These activator RNAs are intranuclear messengers, which migrate to particular receptor genes scattered throughout the chromosomes and combine with these spatially separated replicas of the integrator genes in such a way as to cause them to switch mRNA transcription on or off in adjacent producer genes. The model thus depicts a series of interlocking DNA and RNA sequences which orchestrate the synthesis of enzymes and other proteins by batteries of producer genes in such a way as to enable the organism to respond adaptively to changing stimuli. Since the function of the integrator and receptor genes in this interlocking network would be very sequence-dependent, no point mutations in these genes would be neutral to natural selection; most nucleotide substitutions in such genes would be deleterious and selected against. Now if the marked lengthening of generation time coincident with the remarkable expansion of the cerebrum and other changes of this type in the primate lineage leading to Man were related in some way to the evolution of more complexly interlocking DNA and RNA sequences, then we could conclude that natural selection would henceforth slow down the rate of nucleotide substitutions in such a lineage.

Amino Acid Studies

The trickle of sequence data on proteins which began to gather in the early 1960's is now a stream which promises to become a flood as automation of the sequencing procedures advances. Much of the data already gathered is organized in the "Atlas of Protein Structure and Sequence" (Dayhoff, 1969). The known sequences on certain proteins, in particular the various globins, cytochromes *c*, and fibrinopeptides of different species, are now sufficiently extensive to reveal important insights on primate phylogeny. Computer data processing methods similar to those of numerical taxonomy (Sokal and Sneath, 1963) are being used to extract the maximum amount of information from this sequence data. The general approach is illustrated by the way the globin sequences were interpreted and handled (Barnabas *et al.*, 1971).

The sequences of all the globin chains in present day vertebrates are specified by genes which have lines of descent that are thought to trace back 750 or more million years to a single common ancestral globin gene in a primitive stock of protovertebrates. By a process of gene duplication and intragenic mutational changes involving largely nucleotide base substitutions (point mutations), deletions and additions, the genes for the various types of globin chains arose in different phyletic lines at different stages of vertebrate phylogeny and diverged from each other within and between lineages of organisms with the passage of time.

In a broad sense, vertebrate globin chains are genetic homologs related by their descent from a common ancestral gene. In a more restricted sense, a set of globin genes in a corresponding set of animal species are treated as homologs only if all members of the gene set descended from a single ancestral locus of the most recent common ancestor of the animal species set. For example, the α-chains in the mammals are a set of genetic homologs; the β-chains are another set.

Methods

Once it is decided which protein chains belong to a set of homologs, the amino acid positions in these chains are numbered to correspond to the codon positions of the hypothetical ancestor of the set. The sequences of the chains in the set are so aligned as to maximize codon homologies throughout the set. This is accomplished by judiciously inserting codon gaps in those members of the set with less than the total number of positions of the hypothetical ancestor. These aligned sequences are placed on separate IBM cards using Dayhoff's single symbols for the 20 different amino acids found in proteins. Then the cards are analyzed position by position by a computer program patterned after that of Fitch and Margoliash (1967a). The computer subjects each member of the set to pairwise comparisons with every other member of the set and, in each comparison, calculates the minimum number of nucleotide changes required to account for the particular amino acid replacements observed at homologous codon sites. The mutation distance values, weighted in each case according to the number of homologous amino acid residues examined, are printed out on new IBM cards ("codon" result cards) in a format suited for processing by our "clad" computer program.

This clad program, which treats the different members of a set as separate species, has alternative procedures, divisive and agglomerative ones, for generating from codon result cards a cladogram, i.e., a graph depicting the presumed order of ancestral branching of these species. First, the program arranges the collection of species in a set as a matrix of species to species mutation distances. Then, for the divisive procedure, it divides the collection into 2 sub-collections and calculates the average mutation distance between them. The computer tries all plausible divisions of the total species collection into sub-collection pairs and chooses that division with the largest between-average mutation distance as the correct cladistic division of the collection. It repeats the process for each sub-collection until singleton species are obtained as the terminal branches.

Starting from the postulate that the longer any 2 species have been without a common ancestor the larger is the mutation distance between them, it has been proved (Moore, 1971) by a rigorous mathematical treatment that the divisive procedure always generates a correct cladogram provided that evolution has occurred at roughly equivalent rates in all lines of descent.

The agglomerative procedure used in the clad program is copied after the unweighted pair group method (UWPGM) described by Sokal and Sneath (1963). It builds the tree from the smallest to the largest branches in a series of pairwise clustering cycles. The first cycle joins together the 2 singleton species in the total collection with the smallest mutation distance between them. Then these joined species are treated as a single species (i.e. single member) of the set in the next agglomerative cycle which searches the revised set to again join the 2 members with the smallest average mutation distance between them. This pair-wise grouping continues until all original singleton species are joined into one final cluster.

The agglomerative procedure compared to the divisive assumes less about the structure of the data. The mathematical treatment of Moore (1971) revealed that unequal rates of evolution among species can be better tolerated by the agglomerative procedure than by the divisive. However, if rates of evolution are grossly dissimilar, then the agglomerative as well as the divisive procedure may fail to produce a correct cladogram.

Differences among the species in a set in their relative rates of evolution can be estimated by the procedure of Fitch and Margoliash (1967a) which calculates the leg lengths as mutation distances between every two consecutive branching points of the tree. By summing up these leg lengths from the common ancestor of the set to each descendent species, one can identify differences among the species in their rates of evolution. Suppose in a set of 10 species, species 1 diverged less from species 2 than from each of the other 8 species, and, similarly, species 2 diverged less from species 1 than from any other species; yet, on the average as revealed by the input mutation distance data, species 2 diverged more than did species 1 from the other 8 species, then the Fitch procedure would calculate a correspondingly larger leg length for species 2 then for species 1 in the descent from their most recent common ancestor. The calculation of leg lengths by the Fitch procedure has been incorporated in the clad program. We call these calculated distances between consecutive branching points *patristic mutation lengths*.

For some sets of sequence data, we have further followed the approach of Fitch and Margoliash (1967a) by constructing alternate branching arrangements, i.e. alternate dendrograms, with our computer program DENDR. The DENDR program calculates the patristic mutation length for each pair of adjacent nodes in a dendrogram, sums the patristic lengths through the sequence of nodes between species to calculate a reconstructed mutation distance for each pair of species in the set, and then compares the reconstructed mutation distances to the given mutation distances in the original matrix to calculate a coefficient for the entire dendrogram called "average per cent standard deviation" (APSD). Among the alternate dendrograms, the one with the lowest APSD, i.e. with the least deviation between original and reconstructed mutation distances, is considered the closest approximation of these tried to a true phylogenetic tree.

Results

Figure 1 shows a gene phylogeny of globins generated by the UWPGM using codon result cards from all possible pairings of 37 vertebrate globin sequences. Figures 2 and 3 give more detailed views with many additional species added of the mammalian α- and β-type globin portions of the tree. It is apparent that present-day mammalian myoglobins and hemoglobins descended from a gene which duplicated extremely early in vertebrate evolution, since the branching which separated myoglobin and hemoglobin genes occurred before the separation of the line leading to lamprey globin. Somewhere later, but still early in vertebrate history, near the time the common ancestor of carp and mammals diverged, a hemoglobin gene duplicated to produce the split between the α-chain genes and the γ-β-chain genes. Then, much later in vertebrate history in the lineage leading to

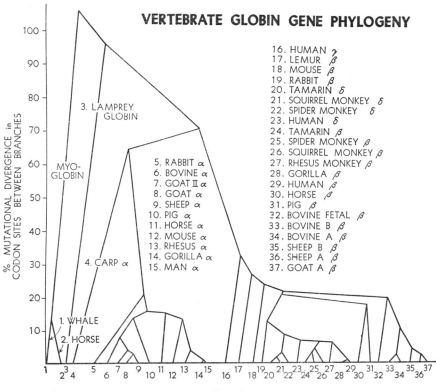

FIG. 1

therian mammals but still earlier than the time of the basal radiation producing the eutherian orders, a γ–β hemoglobin gene duplicated to produce
the first separate ancestor of the human γ-chain gene and the other separate
ancestor of all the mammalian β genes. During the course of eutherian
phylogeny, a number of these β genes duplicated independently in separate
mammalian lines. Thus, the human δ hemoglobin chain is closer to higher
primate β-chains than to any other mammalian β-chains, and the sheep and
goat β–C chains are closer to the other β-chains in the Bovidae than to any
non-bovid β-chains.

Data shown in Figs 2 and 3 and Tables I and II on the patristic mutation
distances demonstrate that rates of evolution have varied markedly between
branches in the descent of the various mammalian globin genes. On the
whole, the β-type genes evolved more rapidly than the α genes, and among
the β-type genes those produced by duplications, such as the ceboid δ's
and sheep and goat C–β's, evolved immediately after the duplicative events
more rapidly than the genes which duplicated. In the case of both α and

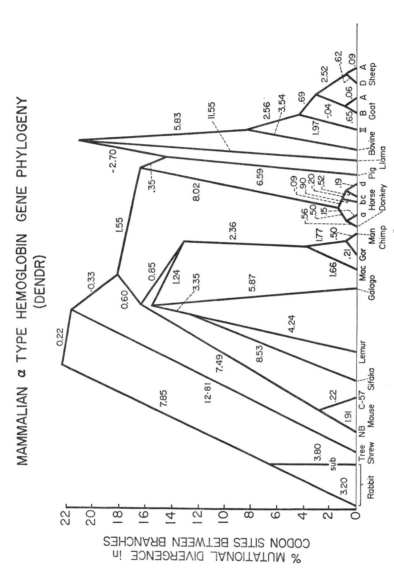

Fig. 2. The best phylogenetic tree of mammalian α-globin genes, according to the APSD criterion, among 30 alternate dendrograms tried out. Horse: a, slow-moving hemoglobin (at alkaline pH), in which 24th residue of α-chain is phenylalanine; b, fast-moving hemoglobin (at alkaline pH), in which 24th residue of α-chain is phenylalanine; c, fast-moving hemoglobin, in which 24th residue of α-chain is tyrosine; d, slow-moving hemoglobin, in which 24th residue of α-chain is tyrosine. Goat: II, goat α-chain, which is the product of duplicated gene; A, common goat α-chain; B, allele of the common goat α-chain. Sheep: A, common sheep α-chain; D, allele of the common sheep α-chain.

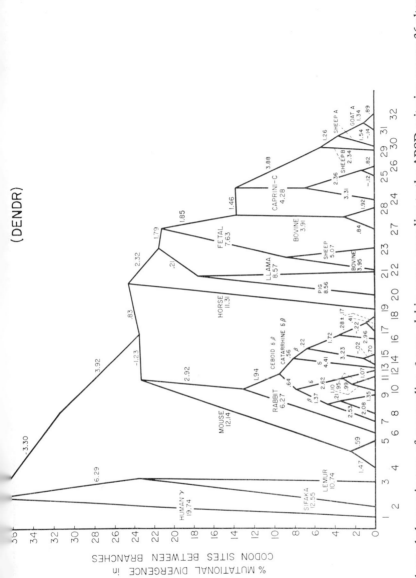

FIG. 3. The best phylogenetic tree of mammalian β-type globin genes according to the APSD criterion among 36 alternate dendrograms tried out. 1, Human gamma; 2, sifaka (*Propithecus verreauxi*); 3, lemur (*Lemur fulvus*); 4, mouse SEC and C57B1; 5, mouse AKR; 6, rabbit (*Oryctolagus cuniculus*); 7, squirrel monkey (*Saimiri scirueus*) β; 8, marmoset (*Saguinus nigri-collis*) β; 9, spider monkey (*Ateles geoffroyi*)β; 10, squirrel monkey (*Macaca fuscata*) δ; 11, Marmoset δ; 12, spider monkey δ; 13, human δ; 14, Japanese macaque (*Macaca fuscata*); 15, rhesus monkey; 16, gibbon (*Hylobates lar*); 17, gorilla; 18, Man and chimpanzee; 19, horse; 20, pig; 21, llama; 22, bovine fetal; 23, sheep fetal; 24, barbary sheep (*Ammotragus lervia*) C; 25, goat C; 26, sheep C; 27, bovine A; 28, bovine B; 29, sheep B; 30, sheep A; 31, goat A; 32, goat A (β allele).

β globin genes the most slowly-evolving lines of descent are those which led to Man and other higher primates.

The slower hemoglobin evolution in higher primates than in prosimians and other mammals was demonstrated in yet another way. Using the rules described by Fitch and Margoliash (1967a) ancestral amino acid sequences were deduced (Barnabas *et al.*, 1971) for mammalian α- and β-chains at the major branching points in their phylogenies. The mutation distances from these ancestral sequences to the contemporary globins descending from them were then determined by the codon program. These results, listed in column 1 of Tables I and II, show that hemoglobin evolution in the higher primates is much slower than in other mammalian lines.

From the branching structure of the globin gene phylogenies (Figs 1, 2 and 3) decisions were made on how to subdivide the various globin species

TABLE I. Percent mutational divergence of mammalian α-chain genes during descent from their common ancestor

α-chain of animal species	From reconstructed ancestral sequences	From patristic mutation lengths
α-Human	7·70	5·97
α-Chimpanzee	7·70	5·97
α-Rhesus monkey	8·40	5·36
α-Japanese monkey	8·40	5·36
α-Gorilla	8·40	5·68
α-Goat A	11·20	8·19
α-Mouse C–57B1	11·30	8·20
α-Bovine	12·00	8·46
α-Goat B (allele)	12·00	8·77
α-Bush Baby	12·70	8·45
α-Brown lemur	12·70	10·17
α-Horse slow 24 Y	12·70	9·36
α-Mouse NB	13·40	9·89
α-Horse fast 24 Y	13·40	9·29
α-Horse slow 24 F	13·40	10·11
α-Sheep A	13·40	10·78
α-Pig	14·10	8·38
α-Horse fast 24 F	14·10	10·27
α-Donkey	14·10	10·52
α-Goat II (duplicated)	14·10	9·45
α-Sheep D	14·10	11·31
α-Rabbit	14·10	11·06
α-Llama	15·60	10·64
α-Tree shrew	17·00	13·03
α-Sifaka	17·00	14·46

TABLE II. Percent mutational divergence of mammalian β-chain genes
during descent from their common ancestor

β-like chains of animal species	From reconstructed ancestral sequences	From patristic mutation lengths
β-Human	7·50	6·36
β-Gorilla	7·50	6·32
β-Spider monkey	7·50	7·20
β-Tamarin	7·50	7·93
β-Chimpanzee	8·20	6·02
β-Japanese monkey	8·20	7·62
β-Rabbit	8·20	9·19
β-Rhesus monkey	8·80	8·34
β-Squirrel monkey	8·80	8·17
β-Human	9·50	8·60
β-Spider monkey	10·20	7·96
β-Tamarin	10·20	8·83
β-Gibbon	10·20	9·09
β-Squirrel monkey	10·90	8·98
β-Pig	14·90	11·92
β-Mouse C–57B1 & SEC	16·40	12·21
β-Mouse AKR	17·12	13·09
β-Bovine A	17·20	11·54
β-Sheep B	17·90	14·47
β-Llama	18·40	11·93
β-Bovine B	18·60	12·62
β-Horse	19·70	12·13
β-Goat A	20·60	14·59
β-Bovinc fetal	20·50	16·52
β-Sheep A	20·60	14·93
β-Goat C	21·30	14·77
β-Goat A^2 (allele)	22·10	15·62
β-Sheep C	23·00	15·71
β-Sheep Barbary C	23·00	15·84
β-Sheep fetal	24·90	17·64

into more restricted sets of homologs which can be used for constructing animal-species phylogenies. The α-chain sequences were treated as one set and adult β-type chains as another set. Homologs of the γ-chain found in the human fetus are found in other primate fetuses and, eventually, when a number of these chains are sequenced, they could also be treated as a set at least for deriving relationships among the primates.

In treating the various β-chains as a set the rule was adopted that duplicated β-chains could be incorporated in the set as long as the gene–species members in each 2 sub-collections descended from a duplicated gene were not cross-compared between sub-collections. Thus, human δ would not be compared to any catarrhine β's but would be compared to lemur β's and all non-primate β's. In the β animal–species set the mutation distance between Man and rhesus monkey was determined solely by the human β and rhesus β codon comparison, whereas the mutation distance between Man and mouse is the average from the codon comparisons of mouse β with both human β and human δ.

The clad program has the capability of combining the codon results from different protein sets, weighting the average by the number of amino acid residues in each sequence comparison. Thus, in addition to obtaining cladograms on the individual sets of protein homologs (α globin chain, β globin chain, cytochrome *c* and fibrinopeptides A and B), animal–species cladograms from combined sets of sequence data were also obtained. Since the cladograms from such combined results encompass a larger number of codon sites, they are more likely to reflect correctly the cladistic relationships of the animal–species than any cladogram from a single set of protein homologs.

Figure 4 shows a cladogram of mammals from combining the codon results of α and β globins and fibrinopeptides A and B. Again, the patristic mutation lengths for this tree of combined blood–protein sequences show that much less mutational divergence occurred from the common ancestor of the set in the lines of descent leading to the higher primates than to the other mammals [mouse–rat (mouse hemoglobin date combined with rat fibrinopeptide data), rabbit, horse, donkey, pig, llama, bovine, goat and sheep].

The cladogram in Fig. 5 of 8 mammalian species (Man, chimpanzee, rhesus monkey, rabbit, horse, pig, bovine and sheep) was produced from codon result cards on cytochrome *c*, α- and β-hemoglobin chains and fibrinopeptide A. These sequence comparisons, encompassing on the average about 407 homologous sites, sample highly conservative (cytochrome *c*) to rapidly evolved (fibrinopeptide A) sections of mammalian genomes and provide a quantitative representation of genetic divergence among the orders Lagomorpha, Primates, Perissodactyla and Artiodactyla, representing the several major phylogenetic branches of the Eutheria which, according to the mammalian paleontologist Malcolm McKenna, trace back to a common ancestral stock in the middle to lower Cretaceous (about 90 million years ago). This cladogram highlights the close genetic relationship between Man and chimpanzee in which no codon divergence at all was found for the particular sequences examined. Furthermore, Man and rhesus

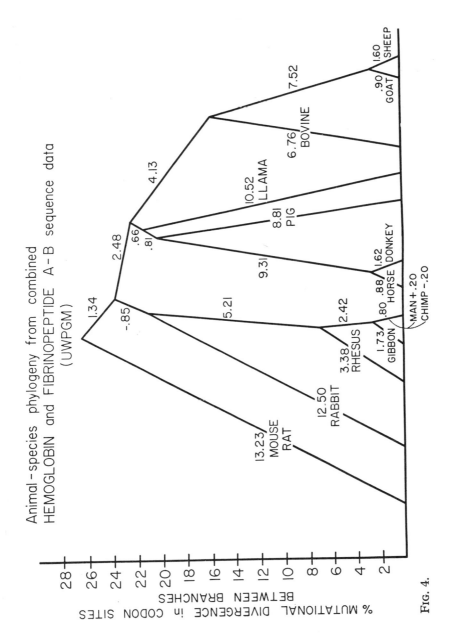

Animal-species phylogeny from combined HEMOGLOBIN and FIBRINOPEPTIDE A-B sequence data (UWPGM)

% MUTATIONAL DIVERGENCE in CODON SITES BETWEEN BRANCHES

FIG. 4.

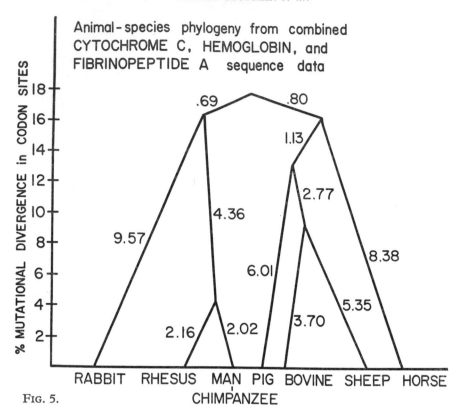

FIG. 5.

monkey showed only about 4% difference in codon sequence compared to about 17–20% between primates and non-primates.

Discussion

Wilson and Sarich (1969) have recently challenged the opinion of paleontologists on the time scale for the branching points in hominoid phylogeny. They contend, from a limited selection of biochemical and immunological evidence on DNA and proteins, that the actual times of divergence between the hominoid branches has been much more recent than the bulk of the fossil evidence indicates (Simons, 1965, 1967; McKenna, 1970). Assuming a uniform rate model of molecular evolution and taking 30 million years ago as the branching point between hominoids and cercopithecoids (an arbitrary assessment of the fossil record on their part), they proceeded to calculate that the Man–African ape and Man–gibbon times of divergence were about five and ten million years ago respectively. Table III shows the mutation distance values from the amino acid sequence

TABLE III. Hypothetical branching times from amino acid sequence data[a]

Proteins examined	Average number of amino acid sites compared	Branching between	Mutational divergence (%)	Calculated branching time 10^6 years ago
1. αHb, βHb, Cyto. Fib. A and B, CAI	549	Man, Chimp–OWM Man–Chimp	4·52 0·17	30 1·1
2. αHb, βHb	287	Man, Gor.–OWM Man–Gor.	5·40 0·70	30 3·9
3. αHb	141	Man, Chimp, Gor.–OWM Catarr.–Prosimians[b]	3·75 13·12	30 105
4. αHb, βHb, Cyto. Fib. A	407	Man, Chimp–OWM Rab.–Catarr.–Ungulates[c] Bovine–Sheep	4·18 17·77 9·05	30 127 65

[a] Calculated using the uniform rate model of molecular evolution with the hominoid—Old World monkey split fixed arbitrarily, after Sarich and Wilson, at 30 million years ago.
[b] Galago, lemur, and sifaka.
[c] Horse, pig, bovine, sheep.

substitution data on α- and β-type globin chains, fibrinopeptides A and B and carbonic anhydrase Is (Tashian and Stroup, 1970) treated in a similar manner. Not only is the Man–African ape divergence time (1–4 million years ago) derived in this way shorter than that estimated from the fossil record, but the higher primate–prosimian and primate–non-primate placental mammal divergence times (105 and 127 million years ago respectively), also the bovine–caprine time (65 million years ago), are greater. Thus, the overall evidence on branching points in mammalian phylogeny from the fossil record are challenged when the uniform rate model is used to analyze the codon sequence mutation distance data.

But is the uniform rate model sound? Even without reference to the fossil record, purely in terms of the sequence divergence data themselves, it was demonstrated by the Fitch calculations, also by the mutation distances from the reconstructed ancestral sequences, that molecular evolution was slower in the descent of higher primates than in the descent of prosimians and various non-primate placental mammals. A model of decelerating protein evolution in the higher primates is more consistent with the sequence divergence data than is the uniform rate model. This result was predicted in the theory discussed earlier in the chapter by the role ascribed to anagenetic evolution as limiting the possibilities for genetic change during cladogenesis.

Malcolm McKenna, from his assessment of the overall fossil evidence, believes one can reasonably use as rough approximations of ancestral branching times: 20 million years ago (Man–African apes); 35 million years ago (Man–gibbon); 45 million years ago (hominoid–cercopithecoid); 55 million years ago (catarrhine–ceboid); 65 million years ago (prosimian–catarrhines and ceboids); 80 million years ago (primates–most other eutherian orders); 90 million years ago (lagomorphs–all other eutherian orders); 65 million years ago (perissodactyl–artiodactyl); 50 million years ago (pig–bovids); and 25 million years ago (bovine–caprine),* as long as one is aware that the fossil evidence is quite incomplete, especially with respect to the Man–African ape and hominoid–cercopithecoid splitting. These branching times from paleontological evidence and the mutation distance data in Tables I and II were used to depict in Figs 6 and 7 the rates of molecular evolution of representative genes and animal–species. This treatment portrays a decelerating rate of evolution in the catarrhine primates, especially so in the descent of Man and the African apes. The recent findings in the Omo basin of Ethiopia of fossil teeth and mandibles of the immediate generic ancestor of *Homo*, *Australopithecus*, dating from

---

*McKenna would tend to choose a more recent time (15 million years ago) for this branching point but would not rule out the later time listed.

RATES OF EVOLUTION OF DIFFERENT MAMMALIAN α HEMOGLOBIN GENES

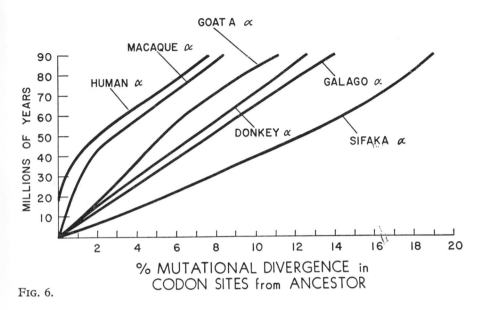

FIG. 6.

RATES OF EVOLUTION OF DIFFERENT MAMMALIAN β-TYPE HEMOGLOBIN

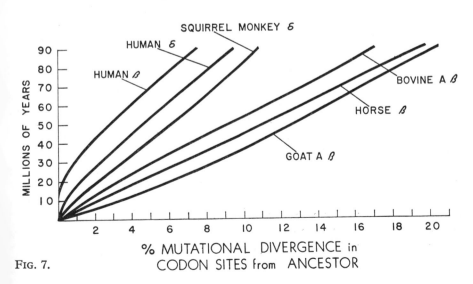

FIG. 7.

as far back as the terminal Pliocene, about 4 million years ago (Howell, 1969), supports, in the context of this discussion, a decelerating rate model for hominoid molecular evolution.

Although the deceleration of molecular evolution is most marked during the final stages of the descent of the lineage leading to Man, the β globin gene results (Fig. 7) depict a relatively slower rate of evolution in this lineage compared to other mammalian lineages, starting from the very beginning of eutherian evolution (about 90 million years ago). This agrees with the idea that during mammalian evolution the progenitors of Man were always in the most anagenetically advanced lineage. Presumably, allelism of β globin genes was present in the basal population, ancestral to all the present-day eutherian mammals, with certain alleles offering the anagenetically evolving portion of this ancient population more advantageous adaptations than the other alleles. As selection favored the spread of such advantageous alleles in this advancing phyletic line, it also acted to eliminate as detrimental alleles most new mutations in these particular genes. It is as though the advancing primates remained in some aspects of their mode of life as their ancestors in the original central population of basal eutherians; thus stabilizing selection would still preserve in these primates certain ancient but adaptive gene complexes. Such selection would be the basis at the molecular–genetic level for the fact that human beings are very generalized mammals.

Aside from depicting differences in relative rates of codon evolution, the amino acid sequence data also provide evidence on the phylogenetic relationships of primates and other placental mammals. The data on fibrinopeptides and α globin chains (but not β) support the paleontological evidence as interpreted by McKenna (1969, 1970) that the most ancient splitting in the Eutheria separates Lagomorpha from all other placentals, also McKenna's view that the rodents evolved out of the early primates. In contrast to immunological data on albumins and several other serum proteins, the α globin data depict a greater distance of tree shrew from primates and ungulates than of primates and ungulates from each other. On the other hand, the α globin data demonstrate that the branching of prosimians (lemur, sifaka and galago) from the line leading to higher primates occurred after the primates had separated from all other mammalian branches. β Globin data show the ceboids to branch away from the line leading to cercopithecoids and hominoids just before these 2 catarrhine groups branched apart. The partial sequence data available on the gibbon β-chain places gibbon closer to other hominoids than to cercopithecoids (macaques) and ceboids. However, in a recent study of catarrhine fibrinopeptide amino acid sequences, Mross *et al.* (1970) found that the divergence of gibbon from chimpanzee and Man approaches that given by cercopithecoids (macaque, vervet, drill) and that the cercopithecoids diverge less from the gibbons than from chimpanzee and Man. While orang utan hemoglobin chains have not been sequenced, amino acid composition data show them to be more divergent from hemoglobins of African apes and Man than these latter hemoglobins are from one another (Buettner-Janusch et al., 1969). Furthermore, partial sequence data on catarrhine carbonic anhydrase I (Tashian and Stroup, 1970) depict first cercopithecoids (macaques, vervet, baboon) splitting from orang utan, chimpanzee and Man, next orang utan splitting from chimpanzee and Man, and then chimpanzee and Man diverging by only one point mutation over the 133 homologous amino acid sites which were examined. Thus these data indicate that the African apes have a much closer genetic relationship to Man than to gibbon or orang utan. This view of hominoid genetic relationships has already been proposed by an extensive body of immunodiffusion data on the divergence between primate species of protein antigenic sites.

Immunological Studies

For many years, beginning long before biochemists had figured out how to sequence proteins, serologists were comparing animal species in terms of the antigenic specificity of proteins and were, thereby, producing data which could be used to decipher phylogeny. However, the significance for taxonomy of the immunological precipitin work of Nuttall (1904), Boyden (1942), and others, which demonstrated the "species specificity" of proteins (Landsteiner, 1947), was not fully grasped in those years, for it was not yet known that DNA contained a genetic code which determined with high precision the detailed structures of proteins. In recent years, the contribution of immunological approaches to the study of phylogeny has gained fairly broad acceptance even among biochemists, some of whom have been impressed with data suggesting that the degree of amino acid sequence difference between proteins correlates with the degree of antigenic difference (e.g. Arnheim et al., 1969; Cocks and Wilson, 1969 and Reichlin et. al., 1970).

Starting with the classic work of Nuttall (1904), a number of investigations of primate relationships utilizing immunological comparisons of proteins, have been carried out (e.g. Mollison, references in Kramp, 1956; Wolfe, 1933, 1939; Boyden, 1958; Paluska and Korinek, 1969; Williams and Wemyss, 1961; Picard et al., 1963) including recent investigations utilizing quantitative precipitin (Hafleigh and Williams, 1966), micro-complement fixation (Sarich and Wilson, 1966, 1967; Sarich, 1968), and radioimmune inhibition of precipitation (Wang et al., 1968) techniques. But by far the most extensive body of immunological data on proteins depicting primate relationships has been obtained by an immunodiffusion technique in which the species comparisons were carried out in trefoil Ouchterlony plates (Goodman, 1960, 1961, 1962a, b, 1963a, b, 1964, 1965, 1967a, b, 1968; Baur and Goodman, 1964; Goodman *et al.*, 1967, 1970; Goodman and Moore, 1971). This technique, which works well with either purified proteins or protein mixtures serving as the immunizing antigens, is being used to sample the degrees of genome divergence among primate species in terms of as many antigenic sites as feasible.

Methods

The trefoil Ouchterlony plate (made by Grafar Corporation, 12613 Woodrow Wilson, Detroit, Michigan 48238) consists of an agar field bounded on 3 sides by wells. An antibody preparation is allowed to diffuse upward from the bottom well toward antigen preparations which diffuse inward from the upper right and upper left wells. Typically, a band of precipitation forms between the antibody and the 2 antigens (Fig. 8).

When the antigen solutions in both upper wells are the same (homologous) as that which was injected into the host to produce the antiserum, a "reaction of identity" occurs, and the precipitation band is that shown in Fig. 8(a). Figure 8(b) illustrates the type of precipitation band resulting when the left well is charged with the homologous antigen and the right well with the heterologous antigen. The line of precipitation extending out from the left side is called a spur and represents all antigenic sites present in the homologous antigen but absent from the heterologous antigen. Spur sizes are graded visually from 0 to 5. The size of the spur is a measure of the antigenic distance of the homologous (A_h) to the heterologous species; the longer the spur, the greater the antigenic distance. Figure 8(c) illustrates the spurs which are sometimes obtained when the antigens in both left and right wells are heterologous. When two heterologous antigens (A_l and A_r) are used, each antigen might develop a spur against the other. The one developing the longer spur shows of the two the least antigenic distance from the homologous species, i.e. it shares more antigenic sites with the homologous antigen than does the other. By subtracting the shorter spur from the longer (in this case A_r from A_l) the net spur size is obtained. This value can be added to the spur size formed by A_h against A_l to give a measure of the antigenic distance of A_h to A_r.

The set theoretical definition of the net spur shown in Figure 8(c), $S_{lr}^h = N(A_h \cap A_l \cap A_r') - N(A_h \cap A_r \cap A_l')$, simply says that with antisera, $As(A_h)$, the net spur formed between heterologous antigens A_l and A_r is equal to the number of antigenic sites which the homologous antigen (A_h) shares with A_l but not with A_r minus the number of antigenic sites which A_h shares with A_r but not with A_l.

In the special case shown in Fig. 8(b), where the homologous antigen is

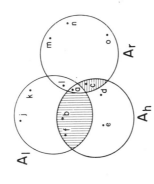
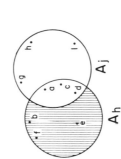

IDENTITY   COMPARISON   COMPARISON

$$S_{hh}^h = N(A_h \cap A_h')$$
$$= N(\emptyset) = 0$$

(a)

$$S_{hj}^h = N(A_h \cap A_j')$$
$$= N(\langle bef \rangle) = 3$$

(b)

$$S_{lr}^h = N(A_h \cap A_l \cap A_r') - N(A_h \cap A_r \cap A_l')$$
$$= N(\langle bf \rangle) - N(\langle c \rangle) = 2 - 1 = 1$$

(c)

FIG. 8. Types of comparisons in trefoil Ouchterlony plates, and the set theoretical definition of the net spurs formed in these comparisons. $As(A_h)$ is the antiserum to the homologous antigen treated as a set of antigenic sites, A_h, from the homologous species h. The antibodies of $As(A_h)$ are directed against antigenic sites: a, b, c, d, e, and f. A_j, A_l, and A_r are heterologous antigens, treated as sets of antigenic sites from heterologous species j, l, and r respectively.

G

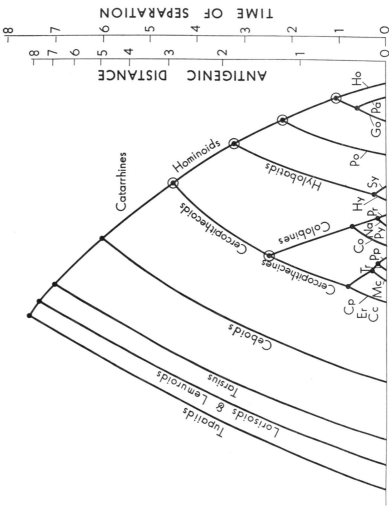

FIG. 9. Cladogenetic tree from computer processing of rabbit anti catarrhine data: Cp, *Cercopithecus*; Er, *Erythrocebus*; Cc, *Cercocebus*; Mc, *Macaca*; Tr, *Theropithecus*; Pp, *Papio*; Co, *Colobus*; Na, *Nasalis*; Py, *Pygathrix*; Pr, *Presbytis*; Hy, *Hylobates*; Sy, *Symphalangus*; Po, *Pongo*; Go, *Gorilla*; Pa, Pan; and Ho, *Homo*. The points surrounded by circles represent the most significant cladogenetic conclusions to be drawn from the rabbit anti catarrhine data.

directly compared to a heterologous antigen (A_j), the formula for the net spur reduces to $S_{hj}^h = N(A_h \cap A_j')$.

Empirical evidence, as well as mathematical deductions from the set theoretical definition of the net spur, demonstrates that any net spur observed as data can be expressed as the difference between 2 homologous spurs (Moore and Goodman, 1968), i.e., $S_{hr}^h - S_{hl}^h = S_{lr}^h$, or, as the sum of the net spurs in a homologous and heterologous comparison, $S_{hr}^h = S_{hl}^h + S_{lr}^h$.

On the basis of this equation, a computer program called "taxon" can take net spur data involving homologous to heterologous and heterologous to heterologous comparisons and solve, via simultaneous linear equations, for the antigenic distance of each member in a particular series of species from a single reference (homologous) species (Moore and Goodman, 1968; Goodman and Moore, 1971). The member species in this series are then arranged in order of ascending antigenic distance in a taxonomic distance table. The clad computer program then depicts the taxonomic distance tables for all homologous species in the various comparison series as the sequence of ancestral branching in the total collection of species.

These taxon and clad computer programs allow results obtained with antisera to single purified proteins to be combined with those obtained with antisera to whole plasma or other protein mixtures, weighting the results used to generate a taxonomic distance table according to the numbers of precipitin lines formed in the comparisons. The results described below were obtained with antisera to 7 highly purified human proteins, to several cruder protein fractions and to whole sera from different primate species. Since the species divergencies were calculated from an averaging of many antigenic sites from a number of proteins, the resulting cladograms were thought to be quite reliable.

Results

About 5600 trefoil Ouchterlony plate comparisons were carried out using 199 monkey, rabbit and chicken antisera to 12 catarrhine species and, in the antigen wells, sera or plasmas from 394 animals belonging to 41 catarrhine primate species, 210 animals belonging to 29 non-catarrhine primate species, and 83 animals belonging to 48 non-primate mammalian species (Goodman and Moore, 1971).

The rabbit and chicken antisera results were used to generate catarrhine cladograms. In general, the two cladograms agreed with each other regarding the separation of lineages. The lineage leading to the catarrhine primates separated first from various non-primate mammals and the tree shrews, then from the lemuroids and lorisoids, and shortly thereafter from the tarsier. The ceboids branched off next, and the catarrhines (cercopithecoids and hominoids) separated as a monophyletic group. Within the catarrhine group, the most ancestral branching was between the hominoids and cercopithecoids.

Within the Cercopithecoidea, the colobines (*Presbytis, Pygathrix, Nasalis, Colobus*) separated from the cercopithecines (*Macaca, Cercocebus, Papio, Theropithecus, Cercopithecus, Erythrocebus*). Within the cercopithecines, a *Macaca–Papio–Theropithecus* branch separated from a *Cercocebus–Cercopithecus–Erythrocebus* branch, and then a *Papio–Theropithecus* branch separated from *Macaca*. The grouping of *Cercocebus* with *Cercopithecus* and *Erythrocebus* was based on rather meager data, and the possibility exists that *Cercocebus* may be cladistically closer to a *Macaca–Papio–Theropithecus* branch as suggested by the chromosome data of Chiarelli (1968) and the comparative anatomical data of Hill (1968) than to a *Cercopithecus–Erythrocebus* branch.

There was some disagreement between the rabbit and chicken antisera results regarding the branching sequence of the Hominoidea. Both depicted the gibbons, *Symphalangus* and *Hylobates*, as separating first and *Pongo* next from *Homo*, *Pan* and *Gorilla*. However, the rabbit antisera cladogram depicted *Gorilla* and *Pan* separating together from *Homo*, whereas in the chicken antisera cladogram, *Homo* and *Gorilla* separated together from *Pan*. These differing results varied only within the error of measurement of the spur technique, and the addition of further data is expected to resolve the conflict.

Both Old and New World monkey anti-Hominoid sera were used to generate cladograms of the Hominoidea. As with the rabbit and chicken antisera results, the data obtained from the two kinds of monkeys agreed except for the branching sequence of *Pan*, *Homo* and *Gorilla*. According to both ceboid and cercopithecoid antisera cladograms, *Pan*, *Homo* and *Gorilla* branched together from the other hominoids. The cercopithecoid antisera cladogram then separates *Pan* and *Homo* together from *Gorilla*, whereas the ceboid antisera cladogram separates *Pan* from a *Homo–Gorilla* branch. Again, as with the rabbit and chicken results, these data disagree only within the limits of error of the spur technique, and thus one cannot be chosen over the other.

Even when the separation of branches within the catarrhine primates was represented in a cladogram by the decelerating rate model of molecular evolution, as shown in the rabbit anti-catarrhine cladogram (Fig. 9), the much closer relationship of the African apes to Man than to the Asiatic apes is evident.

There are now either amino acid substitution or immunologic data on the protein products of at least 17 gene loci which demonstrate the closer relationship of Man to the African apes than to the Asiatic apes (Table IV).

TABLE IV. List of protein products of different gene loci showing closer relationship of Man to African apes than to Asiatic apes

Protein Product	Hominoids Compared	Type of Data
1. α hemoglobin chain	Man, Chimp, Gor, Orang[a], Gib[a]	Amino acid composition and sequence
2. β hemoglobin chain	Man, Chimp, Gor, Orang[a], Gib	Amino acid composition and sequence
3. Fibrinopeptides A and B	Man, Chimp, Gor, Orang, Gib	Amino acid composition and sequence
4. Carbonic anhydrase I	Man, Chimp, Orang	Amino acid composition and sequence
5. Immunoglobulin G	Man, Chimp, Gor, Orang, Gib	Immunological
6. Immunoglobulin A	Man, Chimp, Gor, Orang, Gib	Immunological
7. Immunoglobulin M	Man, Chimp, Gor, Orang, Gib	Immunological
8. Transferrin	Man, Chimp, Gor, Orang, Gib	Immunological
9. Ceruloplasmin	Man, Chimp, Gor, Orang, Gib	Immunological
10. Albumin	Man, Chimp, Gor, Orang, Gib	Immunological
11. Certain minor albumin-like components	Man, Chimp, Gor, Orang, Gib	Immunological
12. α₂ macroglobulin	Man, Chimp, Gor, Orang, Gib	Immunological
13. Prealbumin	Man, Chimp, Gor, Orang, Gib	Immunological
14. Orosomucoid	Man, Chimp, Gor, Orang, Gib	Immunological
15. Thyroglobulin	Man, Chimp, Gor, Orang, Gib	Immunological
16. Certain lipoproteins	Man, Chimp, Gor, Orang, Gib	Immunological
17. A Lens protein	Man, Chimp, Gib	Immunological

[a] The greater divergence of these Asiatic apes from Man and the African apes was demonstrated solely from the amino acid composition data; sequences have not been done as yet, except for the partial sequence on gibbon β hemoglobin chain.

On at least 7 of these protein products and on a number of undefined serum protein antigens, the reciprocal of this, i.e. a closer relationship of the African apes to Man than to the Asiatic apes, has also been demonstrated.

Discussion

The 2 catarrhine cladograms (from rabbit and chicken antisera results) are consistent with major phylogenetic conclusions in the literature on the extant catarrhine primates. They support those formal classifications which depict a closer relationship of cercopithecoids to hominoids than to ceboids by dividing the sub-order of these higher primates into the infra-orders Catarrhini (cercopithecoids and hominoids) and Platyrrhini (ceboids). They also support the generally recognized division of the catarrhine primates into super-families Hominoidea and Cercopithecoidea and the further sub-division of the latter into Colobinae and Cercopiethecinae.

Different authorities have conflicting views on the ancestry network and sub-divisions of Homonoidea. Opinion varies as to whether the African apes are cladistically closer to Man or the orang utan, although the gorilla and chimpanzee are almost always represented taxonomically as if they were closer to the orang utan; i.e. members of the family Pongidae. Some authorities treat the gibbons as a sub-family of the Pongidae, grouping them at the family level with the orang utan and African apes. Other authorities treat them as a separate family, Hylobatidae. The view of certain older primate anatomists reviewed by von Koenigswald (1968) that the hylo-batides are an intermediate form between monkeys and anthropoid apes is carried further by Hammerton *et al.* (1963) and Chiarelli (1963, 1968), who suggest from their studies of chromosome patterns that the gibbons belong in the Cercopithecoidea rather than in the Hominoidea. However, the immunodiffusion findings reaffirm the hominoid classification of the gibbon and support placing *Hylobates* and *Symphalangus* in Hylobatidae as a separate hominoid family.

They also depict a common ancestor for *Homo*, *Pan* and *Gorilla* after the divergence of the line leading to *Pongo*, raising the question of whether to transfer *Pan* and *Gorilla* from the Pongidae to the Hominidae. Further-more, the rather widely-followed proposal of Simpson that the gorilla should be grouped with the chimpanzee at the genus level, thus be classed as *Pan gorilla* rather than *Gorilla gorilla*, is not supported. Indeed, the two branching points separating the ancestral lineages of chimpanzee, Man, and gorilla appear to have been so close to each other in time that the present immunodiffusion data cannot consistently describe the branching sequence of these three hominoids.

In summary, a strictly cladistic classification from the immunodiffusion findings on the catarrhine primates would divide the extant Hominoidea into only two families, Hominidae and Hylobatidae, eliminating the traditional Pongidae; Hominidae would then divide into two sub-families: Ponginae containing *Pongo*, and Homininae containing *Pan*, *Homo* and *Gorilla*.

Another aspect of the comparative immunological data that may be commented on is that these data reveal differences in the rates of evolution of different human proteins which can be related to ontogenetic patterns. For example, albumen, which is synthesized from early embryonic life on, appears to evolve much more slowly (Goodman 1962a, 1965) than the γ-globulins which are not synthesized in significant amounts until well after birth. Baur (1969, 1970) has extended these observations to 28 different purified human proteins. He found that the broader cross-reactions with other mammalian species, due to slowly-evolving antigenic sites on the cross-reacting antigens, were given by antisera to those proteins which are synthesized from early fetal life on and have the ability to cross the placenta, whereas the opposite pattern, due to antigens with many rapidly-evolving antigenic sites, was provided by those proteins which fail to cross the placenta, especially the ontogenetically late-appearing ones. Baur feels that the concept (Goodman, 1961, 1962a, b, 1963a, 1965) that transplacental maternal immunizations have slowed the rate of evolution of certain proteins in the higher primates provides a useful interpretation of his findings.

Review of DNA Data on Primate Phylogeny

The importance of the DNA hybrid reassociation approach (Hoyer *et al.*, 1964, 1965; Hoyer and Roberts, 1967; Martin and Hoyer, 1967; Kohne, 1970; Kohne *et al.*, 1971) in the study of primate phylogeny can hardly be underestimated. Neither amino acid sequencing nor comparative immunologic testing on proteins will be able, in the immediate future, to compare more than a tiny fraction of the coded genetic information in the genomes of primates or other animals. There are just too many thousands and thousands of different gene loci and corresponding thousands of different proteins to be compared in different organisms. While the arguments we have already presented suggest that only the homologs in different species of a handful of proteins have to be compared to be able to depict the ancestral branching sequence of these species, this presupposes that the homologs of each functional protein in every pair of species are coded for by 2 genes, one in each species, which are direct descendants of a gene in the most recent common ancestor of the 2 species. This condition would be met if most proteins were specified by genes made up of non-repeated DNA sequences, and if the mutational differences between intraspecies alleles of these genes had always been minor.

We know, however, that the polypeptide chains of certain proteins, in particular the light and heavy chains of the immunoglobulins, are coded for by populations of genes. There is also evidence that the various genes for the different globin polypeptide chains frequently duplicate, and that in some organisms several similar, but not identical, genes specify the synthesis of the same functional globin chain type, e.g. γ-chain (Shroeder *et al.*, 1968). Indeed, we cannot yet rule out the possibility for any single functional protein which has been studied in vertebrates that a population of genes exists for it, since the evidence that this is not so might merely mean then only one member gene in a population is turned on during the synthesis of the protein. If such a population did exist it would certainly be possible, in some cases, that the particular member genes turned on in 2 different species were not direct descendants of the same gene in the most recent common ancestor of the 2 species. The way out of this dilemma, where we suppose but are not really sure that we are comparing direct protein homologs, is to carry out homology measurements between species over a large portion of the total genetic sequences in the genomes. It is precisely this feat which can be accomplished by the DNA hybrid reassociation techniques.

Two main sets of data on primate DNA's have been gathered so far: the original set utilizing the DNA agar-technique (Hoyer *et al.*, 1964, 1965; Martin and Hoyer, 1967) which dealt with the families of repeated sequence DNA, and a recent set utilizing hydroxyapatite columns which deal with the non-repeated sequences (Kohne, 1970; Kohne *et al.*, 1971). Only the data on the non-repeated DNA sequences meet the condition that any differences present between 2 species arose subsequent to the species branching apart. Nevertheless, it is reassuring that the 2 sets of data not only agree with each other in depicting the same order of relationships among the primates compared but also with the relevant protein data on these primates reviewed in the preceding sections.

When working with either the repeated or non-repeated sequence DNA, fragments of single-stranded, radioactively-tagged DNA from the homologous species are reassociated with non-radioactive single-stranded DNA of the homologous species and each member in a series of heterologous species. The smaller the degree of complementarity between homologous and heterologous DNA, the less reassociation occurs; thus, a measure of genetic relatedness between homologous and heterologous species is provided. Temperature is an important parameter. The higher the temperature, the more perfect must be the complementarity (the matching of base pairs) between the DNA strands for reassociation to occur. The original experiments of Hoyer *et al.* (1964, 1965) by the DNA–agar technique were conducted at 60°C with radioactive DNA from Man, rhesus monkey, green monkey, capuchin monkey, and galago as homologous species and non-radioactive DNA from these animals and from chimpanzee, gibbon, owl monkey, tarsier, slow loris, potto, lemur, tree shrew, hedgehog, mouse and several other animals as the heterologous species used in one or another series of comparisons.

The results of these early experiments showed extensive similarities among higher primates and agreed with protein data in placing hominoids and cercopithecoids closer to each other than to ceboids. In further agreement with protein data, gibbon DNA was not quite as similar to human DNA as was chimpanzee DNA. In a series of comparisons which demonstrated that latter point, the homologous DNA was human, and the degrees of correspondence of heterologous DNAs to the human calculated in the test system were as follows: chimpanzee, 100%; gibbon, 94%; rhesus monkey, 88%; capuchin monkey, 83%; tarsier, 65%; slow loris, 58%; galago, 58%; lemur, 47%; tree shrew, 28%; hedgehog, 21%; mouse, 19%; and chicken, 10% (Hoyer and Roberts, 1967). The later experiments of Martin and Hoyer (1967) utilizing AT-rich and CG-rich DNA fractions as well as whole DNA were conducted at 69°C, again with human DNA as the homologous one. At the higher temperature more divergence between homologous and heterologous DNA was observed. The average chimpanzee cross-section was 91% compared to 76% for gibbon and 66% for rhesus monkey.

It appears that what was largely being measured in these hybrid reassociation experiments with the repeated sequence DNA was a divergence in genomes due, on the one hand, to new DNA arising in an evolving species line after its ancestral separation from other species, and, on the other hand, to the decay of pre-existing DNA families in the evolving lineage. Many examples were found in which a family of repeated sequences exists in the genome of one species while a closely-related species does not possess that family. Such is the case with mouse and rat; a large portion of mouse satellite DNA is composed of repeated sequences not matched by any repeated-sequence DNA in the rat. This portion of the mouse DNA is therefore thought to have been formed by a saltatory replication in the mouse after separation from the rat which occurred about 10 million years ago. Similarly, the green monkey has a particular family of repeated sequences which is not found in the rhesus monkey, a closely-related species, and Man has a family which the chimpanzee does not possess.

The recent work on non-repeated DNA sequences (Kohne, 1970; Kohne et al., 1971) is yielding much more precise information on the rate of nucleotide substitution during evolution and is providing the most exact measurements yet available on the actual percentages of nucleotide sequence divergence among primate species. In this work, the radioactive DNAs were from Man, green monkey and mouse, and the non-radioactive

G*

DNAs were from these animals and from chimpanzee, gibbon, rhesus monkey, colobus, capuchin monkey, galago and rat. The criterion of thermostability was employed to measure the extent of nucleotide pair mismatching on forming, from single-stranded non-repeated DNA sequences, double-stranded interspecies hybrids. The nucleotide difference between Man and chimpanzee was found to be 2·5%, between Man and gibbon 6·2%, between hominoids (Man, chimpanzee and gibbon) and cercopithecoids (rhesus and green monkey) 10·5%, between these catarrhines and capuchin monkey (a ceboid) 18%, and between these members of the Anthropoidea and galago (a lorisoid) about 54%. Furthermore, the difference between rhesus and green monkey is 2·9%, between green monkey and colobus 4·5% (Hoyer, 1970) and between mouse and rat 33·4%. Thus, the small divergence between Man and chimpanzee is equivalent to that between 2 Old World monkeys closely related at the sub-family level, and the divergence between mouse and rat is 3 times greater than that between Man and Old World monkey. Kohne and Hoyer plan on extending these studies to include all principal members of the Hominoidea (gorilla, Man, chimpanzee, orang utan and gibbon) using as homologous DNAs preparations from gorilla and chimpanzee, as well as from Man. The results should provide a decisive answer to the question of whether the African apes are closer genetically to Man or to the orang utan, and, if found to be closer to Man, should answer the further question of which 2 of the 3 species are closest to each other: Man and chimpanzee, Man and gorilla, or chimpanzee and gorilla?

Kohne *et al.* (1971) have also studied the question of whether the rate of nucleotide evolution has been constant over time in different lineages, or whether it has varied in these lineages in accordance with some independently determinable parameter other than time. When the model of uniform rate of evolution over time is employed, and 30 million years ago is taken for the hominoid–cercopithecoid split after Sarich and Wilson (1967), absurdly ancient divergence times of 94 million years ago for the mouse–rat split and 152 million years ago for the catarrhine–galago split are calculated from the DNA data. These so violate the weight of fossil evidence that they cannot be accepted. Instead, Kohne *et al.* (1971) dated all branching points, not just that of the hominoid–cercopithecoid split, from paleontological evidence. When this was done, the results showed that the rate of nucleotide change was not constant with absolute time, but, instead, had been decreasing during the last 65–70 million years, especially in the primate line leading to Man. Evidence from amino acid sequence data that molecular evolution has decelerated in the higher primates was presented earlier in this chapter, and a theory concerning the process of anagenesis predicting this result was also reviewed (Goodman, 1961, 1963a, 1965).

Kohne *et al.* (1971) proposed that rates of nucleotide change are constant with generation time rather than with absolute time. Generation times are not ascertainable from the fossil record, but it is assumed that present-day generation times provide reasonable estimates of those characteristic of the species during evolution. In some cases, generation times are known to have increased during evolution, and this is assumed to have occurred gradually. Using the shortest generation time of the pair of species being compared and the divergence times proposed by paleontologists, they calculated the rates of nucleotide change on a per generation basis rather than absolute time basis.

Using their own DNA data on primates and rodents and that of Laird *et al.* (1969) on bovids, they found a remarkable similarity in the results. That is, primates, rodents and bovids all showed similar rates of nucleotide sequence change per generation. If the mutation rate itself is generation–time dependent, then the marked deceleration of molecular evolution in the lineage leading to Man could simply be attributed to the progressive lengthening of generation time in this lineage. We have already argued why selection against new mutations and not just a slower rate for the occurrence of neutral mutations must be considered an important factor in this process.

ADAPTIVE EVOLUTION—ENZYMES

Because of their nature and function, enzymes offer a unique opportunity for the study of evolution at the molecular level. First, since they are proteins, they are direct gene products, and, as such, their amino acid sequences reflect fairly accurately the nucleotide sequences of DNA. Secondly, although they are usually present in the body in very small amounts, their presence can be detected easily by means of their catalytic activity. Finally, the occurrence of many enzymes as isozymes (different molecular forms with the same catalytic activity) and the relative concentrations of these isozymes seems to be significant with regard to adaptive changes during evolution. It is in the study of isozymes, particularly, that the influence of control mechanisms is evidenced.

Instances of repression and derepression of genes in the same animal are plentiful and well known. For instance, a "switchover" from fetal to adult hemoglobin takes place during the first few months of life, even though both genes were present from the beginning. Similarly, tissue-specific reactivity to activating genes is exemplified by the finding that in rats one form of alcohol dehydrogenase is present in the liver while another form is present in the eye (Koen and Shaw, 1966).

The study of enzymes and their isozymes in primates can yield two different kinds of information: (i) the incidence of amino acid substitutions in the protein molecule, as evidenced by changes in electrophoretic migration, and (ii) grade relationships among members of the order Primates, as evidenced by relative concentrations of isozymes of an enzyme. Usually, both kinds of information can be extracted from a single electrophoretic analysis, the first by the distance and direction of migration of the charged enzyme molecules in an electric field, the second by the amount of catalytic activity revealed by the specific staining technique used.

Only a few amino acid substitutions which alter the electrophoretic migration of an enzyme have been shown to be associated with any alteration in functional efficiency; for the most part, at least *in vitro*, a change in electric charge of an enzyme does not seem to make any discernable difference in its catalytic function. This type of enzyme variation might therefore be tentatively regarded as due to "neutral mutations", except for the cases where it has been proven not to be so.

A shift in isozyme concentration, on the other hand, is, as shown by work in our laboratory, much more likely to reflect grade relationships among the primates. As we interpret our findings, shifts in isozyme concentrations are under the control of regulatory mechanisms which are more deeply rooted in the evolutionary history of the order.

Amino acid substitutions in enzymes may, when enough data has been gathered, yield the same kind of information that is now being obtained from analyses of cytochrome *c*, hemoglobins, etc. Amino acid substitution data, combined with electrophoretic data, may yield insights into adaptive evolution which could not be drawn from either one alone. This is because the manner in which a protein molecule migrates in an electric field gives us information about its surface charges and conformation. If, for example, these have remained the same through large stretches of evolutionary time while a variety of amino acid substitutions have occurred in the molecule, one may assume that retention of this particular surface conformation and charge conferred some advantage on the organisms possessing it. Furthermore, isozyme shifts may reveal the level of adaptive improvements which a primate has achieved. This latter type of information is exemplified by our data on lactate dehydrogenase in primates.

There is a growing body of evidence that, at least for some enzymes, one isozymic form may function more efficiently under one set of conditions while another is better suited to other conditions. Such is the case with hemoglobin (which may be broadly classified as an enzyme) and for lactate dehydrogenase. Dawson *et al.* (1964) have presented kinetic evidence that the B sub-unit of lactate dehydrogenase is more suited to aerobic metabolism and the A sub-unit is more efficient under anaerobic conditions. This is supported by the tissue distribution of the lactate dehydrogenase isozymes: aerobic tissues (e.g. heart) have a larger proportion of the B sub-unit, and anaerobic tissues (e.g. skeletal muscle) have predominantly the A sub-unit. Aerobic metabolism yields more energy than does anaerobic, therefore, it appears that tissues or systems which are rich in the B sub-unit would be supplied with more sustained energy than those having mostly the A sub-unit. It is well known that this kind of division exists in tissues of the same animal—called tissue specificity of isozymes. We have found that among primates there is an increasing amount of the B sub-unit in homologous tissues of different animals as the evolutionary grade increases. That is, not only do all primate hearts have a greater proportion of the B sub-unit than do skeletal muscles, but both heart and skeletal muscle (and all other tissues) of higher primates contain proportionately more B sub-units than do the same tissues of lower primates.

Methods

Enzymes can be separated from other materials and from each other according to their physical characteristics (solubility, molecular weight, electric charge, etc.) and assayed according to their enzyme activity. In much of the work on comparative enzymology in primates, gel electrophoresis has been used as the means of separation. An extract of the sample to be tested is inserted into a gel slab, and the gel is placed in an electric field (direct current). The protein molecules migrate through the gel; the direction and distance covered is governed by the net charge, size and shape of the molecules. After electrophoresis the gel is immersed in a developing solution designed to reveal visually only the component of interest while leaving all other components of the original solution invisible. A general protein stain may be used by which any protein of whatever type will be detected, as long as it is present in sufficient amount. Alternatively, the gel may be developed for enzyme activity. In this case, the incubation mixture is composed of the enzyme's substrate (the specific compound on which the enzyme acts, e.g., lactic acid for lactate dehydrogenase) and such other compounds as are necessary to produce an insoluble, visible product. Enzymes are very substrate-specific so that when lactic acid is used only lactate dehydrogenase will act upon it, all other enzymes remain invisible. Many refinements, adaptations and variations of these methods are available in the literature (Poulik, 1957; Shaw and Koen, 1968; Nakamura, 1966; Jordan, 1967; Russell et al., 1964).

By using gel electrophoretic techniques, it is possible to survey many samples for homologous enzymes from different animals or species in a short period of time. One reservation that must be kept in mind is that an amino acid substitution may have occurred which is not revealed upon electrophoresis; if neither the original nor the substitute amino acid was charged, or if the site of substitution is not on the surface of the molecule, then it is unlikely that a difference in electrophoretic mobility will be observed.

Results

Evidences of adaptive evolution

Lactate dehydrogenase. (i) Electrophoretic changes during evolution. All mammals studied thus far synthesize lactate dehydrogenase in 2 basic forms designated the A and B polypeptides. Most tissues contain both forms, but in some tissues (e.g. heart) the B form is most abundant, while in others (e.g. skeletal muscle) the A form predominates.

Lactate dehydrogenase occurs as 5 isozymes, each of which is a tetramer composed of complementary amounts of the 2 sub-units, A and B (Markert and Møller, 1959; Appella and Markert, 1961). After starch gel electrophoresis and staining of the gel, 5 areas of enzyme activity corresponding to the 5 isozymes are revealed. Numbered from anode to cathode, they are designated LDH 1 (BBBB) (heart type), LDH 2 (BBBA), LDH 3 (BBAA), LDH 4 (BAAA), and LDH 5 (AAAA) (muscle type). Synthesis of the B and A sub-units is directed by 2 separate genes which may vary independently of each other (Shaw and Barto, 1963; Syner and Goodman, 1966). Electrophoretic variation of either usually affects the mobility of all isozymes containing that sub-unit. Fritz *et al.* (1969) have demonstrated that in the rat both synthesis and degradation of the A form of lactate dehydrogenase proceed at different rates in different tissues; thus, the predominance of one form over the other is dependent upon both these factors, and both are basically under genetic control.

LDH 1 has retained a remarkable constancy of electrophoretic mobility throughout primate evolution (Koen and Goodman, 1969a), indicating that the charges on the surface of the molecule have probably not changed. Out of 41 primate species investigated from *Tupaia* to *Homo*, 38 had LDH 1's with the same electrophoretic mobility. LDH 5 was observed in 2 positions, a more cathodal one in Tupaioidea, Lemuroidea, Cercopithecoidea and Hominoidea and a more anodal one in Lorisoidea, Tarsioidea and Ceboidea (Fig. 10).

Both LDH 1 and LDH 5 had the same electrophoretic mobility in Tupaia and Lemur (Fig. 10) and among all catarrhine primates, including *Homo*. However, the intermediate bands (LDH's 2, 3 and 4) of *Homo* and *Tupaia* migrated to a more anodal position than those of the lemur, the cercopithecoids and the other hominoids (Fig. 10). Since all the bands are composed of B and A sub-units, the difference in the intermediate bands must necessarily derive from some difference in one or the other polypeptide even though this was not observable in the parent tetramers themselves. By extraction of LDH 1 and LDH 5 from *Pan* and *Homo* and application of the dissociation–reassociation procedure of Markert (1963), this difference was traced to the A sub-unit. It was concluded that a mutation took place in this polypeptide between *Pan* and *Homo*, and that the site of the amino acid substitution does not lie on the surface of the AAAA molecule when it is in the tetrameric form.

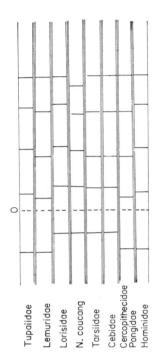

FIG. 10. Lactate dehydrogenase isozyme bands of representative primates. From top to bottom, bands are numbered LDH 1, 2, 3, 4, and 5.

(ii) Evidence for 2 LDH B's produced at 2 gene loci. It is generally assumed that in most animals only one gene locus controls the synthesis of the LDH B polypeptide. However, in *Nycticebus coucang* 2 different LDH B polypeptides are produced, each apparently at a separate locus (Koen and Goodman, 1969b). *N. coucang* is one of the 3 primate species possessing an LDH 1 which is electrophoretically different from that of the other 38. In a series of 25 *N. coucang*, several were observed to have in their tissues a small amount of an LDH 1 which corresponded in mobility to that possessed by most other primate species, in addition to a normal amount of the *N. coucang* LDH 1. One of the *N. coucang* exhibiting this extra band also possessed an LDH pattern entirely consistent with heterozygosity at the LDH B locus. The heterozygous pattern consists of 5 bands in LDH 1, 4 bands in LDH 2, 3 bands in LDH 3, 2 bands in LDH 4, and 1 band in the unaffected LDH 5 (AAAA). The heterozygous condition means that 2 different alleles occupy that gene locus. The appearance of another LDH 1 band (making a total of 6) in the heterozygous *N. coucang* referred to above indicates that a second LDH B gene locus is functioning, albeit at a very low level.

More recently we have observed in another series of *N. coucang*, several who have a second LDH 5 band at the same position as that of the Tupaia, Lemur and catarrhine LDH 5, in addition to the normal *N. coucang* LDH 5. Some of these animals also have the common primate LDH 1. These latter animals, then, have a total of 7 LDH bands, the 5 major bands for slow loris LDH's 1, 2, 3, 4 and 5 and the 2 additional bands, one in the common primate LDH 1 position and the other in the position of LDH 5 of Tupaia, Lemur and the catarrhines (Fig. 10).

(iii) Quantitative increase in B/A ratios. We have shown (Koen and Goodman, 1969a) that the proportion of B to A sub-unit in erythrocytes of primate species increases as the evolutionary scale is ascended (Table V). Additionally, in homologous tissues from *P. potto* (prosimian) and *S.*

TABLE V. Erythrocyte lactate dehydrogenase B/A ratios among primates

PROSIMII

Family	B/A	% B
Tupaiidae	0·63 (0·15)[a]	38·7
Lemuridae	0·66 (0·34)	39·8
Lorisidae	0·83 (0·25)	45·4
Tarsiidae	0·96 (0·02)	49·0

ANTHROPOIDEA

Cebidae	1·81 (0·35)	64·4
Cercopithecedae	1·80 (0·39)	64·3
Pongidae	2·13 (0·35)	68·1
Hominidae	2·92 (0·38)	74·5

[a] standard deviation in parentheses.

sciurea (a member of the Anthropoidea) the ratios of B sub-unit to A sub-unit were higher in the higher primate than in the lower (Table VI). All tissues except liver evidenced this increase, but the tissues showing the most difference between the lower and higher primate were erythrocytes and brain. Within the brain, the areas showing the most difference between *P. potto* and *S. sciurea* were those concerned with the cognitive and integrative functions, while those concerned with basic visceral functions, such as breathing and heartbeat, showed much smaller differences. In *P. potto*, the ratios were substantially lower in the cognitive and integrative brain regions than in the "visceral" regions, whereas in *S. sciurea* the ratios were more uniformly high.

TABLE VI. Tissue LDH B/A ratios of *Saimiri sciurea* and *Perodicticus potto*

Tissue	B/A S.s[a]	B/A P.b[b]	S.s/P.p (B/A)	% B S.s	% B P.p
Brain					
Neocortex					
Occipital pole	2·54	0·64	3·96	71·8	39·0
Calcarine fissure	2·01	0·52	3·87	66·8	34·2
Temporal pole	0·96	0·29	3·31	49·0	22·5
Remaining parts	1·59	0·47	3·38	61·4	32·0
Corpus callosum	1·49	0·33	4·52	59·8	24·8
Limbic system					
Olfactory bulbs	1·52	0·55	2·76	60·3	36·1
Hippocampus	1·01	0·31	3·26	50·2	23·7
Remaining parts	1·40	0·36	3·89	58·3	26·5
Thalamus	2·14	0·35	6·11	68·1	25·9
Midbrain					
Optic chiasma	1·67	0·71	2·35	62·5	41·5
Anterior colliculus	1·49	0·78	1·91	60·0	43·8
Posterior colliculus	2·47	0·80	3·09	71·2	44·4
Cerebellum					
Lateral lobes	2·48	0·77	3·22	71·3	43·5
Vermis	1·78	1·05	1·70	64·2	51·2
Posterior brain stem					
Pons	2·08	0·99	2·10	67·5	49·7
Medulla	2·64	1·01	2·61	72·5	50·2
Spinal cord (adjacent to brain)	1·88	0·66	2·85	65·3	39·8
Erythrocytes	1·97	0·75	2·63	66·3	42·9
Muscle	0·50	0·20	2·50	33·3	16·7
Kidney	1·60	0·65	2·46	61·5	39·4
Heart	3·53	1·79	1·97	77·9	64·2
Skin	0·27	0·15	1·80	21·3	13·0
Spleen	0·61	0·41	1·49	37·9	29·1
Testis	1·03	0·80	1·29	50·7	44·4
Adrenal	1·34	1·14	1·18	57·3	53·3
Liver	0·29	0·30	0·97	22·5	23·0

[a] *Saimiri sciurea*
[b] *Perodicticus potto*

Aconitate hydratase and Isocitrate dehydrogenase. In the citric acid cycle, isocitrate dehydrogenase is sequential to aconitase (aconitate hydratase);

i.e. the product of the aconitase reaction is the substrate for isocitrate dehydrogenase. Both exist in 2 sub-cellular forms, mitochondrial and cytoplasmic (Henderson, 1965; Koen and Goodman, 1969c). Distribution of these 2 isozymes is tissue-specific, some tissues possessing more of the mitochondrial form, others of the cytoplasmic form. Not surprisingly, the relative distribution of the 2 forms of both enzymes was the same in most tissues; those tissues which were richest in the mitochondrial isozyme of aconitase were also richest in the mitochondrial isozyme of isocitrate dehydrogenase and vice versa. However, comparing primates with lower animals, the brain was an exception to this rule. In mouse, bovine and dog brain, the mitochondrial form of aconitase predominated, but the cytoplasmic form of isocitrate dehydrogenase. In primates the 2 forms of each enzyme were more nearly equal, with the mitochondrial form of isocitrate dehydrogenase apparently becoming stronger as the evolutionary scale was ascended; in human brain especially, this isozyme of isocitrate dehydrogenase was quite prominent. No differences in the relative concentrations of the 2 forms of aconitase and isocitrate dehydrogenase have been observed in different areas within the brain, as was noted with lactate dehydrogenase. However, human brain white matter was found to be richer in the mitochondrial form of isocitrate dehydrogenase than was gray matter.

Unfortunately, primate erythrocytes do not exhibit either form of aconitase and only the cytoplasmic form of isocitrate dehydrogenase. Since many fewer primate tissue samples are available than primate blood samples, a complete survey of electrophoretic mobilities could not be accomplished. However, comparative studies of tissue enzymes were made between *T. glis, N. coucang, P. potto, S. sciurea, C. patas, C. aethiops, Pongo, Pan* and *Homo*. From this small sample definite conclusions could not be drawn, but it appeared that aconitase and isocitrate dehydrogenase both show variation at the genus level. In aconitase, *Pan* and *Homo* had the same electrophoretic mobility, while *Pongo* was different. In isocitrate dehydrogenase, each of the 3 hominoids was different from the others.

Intraspecies polymorphism of cytoplasmic isocitrate dehydrogenase appears to be fairly common among primates. Many examples have been observed in this laboratory, including several humans. On the other hand, in a survey of tonsils from 60 children, no polymorphism of aconitase was seen. One was, however, observed in several *N. coucang* (Koen and Goodman, 1969b).

Malate dehydrogenase. This enzyme catalyzes the formation of oxaloacetic acid from malic acid. It occurs in 2 forms, cytoplasmic and mitochondrial. Since mammalian erythrocytes have no mitochondria, it might be expected that no mitochondrial isozyme of malate dehydrogenase would be visible in hemolysates. However, among the non-human primates, the mitochondrial isozyme is quite often observed in erythrocyte extracts. Interestingly, a gradation exists among primates with respect to mitochondrial malate dehydrogenase similar to that seen for erythrocyte LDH 5. That is, lower primates possess more mitochondrial malate dehydrogenase than do higher primates, and the decrease is gradual throughout the evolutionary scale. In Man, the mitochondrial form is not visible and the electrophoretic migration was determined from samples of liver and kidney.

Polymorphism of either form in all species examined is quite rare, and the general incidence of variation among primates occurs at the sub-family level. All the hominoids yield the same pattern in both forms, while the ceboids show the same electrophoretic mobility in the cytoplasmic form, but *Cebus* and *Saimiri* differ in the mitochondrial form. Tupaiinae (*Tupaia* and *Urogale*), Lemurinae (catta, fulvus, macaco, mongoz), Lorisinae (*Galago, Nycticebus, Perodicticus* and *Loris*), and Tarsius are distinct from each other but homogeneous within each sub-family. Within the Cercopithecoidea, on the other hand, *Cercopithecus, Cynopithecus* and *Macaca* species possess similar patterns in both cytoplasmic and mitochondrial forms, while *Papio* and *Theropithecus* species have in common different mobilities of each.

Electrophoretic variants of human cytoplasmic malate dehydrogenase are extremely rare, only one family possessing a variant allele having been reported, to our knowledge, in the literature (Davidson and Cortner, 1967a). In our laboratory, among several hundred primate samples, one has been detected in a gorilla. Mitochondrial malate dehydrogenase variants have been reported in 3 families (Davidson and Cortner, 1967b). We have seen only one in a slow loris. All these variants have so far been observed only in the heterozygous form; no homozygous variants have been found.

Phosphoglucomutase (PGM). PGM catalyzes the conversion of glucose-1-phosphate to glucose-6-phosphate. Human PGM is under the control of 3 separate loci, PGM_1, PGM_2 and PGM_3, (Harris *et al.*, 1968); the PGM of higher primates appears to be analogous to that of human (Goodman and Tashian, 1969). The PGM_1 locus commonly is expressed as 3 electrophoretic phenotypes, PGM_1 1, PGM_1 2 and PGM_1 2–1. The PGM_1 1 and PGM_1 2 phenotypes are homozygotes, each always exhibiting 2 bands, while PGM_1 2–1 is the heterozygote exhibiting a 4-banded pattern. Other variant patterns have been observed, but are rare. All members of the Anthropoidea so far examined in this laboratory and by others (Barnicot and Cohen, 1970) possess the same type of PGM_1 double-banded pattern as humans. Variation of this enzyme occurred at the species level.

The circumstance of 2 electrophoretically separable bands appearing as a homozygous pattern could be explained if they were conformational iso-mers of each other as has been suggested for malate dehydrogenase by Kaplan (1968). It could also occur if the 2 bands were controlled by 2 sepa-rate but closely-linked genes. In our laboratory we have observed that some lower primates possess only a single band in what we assume is the analo-gous PGM_1 pattern, while others have double bands.

Among chimpanzees, a distinct geographical distribution of the PGM_1 1 and PGM_1 2 phenotypes occurs. West African chimpanzees have almost exclusively the PGM_1 2 allele, whereas those from East Africa have a very high frequency of the PGM_1 1 allele, with some polymorphism in both (Goodman and Tashian, 1969). Moreover, even in such a small country as the Cameroons, the division is very sharp, those animals from the West Cameroons possessing the PGM_1 2 pattern, those from the East Cameroons the PGM_1 1. Throughout the Anthropoidea there is evidence of a distri-bution into PGM_1 1 and PGM_1 2 types, but due to the frequent uncertainty of geographical origin of the samples, they cannot be sorted according to this criterion as was done with the chimpanzees.

Phosphohexose isomerase. Phosphohexose isomerase catalyzes the inter-conversion of glucose-6-phosphate and fructose-6-phosphate in the gly-colytic pathway. Electrophoretic variation of this enzyme occurred at the genus level, although none was observed between *Tupaia* and *Urogale* or among most members of the Cercopithecidae. However, there was a dif-ference between *Presbytis obscurus* and *P. melalophus* on the one hand and *Pygathrix nemaeus* on the other.

In general, no significance is attached to the direction of electrophoretic variation (whether the protein becomes more- or less-positively charged). However, the gradation from lower to higher primates in direction of mig-ration of phosphohexose isomerase is quite striking and may be found in the future to bear some relationship to the level of organization of the animal.

The lorisoids are the most anodal, the cercopithecoids are slightly anodal, the ceboids become slightly cathodal and the hominoids are very cathodal. Among the Hominoidea, *Hylobates*, *Pongo* and *Gorilla* have the same catho-dal mobility, *Pan* is more cathodal and *Homo* is most cathodal of all.

Several electrophoretic variants of this enzyme have been observed among humans. Also, in a survey of macaques, we found that *M. irus*, *M. fuscata* and *M. cyclopis* were polymorphic at the phosphohexose isomerase locus. About two-thirds of the species members examined possessed a phosphohexose isomerase electrophoretically the same as that of *M. mulatta*, *M. nemestrina* and *M. speciosa*, but the other one-third of these had a different form of the enzyme. It is not yet known whether or not this polymorphism has a geographical basis.

Glucose-6-phosphate dehydrogenase. This enzyme converts glucose-6-phosphate to 6-phosphogluconate for its entrance into the pentose-phosphate shunt. The shunt is a mechanism for the production of extra energy in addition to that yielded by the glycolytic pathway. It is utilized more in certain tissues than in others and more in young animals than in older ones.

The enzyme is linked to the X chromosome (Boyer *et al.*, 1962; Kirkman and Hendrickson, 1963). Although earlier studies (Chung and Langdon, 1963; Beutler and Collins, 1965) had indicated that the enzyme molecule contains at least 2 different polypeptide chains, Yoshida (1968) has presented evidence that it is made up of several identical sub-units, and that only one structural gene is concerned in their formation.

Among humans, at least 20 different types of glucose-6-phosphate dehydrogenase have been described. The normal form is referred to as the B type, the most common variant as the A type, which migrates faster than the B. Erythrocyte glucose-6-phosphate dehydrogenase deficiency has been linked with malaria (Flatz and Sringam, 1963) and hemolytic anemia (Wong *et al.*, 1965; Tanaka and Beutler, 1969), and a variant characterized by an increase in enzyme activity has been reported (Dern, 1966).

Barnicot and Cohen (1970), investigating this enzyme in the Anthropoidea, found only single bands in all the non-human species examined and no intraspecies differences were noted. Most species exhibited bands with a mobility very similar to that of the human B+ enzyme. Most ceboids were found to have high glucose-6-phosphate dehydrogenase activity. In our laboratory a survey of the lower and higher primates revealed that electrophoretic variation occurred at the genus level.

Electrophoretic variations with no apparent functional significance
6-Phosphogluconate dehydrogenase. This is the second enzyme in the pentose–phosphate shunt; it decarboxylates 6-phosphogluconic acid to yield ribulose-5-phosphate. Throughout the primates, electrophoretic variation occurs in general at the genus level. Interestingly, *Papio* and *Theropithecus* deviate together from *Cercopithecus* and *Cynopithecus* with regard to 6-phosphogluconate dehydrogenase, while the drills yield still a third pattern. *Presbytis melalophus* and *Pygathrix nemaeus* are similar to each other but different from *Presbytis obscurus*. Among the Hominoidae, *Gorilla* and *Homo* have similar mobilities, but *Hylobates*, *Pongo* and *Pan* are each different from the other and from *Gorilla* and *Homo*.

Polymorphism within species is common. Three bands are present in the heterozygote pattern, indicating that this enzyme is composed of 2 sub-units. In a series of macaques, several different 3-banded heterozygote patterns were found, as well as the corresponding homozygous variants.

Acid phosphatase. Barnicot and Cohen (1970) also surveyed acid phosphatase, which consists of 2 components of approximately equal strength, whose mobilities were the same in all the hominoid species examined. There were electrophoretic differences among the cercopithecoids and the ceboids; the variation apparently occurred at the genus level. In addition, there was some intraspecies polymorphism.

Adenylate kinase. Barnicot and Cohen (1970) found intraspecies variation of adenylate kinase to be quite common. Electrophoretic patterns were very similar in the hominoids *Pongo*, *Gorilla* and *Homo*, but more anodal in *Hylobates* as it was in the cercopithecoids.

Esterases. Tashian (1965) investigated by electrophoresis the erythrocyte esterases of 23 primate species. Similar patterns were obtained in *Homo*, *Pan*, *Gorilla* and *Pongo*, but *Hylobates* was different. On the other hand, Syner and Goodman (unpublished data) found brain esterases of *Hylobates* similar to those of *Homo* and *Pan*. The hominoids and cercopithecoids showed rather small interspecies variations; however, among the ceboids and prosimians patterns varied sharply from species to species.

Carbonic anhydrase. This enzyme catalyzes the reversible hydration of CO_2 to form carbonic acid, reversible hydration of aldehydes (Pocker and Meany, 1964) and the splitting of esters (Tashian *et al.*, 1964). There are 3 distinct forms of the human enzyme denoted as A, B and C (Edsall, 1968). Each consists of a single polypeptide chain. Carbonic anhydrases can be visualized on a starch gel after electrophoresis by means of their esterase activity or by the direct hydration of CO_2 in the presence of a pH indicator. If an esterase stain is used, carbonic anhydrase can be distinguished from other esterases either by specific inhibition by sulfonamides or by their preference for β-naphthyl acetate, while other esterases react more strongly with α-naphthyl acetate.

Tashian and Shreffler (1968) investigated genetic and phylogenetic variation in the B and C isozymes of carbonic anhydrase among primates. They found that the B form showed considerably more intraspecies variation than the C form. Some primate species have only one form of the enzyme; when this occurs, the form seems to correspond more often with the C form than with the B form. These authors have theorized from comparison of tryptic peptide patterns, amino acid compositions, and immunodiffusion data that the C isozyme is the older evolutionary form, that the B and C forms had a common genetic origin, and that a gene duplication gave rise to B.

Discussion

Even a cursory examination of enzyme relationships reveals that electrophoretic differences *per se* may be irrelevant to the course of adaptive or

Darwinian evolution. Of the many electrophoretic differences observed, only a handful impart an obvious functional change to the enzyme. On the other hand, enzyme assay *in vitro* is usually carried out under unphysiological conditions so that slightly deleterious or slightly advantageous functional changes may be easily missed due to the limitations of the method. Again, as pointed out by King and Jukes (1969), deleterious changes may only be revealed as such when an "artificial" change in the environment occurs, as with inability of primates and guinea pigs to synthesize ascorbic acid.

This latter type of change is that hypothesized by Dawson *et al.* (1964) for the 2 types of lactate dehydrogenase. Although synthesis of the B and A sub-units is directed by 2 separate structural genes, the relative concentrations of these sub-units in tissues is obviously under a common genetic control. In the context of the proposal of Dawson *et al.* (1964) that LDH 1, composed of 4 B sub-units, is more suited for aerobic metabolism, while LDH 5, composed of 4 A sub-units, is more suited for anaerobic metabolism, our observation of the relative proportions of the B and A sub-units in tissues of primates assumes importance. The ratio of B to A increased in all tissues (except liver) as the primates approached Man. In terms of the theory of Dawson *et al.*, this implies increasing metabolic efficiency. These differences were most striking in brain; the next largest differences were in erythrocytes, other tissues exhibited smaller differences, and in liver the isozyme distribution was the same. In all of 17 homologous brain areas compared between *P. potto*, a prosimian and *S. sciurea*, a ceboid, the higher primate had a higher ratio of B to A sub-units. Furthermore, the areas of the brain concerned with cognitive functions showed more differences than those concerned with the vital functions.

The structural modifications of lactate dehydrogenase revealed by electrophoresis were of 3 types: mutation in the B sub-unit, reflected in LDH's 1, 2, 3 and 4; mutation in the A sub-unit, reflected in LDH's 2, 3, 4 and 5, and another mutation in the A sub-unit, reflected in LDH's 2, 3 and 4 only. The last type of mutation was definitely shown to reside in the A sub-unit only between *Pan* and *Homo*; for the several other incidences wherein only the 3 intermediate LDH bands were affected, we have inferred that those were also due to the A sub-unit.

At first glance, the several electrophoretic variations of lactate dehydrogenase in primates might convince us that these represent neutral mutations which have become fixed in populations by genetic drift (non-Darwinian evolution according to King and Jukes, 1969). However, with regard to the 2 different positions of LDH 5 (that assumed by the Tupaiidae, Lemuridae, Cercopithecidae and Hominoidea on the one hand, and the less cathodal one assumed by the Lorisidae, Tarsius and Cebidae on the other), there is some inferential evidence that this may be an adaptive mutation. In a survey of squirrel erythrocytes supplied by Dr Mary Etta Hight, we have observed that the same type of segregation into more- and less-cathodal LDH 5's occurs between ground squirrels and flying squirrels. All squirrel species surveyed (8) had an LDH 1 which was identical in electrophoretic mobility to the common primate LDH 1. It has been suggested that the branching of rodents, of which squirrels represent a generalized form, from the primate line was subsequent to the branching of primates from other mammals. The possibility exists, therefore, that the 2 different positions occupied by LDH 5 in primates are related to functional adaptation according to Darwinian evolution.

The observation that the common primate LDH 1 and the LDH 5 of Tupaia, Lemur and the catarrhini are also present in some members of *N. coucang* is additional evidence that a control mechanism is operative with regard to the synthesis of the different sub-units of the enzyme. In this case, not only are the relative concentrations of the B and A sub-units specified, but also each of these sub-units is produced in 2 forms. This indicates to us that the control mechanism in *N. coucang* is not so rigidly adjusted as it became in higher primates, thus allowing more than one structural gene to be expressed in the formation of homologous enzymes.

According to Nei (1969), Kohne (1970) and others, when gene duplication occurs and one of the pair subsequently mutates, either both are fixed in the population, or one becomes fixed and the other becomes "silent". The silent partner then is free to undergo further mutation without being subject to selective forces. It is therefore reasoned that more mutations would be found in the silent gene partner after sufficient lapse of time than in the active partner. In *N. coucang* it is obvious that the structural genes for the common primate LDH B and Lemuriform LDH A are the silent genes, since they are expressed only in some tissues of some members of the species and then only at a fraction of the activity of the *N. coucang* genes. Considering the length of time since separation of the lorisoids and lemuroids, one would expect to find these "silent" genes by this time to be so changed as to be unrecognizable, if they could be identified as lactate dehydrogenase at all. Yet, when expressed even today, they are the same form. Even taking into account the constancy of lactate dehydrogenase surface charges throughout primate evolution, this seems remarkable.

With regard to the findings on aconitase and isocitrate dehydrogenase, it is obvious that in terms of cellular efficiency in the production of energy, the situation of 2 sequentially related enzymes in the same part of the cell would be the most favorable arrangement. Availability of substrate is an important factor in the efficiency with which a given enzyme system may operate. If, as in the case of lower mammalian brain, isocitrate is produced within the mitochondrion by aconitase while isocitrate dehydrogenase is outside in the cytoplasm, a considerable slowing down of the cycle could result. The human brain, in contrast to lower animals, has a fairly high concentration of mitochondrial isocitrate dehydrogenase as well as mitochondrial aconitase. Because this would provide a better spatial relationship between the sequential enzymes, an uninterrupted flow of energy would be assured.

Our observation of the progressive decrease of the mitochondrial form of the malate dehydrogenase in erythrocytes as lower primates are compared to Man cannot be interpreted in terms of cellular efficiency, because erythrocytes have no mitochondria. This may, therefore, be a molecular example of gradual loss of a function which is no longer useful in this tissue, such as is theorized for the loss of body hair, prehensile toes, and the ability to synthesize ascorbic acid.

Phosphoglucomutase polymorphism among chimpanzees is quite sharply divided between the Eastern and Western zones of Africa. Although no functional differences in the 2 forms of the enzyme have yet been observed, the geographic distribution of the polymorphism leads one to suspect that there may be this type of difference. We are investigating this further in our laboratory.

Phosphohexose isomerase has not been investigated sufficiently in the primates to draw any conclusions as to the relative functional adequacy of the different forms. However, the regular progressive increase in positive charge (as evidenced by the more cathodal electrophoretic mobilities) as the primates ascend toward Man is most striking. Furthermore, the similarity of the polymorphism of this enzyme among macaques to that of phosphoglucomutase among chimpanzees suggests an adjustment to environmental forces, even though our sampling method did not permit us to differentiate the macaques as to geographical origin.

Glucose-6-phosphate dehydrogenase must be classed as a functionally adapting enzyme, since so many of its known forms are associated with some functional alteration. Although these alterations are usually deleterious, they are assuredly not neutral.

While we have attempted to classify the enzyme mutations into those which are functionally adaptive and those which are not, it is quite possible that the distinction between these categories will prove to be flexible. A mutation which is adaptive in one species may be carried by a related species in which it is neutral or, according to the thesis developed earlier under "Divergent Codon Evolution", an originally neutral mutation may become adaptively valuable when coupled with another neutral mutation.

Investigation of those enzymes whose mutations are classed as neutral promises, when much more work is done, to contribute a great deal to the unravelling of phyletic relationships. Our classification of enzyme-surface mutations in primates into those which occur at the family, generic, species, or individual level is analogous to the differentiation of proteins into slowly-evolving and quickly-evolving ones, as has been done by investigators using amino acid sequence data, except that, by surveying several enzymes across a broad range of primate species, gradations in the rate of change come to light.

REFERENCES

Appella, E. and Markert, C. L. (1961). *Biochem. Biophys. Res. Commun.* **6**, 171.
Arnheim, N., Prager, E. M. and Wilson, A. C. (1969). *J. Biol. Chem.* **244**, 2085.
Barnabas, J., Goodman, M. and Moore, G. W. (1971). *Comp. Biochem. Physiol.* **39**, 455.
Barnicot, N. A. and Cohen, P. (1970). *Biochem. Genet.* **4**, 41.
Baur, K. (1969). *Humangenetik* **7**, 225.
Baur, K. (1970). *Humangenetik* **10**, 8.
Baur, S. and Goodman, M. (1964). *J. Immunol.* **93**, 183.
Benesch, R. E. and Benesch, R. (1970). *Fed. Proc.* **29**, 1101.
Beutler, E. and Collins, Z. (1965). *Science, N.Y.* **150**, 1306.
Bolton, E. T. and McCarthy, B. J. (1962). *Proc. Nat. Acad. Sci. U.S.A.* **48**, 1390.
Bonner, J., Dahmus, M. E., Fambrough, D., Huang, R. C., Marushige, K. and Yuan, Y. J. (1968). *Science, N.Y.* **159**, 47.
Boyden, A. A. (1942). *Physiol. Zool.* **15**, 109.
Boyden, A. A. (1958). *In* "Serological and Biochemical Comparisons of Proteins" (W. Cole, ed.), pp. 3–24. Rutgers University Press, New Brunswick.
Boyer, S. H., Porter, I. H. and Weilbacher, R. G. (1962). *Proc. Nat. Acad. Sci. U.S.A.* **48**, 1868.
Britten, R. J. and Davidson, E. H. (1969). *Science, N.Y.* **165**, 349.
Britten, R. J. and Kohne, D. E. (1968). *Science, N.Y.* **161**, 529.
Buettner-Janusch, J., Buettner-Janusch, V. and Mason, G. A. (1969). *Arch. Biochem. Biophys.* **133**, 164.
Chiarelli, B. (1962a). *Caryologia* **15**, 99.
Chiarelli, B. (1962b). *Caryologia* **15**, 401.
Chiarelli, B. (1963). *Caryologia* **16**, 637.
Chiarelli, B. (1968). *In* "Taxonomy and Phylogeny of Old World Primates with References to the Origin of Man" (B. Chiarelli, ed.), pp. 151–186. Rosenberg and Sellier, Turin.
Chung, A. E. and Langdon, R. G. (1963). *J. Biol. Chem.* **238**, 2317.
Cinader, B. (1963). *Ann. N.Y. Acad. Sci.* **103**, 495.
Cocks, G. T. and Wilson, A. C. (1969). *Science, N.Y.* **164**, 188.
Corbin, K. W. and Uzzell, T. (1970). *Amer. Natur.* **104**, 37.
Davidson, R. G. and Cortner, J. A. (1967a). *Nature, Lond.* **215**, 761.
Davidson, R. G. and Cortner, J. A. (1967b). *Science, N.Y.* **157**, 1569.

Dawson, D. M., Goodfriend, T. L. and Kaplan, N. O. (1964). *Science, N.Y.* **143**, 929.

Dayhoff, M. O. (1969). "Atlas of Protein Sequence and Structure". Vol. 4. Nat. Biochem. Res. Fndn., Silver Spring.

Dern, R. J. (1966). *J. Lab. Clin. Med.* **68**, 560.

Edsall, J. T. (1968). *Ann. N.Y. Acad. Sci.* **151**, 41.

Fitch, W. M. and Margoliash, E. (1967a). *Science, N.Y.* **155**, 279.

Fitch, W. M. and Margoliash, E. (1967b). *Biochem. Genet.* **1**, 65.

Flatz, G. and Sringam, S. (1963). *Lancet* **II**, 1248.

Fritz, P. J., Vesell, E. S., White, E. L. and Pruitt, K. M. (1969). *Proc. Nat. Acad. Sci. U.S.A.* **62**, 2.

Goodman, M. (1960). *Amer. Natur.* **94**, 184.

Goodman, M. (1961). *Human Biol.* **33**, 131.

Goodman, M. (1962a). *Human Biol.* **34**, 104.

Goodman, M. (1962b). *Ann. N.Y. Acad. Sci.* **102**, 219.

Goodman, M. (1963a). *In* "Classification and Human Evolution" (S. L. Washburn, ed.), pp. 204–234. Aldine Press, Chicago.

Goodman, M. (1963b). *Human Biol.* **35**, 277.

Goodman, M. (1964). *In* "Taxonomic Biochemistry and Serology" (C. A. Leone, ed.), pp. 467–486. Ronald Press, New York.

Goodman, M. (1965). *In* "Protides of the Biological Fluids" (H. Peeters, ed.), pp. 70–86. Elsevier, Amsterdam.

Goodman, M. (1967a). *Amer. J. Phys. Anthropol.* **26**, 255.

Goodman, M. (1967b). *Primates* **8**, 1.

Goodman, M. (1968). *Prim. Med.* **1**, 10.

Goodman, M. and Moore, G. W. (1971). *System. Zool.* **20**, 19.

Goodman, M. and Tashian, R. E. (1969). *Human Biol.* **41**, 237.

Goodman, M., Farris, W. and Poulik, E. (1967). *In* "The Baboon in Medical Research" (H. Vagtbord, ed.), Vol. 2, pp. 283–295. University of Texas Press, Austin.

Goodman, M., Moore, G. W., Farris, W. and Poulik, E. (1970). *In* "The Chimpanzee" (C. H. Bourne, ed.), Vol. 2, pp. 318–360. Karger, New York.

Hafleigh, A. L. and Williams, C. A. Jr. (1966). *Science, N.Y.* **151**, 1530.

Hammerton, J. L., Klinger, A. P., Mutton, D. E. and Lange, E. M. (1963). *Cytogenetics* **2**, 240.

Hamilton, W. J., Boyd, J. D. and Mossman, H. W. (1952). "Human Embryology", 2nd edition. Williams and Wilkins, Baltimore.

Harris, H., Hopkinson, D. A. and Luffman, J. (1968). *Ann. N.Y. Acad. Sci.* **151**, 232.

Henderson, N. W. (1965). *Exptl. Zool.* **158**, 263.

Hill, W. C. O. (1968). *In* "Taxonomy and Phylogeny of Old World Primates with Reference to the Origin of Man" (B. Chiarelli, ed.), pp. 7–15. Rosenberg and Sellier, Turin.

Howell, F. C. (1969). *Nature, Lond.* **223**, 1234.

Hoyer, B. H. (1970). Personal communication.

Hoyer, B. H., McCarthy, B. J. and Bolton, E. G. (1964). *Science, N.Y.* **144**, 959.

Hoyer, B. H., Bolton, E. T. and McCarthy, B. J. (1965). *In* "Evolving Genes and Proteins" (V. Bryson and H. J. Vogel, eds), pp. 581–590. Academic Press, New York.

Hoyer, B. H. and Roberts, R. B. (1967). *In* "Molecular Genetics", Part 2, p. 477 ff. Academic Press, New York.

Huxley, H. E. (1964). "Evolution: The Modern Synthesis", Science Edition. John Wiley and Sons, Inc., New York.

Ingram, V. M. (1958). *Biochim. Biophys. Acta* **78**, 539.

Ingram, V. M. (1961). *Nature, Lond.* **189**, 704.

Jordan, E. M. (1967). *Nature, Lond.* **216**, 78.

Kaplan, N. O. (1968). *Ann. N.Y. Acad. Sci.* **151**, 382.

Kimura, M. (1968). *Nature, Lond.* **217**, 624.

King, J. L. and Jukes, T. H. (1969). *Science, N.Y.* **164**, 788.

Kirkman, H. N. and Hendrickson, E. G. (1963). *Amer. J. Hum. Genet.* **15**, 241.

Koen, A. L. and Goodman, M. (1969a). *Biochem. Genet.* **3**, 457.

Koen, A. L. and Goodman, M. (1969b). *Biochem. Genet.* **3**, 595.

Koen, A. L. and Goodman, M. (1969c). *Biochim. Biophys. Acta* **128**, 48.

Kohne, D. E. (1970). *Quart. Rev. Biophys.* **3**, 327.

Kohne, D. E., Chiscon, J. A. and Hoyer, B. H. (1971). "Carnegie Institution Year Book" **69**, 4886.

Kramp, V. P. (1956). *In* "Primatologia I. Systematik Phylogenie Ontogenie" (H. Hofer, A. H. Schultz and D. Stark, eds). Karger, Basle.

Laird, C. D., McConaughy, B. L. and McCarthy, B. J. (1969). *Nature, Lond.* **224**, 149.

Landsteiner, K. (1947). "The Specificity of Serological Reactions". Harvard University Press, Cambridge.

Lengyel, P., Speyer, J. F. and Ochoa, S. (1961). *Proc. Nat. Acad. Sci. U.S.A.* **47**, 1936.

Lengyel, P., Speyer, J. F., Basilio, C. and Ochoa, S. (1962). *Proc. Nat. Acad. Sci. U.S.A.* **48**, 282.

LeGros Clark, W. E. (1959). "The Antecedents of Man". Edinburgh University Press, Edinburgh.

Littau, V. C., Allfrey, V. G. and Mirsky, A. E. (1964). *Proc. Nat. Acad. Sci. U.S.A.* **52**, 93.

Margoliash, E., Fitch, W. M. and Dickerson, R. E. (1968). *Brookhaven Symp. Biol.* **21**, 259.

Markert, C. L. (1963). *Science, N.Y.* **140**, 1329.

Markert, C. L. and Mlleør, F. (1959). *Proc. Nat. Acad. Sci. U.S.A.* **45**, 753.

Martin, M. A. and Hoyer, B. H. (1967). *J. Molec. Biol.* **27**, 113.

McCarthy, B. J. and Bolton, E. T. (1963). *Proc. Nat. Acad. Sci. U.S.A.* **50**, 156.

McKenna, M. G. (1969). *Ann. N.Y. Acad. Sci.* **167**, 217.

McKenna, M. G. (1970). Personal communication.

Moore, G. W. (1971). Ph.D. thesis. North Carolina State University at Raleigh.

Moore, G. W. and Goodman, M. (1968). *Bull. Math. Biophys.* **30**, 279.

Mross, G. A., Doolittle, R. F. and Roberts, B. F. (1970). *Science, N.Y.* **120**, 468.

Nakamura, S. (1966). "Cross Electrophoresis, Its Principles and Applications". Elsevier, New York.

Nei, M. (1969). *Nature, Lond.* **370**, 40.

Noland, C. and Margoliash, E. (1968). *Ann. Rev. Biochem.* **32**, 727.

Nuttall, G. H. F. (1904). "Blood Immunity and Blood Relationship". Cambridge University Press, Cambridge.

Paluska, E. and Korinek, J. (1960). *Z. Immun. Forschung und exper. Therapie* **119**, 244.

Picard, T., Heremans, J. and Vandebroek, G. (1963). *Mammalia* **27**, 285.

Perutz, M. F., Kendrew, J. C. and Watson, H. C. (1965). *J. Molec. Biol.* **13**, 669.

Pocker, Y. and Meany, E. (1964). *Biochemistry* **4**, 2535.

Poulik, M. D. (1957). *Nature, Lond.* **180**, 1477.

Reichlin, M., Nisonoff, A. and Margoliash, E. (1970). *J. Biol. Chem.* **245**, 947.

Rensch, B. (1959). "Evolution above the Species Level". Methuen, London.

Russell, B., Levitt, J. and Polson, A. (1964). *Biochem. Biophys. Acta* **79**, 622.

Sarich, V. M. (1968). *In* "Perspectives on Human Evolution" (S. L. Washburn and P. C. Jay, eds), pp. 97–121. Holt, Rinehart and Winston, New York.

Sarich, V. M. and Wilson, A. C. (1966). *Science, N.Y.* **154**, 1563.

Sarich, V. M. and Wilson, A. C. (1967). *Science, N.Y.* **158**, 1200.

Schildkraut, C., Marmur, J. and Doty, P. (1961). *J. Molec. Biol.* **5**, 595.

Shaw, C. R. and Barto, E. (1963). *Proc. Nat. Acad. Sci. U.S.A.* **50**, 211.

Shaw, C. R. and Koen, A. L. (1968). *In* "Chromatographic and Electrophoretic Techniques", 2nd edition (I. Smith, ed.), Vol. 5, pp. 325–364. W. Heinemann Ltd., London.

Shroeder, W. A., Huisman, T. H. J., Shelton, J. R., Shelton, J. B., Kleihauer, E. F., Dozy, A. M. and Robberson, B. (1968). *Proc. Nat. Acad. Sci. U.S.A.* **60**, 537.

Simons, E. L. (1965). *Nature, Lond.* **205**, 135.

Simons, E. L. (1967). *Sci. Amer.* **217**, 28.

Sokal, R. R. and Sneath, P. H. A. (1963). "Principles of Numerical Taxonomy". W. H. Freeman and Company, San Francisco and London.

Speyer, J. F., Lengyel, P., Basilio, C. and Ochoa, S. (1962a). *Proc. Nat. Acad. Sci. U.S.A.* **48**, 63.

Speyer, J. F., Lengyel, P., Basilio, C. and Ochoa, S. (1962b). *Proc. Nat. Acad. Sci. U.S.A.* **48**, 441.

Syner, F. N. and Goodman, M. (1966). *Nature, Lond.* **151**, 206.

Tanaka, K. R. and Beutler, E. (1969). *J. Lab. Clin. Med.* **73**, 657.

Tashian, R. E. (1965a). *Amer. J. Hum. Genet.* **17**, 257.

Tashian, R. E. (1965b). *Amer. J. Hum. Genet.* **17**, 137.

Tashian, R. E. and Stroup, S. R. (1970). *Biochem. Biophys. Research Communications* **41**, 1457.

Tashian, R. E., Douglas, D. P. and Yu, Y. L. (1964). *Biochem. Biophys. Res. Commun.* **14**, 256.

Tashian, R. E. and Shreffler, D. C. (1969). *Ann. N.Y. Acad. Sci.* **151**, 64.

Tyuma, I. and Shimizu, A. (1970). *Fed. Proc.* **29**, 1112.

von Koenigswald, G. H. R. (1968). *In* "Taxonomy and Phylogeny of Old World Primates with References to the Origin of Man" (B. Chiarelli, ed.), pp. 271–276. Rosenberg and Sellier, Turin.

Wang, A. C., Schuster, J., Epstein, A. and Fudenberg, H. H. (1968). *Biochem. Genet.* **1**, 347.

Waring, M. and Britten, R. J. (1966). *Science, N.Y.* **154**, 791.

Williams, C. H. and Wemyss, C. T. Jr. (1961). *Ann. N.Y. Acad. Sci.* **94**, 77.

Wilson, A. C. and Sarich, V. M. (1969). *Proc. Nat. Acad. Sci. U.S.A.* **63**, 1088.

Wolfe, H. R. (1933). *Physiol. Zool.* **6**, 55.

Wolfe, H. R. (1939). *Biol. Bull.* **76**, 108.

Wong, P. W. K., Shih, L. Y., Hsia, D. Y. -Y. and Tsao, Y. C. (1965). *Nature, Lond.* **208**, 1323.

Yoshida, A. (1968). *Biochem. Genet.* **2**, 237.

Zuckerkandl, E. and Pauling, L. (1965). *In* "Evolving Genes and Proteins" (Bryson and Vogel, eds), pp. 97–166. Academic Press, New York.

Comparison of the Hemoglobins in Non-human Primates and their Importance in the Study of Human Hemoglobins

BOLLING SULLIVAN

*Department of Biochemistry, Duke University Medical Center,
Durham, North Carolina, U.S.A.*

INTRODUCTION

The comprehensive review of animal hemoglobins by Gratzer and Allison (1960) not only summarized the existing data, but also served to stimulate additional comparative studies of hemoglobins. The output of scientific papers since then has made a similar review almost impossible. Certain aspects of hemoglobin biochemistry have been reviewed (Antonini, 1964, 1965, 1967; Antonini and Brunori, 1970; Braunitzer, 1966; Braunitzer *et al.*, 1964; Riggs, 1965; Rossi-Fanelli *et al.*, 1964; Schroeder and Jones, 1965 and Wyman, 1964). Each of these reviews covers part of the overall picture. Similarly, the order Primates contains only a small number of the total species which have hemoglobin. Primate hemoglobins were last reviewed by Buettner-Janusch and Buettner-Janusch (1964) and Barnicot (1969a, b) has discussed certain aspects of the subject. This review will describe the hemoglobin components discovered in primate red cells and the variations which these protein components show. Partial and complete structural studies of hemoglobin polypeptide chains have been undertaken, and these will be examined in light of the functional properties of primate hemoglobins. Finally, variation in structure and function will be compared and related to primate evolution. Knowledge of primate hemoglobins will broaden our perspectives and increase our understanding of the structure, function and evolution of hemoglobins in *Homo sapiens*. Human hemoglobins will be used as points of reference, but no attempt will be made to exhaustively document them since they are often reviewed. Although the focus is on human and other primate hemoglobins, this is a restricted view imposed by convenience. One should always associate and relate *Homo sapiens* to the other hominoids, the hominoids with other primates, primates with other mammals, mammals with other vertebrates, and so on. Regardless of which character is under study, each exists in an evolutionary continuum and should be viewed in an evolutionary context.

NORMAL PRIMATE HEMOGLOBIN COMPONENTS

Inside the red cell there are few proteins other than hemoglobin. The ease with which pure hemoglobin can be obtained has made it a popular subject for study. Nevertheless, it is still unclear exactly how many types of hemoglobins occur in any one species. The hemoglobins of *Homo sapiens* are the best known. When normal adult human red cells are lysed and the resulting hemolysate (about 95% hemoglobin) examined by electrophoresis, 3 heme fractions are resolved. The major fraction, designated hemoglobin A, accounts for 97% of the hemoglobin. The minor fraction, designated hemoglobin A_2, together with small amounts of fetal hemoglobin account for the remaining 3%. Another minor fraction (Hb A_3) is often present in aged red cells. Presumably it results from the covalent bonding of one or more glutathione molecules to normal, adult hemoglobin by disulfide bridges. When the major component, Hb A, is chromatographed on ion-exchange resins (usually CM or DEAE cellulose or Sephadex; or IRC–50) at least 5 additional minor components can be resolved. The genetic origin, sub-unit structure and biological function of most of these minor fractions remains unknown (see review by Schroeder and Holmquist, 1968).

The major component in *Homo sapiens* is a tetramer composed of equal amounts of 2 types of polypeptide chains. Each chain has a molecular weight (m.w.) of about 16,000 and the tetramer m.w. is 64,500. The 2 types of polypeptide chain (designated α and β) have heme at the active site (which binds one molecule of molecular oxygen) and differ in length by 5 residues (141 residues in the α chain and 146 residues in the β chain). The α and β chains in *Homo sapiens* are coded at separate, non-linked loci. Similarities in the amino acid sequences of the α and β chains indicate a genetic relationship between the 2 loci (Ingram, 1961). One chain locus (presumed to be the β) arose from the α chain locus by complete gene duplication followed by translocation and differentiation (for discussions of gene duplication see Dixon, 1966, and Watts and Watts, 1968). Discovery of individuals with 3 types of α chains increases the possibility of 2 loci coding for α chains in some *Homo sapiens* (Abramson *et al.*, 1970; Brimhall *et al.*, 1970). There is no evidence that the human β chain is coded for by more than one locus.

The minor component (HbA_2) is a 64,500 M.W. tetramer containing 2 α and 2 δ chains. The δ chains are very similar to β chains in amino acid sequence, differing at only 10 positions. The δ-chain locus is closely linked to the β-chain locus and crossovers during mispairing are known. The δ-chain locus arose by total gene duplication of the β-chain locus (Ingram, 1961).

Red cells from fetal blood contain a different major hemoglobin (designated hemoglobin F). Fetal hemoglobin (Hb F) is a 64,500 M.W. tetramer containing 2 α and 2 γ chains. The γ chain, like the β and δ chains, contains 146 amino acids. Although the γ chain shares sequence homologies with both the α and β chains, it is believed to have derived from the β-chain locus by gene duplication (Ingram, 1961). *Homo sapiens* has at least 2 loci coding for γ chains (Schroeder *et al.*, 1968; Huisman *et al.*, 1969). The loci differ in coding properties at amino acid residue site 136; glycine at this position in one chain is replaced by alanine in the second chain. Additional hemoglobin heterogeneity in fetal hemolysates (other than the presence of some HbA which replaces Hb F after birth) is largely accounted for by the presence of Hb F_1. This tetramer differs from Hb F by the presence of an acetyl group on the N-terminus of 1 of its 2 gamma chains. It is not known whether or not a separate locus generates the acetylated gamma chain.

Several other tetramers containing 2 types of polypeptide chains occur in very young human embryos. Apparently these are the major components prior to 3 months gestation (Huehns *et al.*, 1964), although one component, Gower 1, is present in low amounts at birth in normal infants (Hecht *et al.*, 1967). Gower 1 apparently contains 2 γ chains and 2 chains designated ϵ. The ϵ chain contains 9 unique tryptic peptides not found in α, β, γ or δ chains (Capp *et al.*, 1967). Gower 2, another component found early in gestation, but not at birth, seems to have 2 α and 2 ϵ chains. Because so little material has been available from human embryos, some confusion naturally exists concerning the sub-unit structure and distribution of these embryonic hemoglobins. The genetic relationships of the ϵ chain to the other chains are not known, but it is interesting that it seems to pair with both α-like and γ-like chains.

Although *Homo sapiens* has a single major adult hemoglobin component, most species of vertebrates have multiple components. Presumably these result from duplications of the α and β chain loci and recombination (usually) into all possible heterotetramers. These multiple components are found in all normal individuals. In contrast to *Homo sapiens*, some fish species have over 15 electrophoretically separable hemoglobin components! For these species the designations major and minor may have little meaning. In primates, however, the hemoglobins do fall into 2 categories, major and minor. Until adequate proof to the contrary is available, one should be aware that not all major components (nor minor components) are strictly homologous. For example, in some cases a minor component may have become the major component with the suppression or elimination of the previous major component. Furthermore, the electrophoretic techniques used to determine the distribution of hemoglobin components suffer some

H

serious drawbacks. Molecules with similar or identical charges may not be resolved. Only 3 of 7 or more minor components are resolved when human hemoglobin is electrophoresed. The products of duplicated α and γ loci in *Homo sapiens* were discovered some 5–9 years after the primary sequences of α and γ chains had been determined. Better electrophoretic, chromatographic and immunologic techniques will undoubtedly uncover additional components in primate hemolysates.

Amino acid sequence studies have shown that gene duplication and point mutation generate most genetic variation in protein structure. We will examine the hemoglobin loci in the primates in order to determine the frequency and significance of these 2 processes. Usually duplicated genes manifest themselves as multiple, non-segregating components. Point mutations segregate. These are "rules of thumb" and of course each occurrence should be thoroughly investigated and if possible proven. Several authors have used the term polymorphic to describe any species or individual that has more than one major hemoglobin component. Usage should be restricted to those individuals or species showing segregating variants.

In the following discussions, the number preceding or following the species (or with the reference) indicates the number of individuals examined. Where pertinent and available the source of the sample(s) has been recorded. If the number of animals examined was not reported, a question mark so indicates.

Adult Components

Buettner-Janusch and Buettner-Janusch (1964) and Hoffman and Gottlieb (1969) both examined single individuals of *Tupaia glis* and found only 1 major hemoglobin.

The lemurs examined by Nute (1968) appear to fall into 2 distinct groups. *Lemur fulvus* (24, including *L. f. fulvus*, *L. f. collaris* and *L. f. rufus*) had a single component. *L. catta* (5) and *L. variegatus* (3) hemolysates were all 2-banded. Buettner-Janusch and Beuttner-Janusch (1964) report a single banded hemolysate for *L. variegatus* that was "difficult to resolve clearly". Probably 2 bands were present. Two bands were observed in hemolysates from *L. catta* and *Hapalemur griseus* (number of animals examined not given in this reference). Single bands were observed in hemolysates from *L. mongoz*, *L. macaco*, *L. f. fulvus*, *L. f. albifrons*, *L. f. collaris* and *L. f. rufus*. A single band was found also in *Propithecus v. verreauxi* and *P. v. coquereli*. It is curious that the apparent duplication includes 2 genera, *Lemur* and *Hapalemur*, but not all members of *Lemur*. Unless *Lemur* is a polyphyletic genus, the *Hapalemur* examined was a genetic variant (heterozygote), or an error exists in the phenotyping, there must have been

2 gene duplications or loss of a second locus in several species of *Lemur*. It is not known which chain locus has duplicated in *L. catta* and *L. variegatus*.

Quite a few species of the Lorisidae have been examined and all are consistent with a duplication of the β chain locus. These show 2 hemoglobin bands in all individuals and in some cases the hemoglobins are able to polymerize. The species examined include *Loris tardigradus* and *Nycticebus coucang* (?-Bucttner-Janusch and Beuttner-Janusch, 1964; 5-Nute, 1968), *Galago crassicaudatus* (?-Buettner-Janusch and Buettner-Janusch, 1964), *Galago c. crassicaudatus* (31-Nute, 1968), *G. c. panganiensis* (36-Stormont and Cooper, 1965), *G. senegalensis* (?-Buettner-Janusch and Buettner-Janusch, 1964; 9-Nute, 1968), *Galago* (*crassicaudatus* and *senegalensis*) (6-Hoffman and Gottlieb, 1969), *Galago* (*Galagoides*) *demidovii* (12-Nute, 1968). Nute *et al.* (1969) suggest on electrophoretic evidence that the duplicated locus is the β chain locus. It will be interesting to trace the distribution of this duplication in other prosimians, since it could be a good phylogenetic marker.

To my knowledge hemolysates from tarsiers have not been examined.

Most hemolysates from New World monkeys have shown a single major hemoglobin band upon electrophoresis. These include *Oedipomidas* (*Saguinus*) *oedipus* (3-Hoffman and Gottlieb, 1969), *Saguinus* (*Tamarinus*) *mystax* (35-Boyer *et al.*, 1969b), *S. nigricollis* (10-Boyer *et al.*, 1969b), *Cebus albifrons* (?-Buettner-Janusch and Buettner-Janusch, 1964), *Saimiri sciureus* (100-Boyer *et al.*, 1969b); (?-Buettner-Janusch and Buettner-Janusch, 1964), *Saimiri* sp. (17-Hoffman and Gottlieb, 1969), *Aotes trivirgatus* (65-Boyer *et al.*, 1969b), *Ateles* sp. (1-Hoffman and Gottlieb, 1969), *Ateles* (*geoffroyi* and *fusciceps*) (106-Boyer *et al.*, 1969b) and *Callicebus moloch* (27-Boyer *et al.*, 1969b).

Buettner-Janusch and Buettner-Janusch (1964) noted 1 and sometimes 2 bands in *Ateles ater* and 2 bands in *Lagothrix lagotricha* and *Cacajao rubicundus*. Boyer *et al.* (1969b) described heterogeniety in the major component of 1 of 4 *Saimiri sciureus* (by compositional analysis) and Nute and Sullivan (submitted for publication) found 2 amino acids at 1 position in a single sample of *Cebus albifrons* β chain. Thus, no major chain duplications in New World monkeys have been proved. Duplications may have occurred in *Cebus albifrons*, *Lagothrix* and *Cacajao*, but these require further study.

Electrophoreses of *Cercopithecus* hemoglobins have shown no evidence for gene duplication. The species examined include *C. aethiops* (3-Jacob and Tappen, 1957; ?-Buettner-Janusch and Buettner-Janusch, 1964; 1-Nute, 1968; 63-Hoffman and Gottlieb, 1969), *C. l'hoesti* (1-Jacob and Tappen, 1958), *C. mitis* (8-Jacob and Tappen, 1957, 1958; ?-Buettner-Janusch and Buettner-Janusch, 1964), *C. mona* (7-Jacob and Tappen,

1958; ?-Buettner-Janusch and Buettner-Janusch, 1964), *C. nictitans* (*ascanius*) (40-Jacob and Tappen, 1957, 1958), *C. patas* (?-Buettner-Janusch and Buettner-Janusch, 1964; Barnicot *et al.*, 1965; 13-Nute, 1968; 6-Hoffman and Gottlieb, 1969), *C. talapoin* (1-Nute, 1968; 3-Hoffman and Gottlieb, 1969). A single *Cercocebus* sp. hemolysate was examined by Hoffman and Gottlieb (1969) and found to be single-banded. In contrast Jacob and Tappen (1957, 1958) examined 18 *C. albigena* and 4 *C. galeritus* hemolysates and found all were 2-banded. Which locus is involved in the apparent duplication is not known.

The genus *Macaca* (including *Cynopithecus*) has been examined in some detail. Those species showing no evidence of gene duplication are *M. niger* (10-Jolly and Barnicot, 1966; 5-Nute, 1968; 1-Ishimoto and Prychodko, in press), *M. cyclopis* (101-Ishimoto *et al.*, in press), *M. fuscata* (217-Ishimoto *et al.*, in press), and *M. mulatta* (300-Buettner-Janusch and Buettner-Janusch, 1964; 524-Gottlieb and Van Lancker, 1963; 589-Hoffman and Gottlieb, 1969; 72-Ishimoto *et al.*, in press). The other species of *Macaca* which have been examined in number require further explanation.

The cynomolgus monkey, *Macaca fascicularis*, presents a most confusing situation. In this discussion it is impossible to divorce the normal phenotype from polymorphism simply because it is almost impossible to determine what the normal phenotype really is! The excellent paper by Barnicot *et al.* (1966) did much to clarify the situation. However, since then additional data, primarily obtained by Ishimoto *et al.* (in press) have confused the picture again.

In order to show the similarities in hemoglobin patterns among species of *Macaca* (no identity of components in different species is implied) I have chosen to label the 3 major hemoglobin components in *M. fascicularis* as fast (F) ($=Q^{mi}$), slow (S) ($=A^{mi}$) and polymerizing (P) ($=P^{mi}$) and those in other species as fast (F) and slow (S). Barnicott *et al.* (1966), who used the mi superscripts, showed that all 3 major components differed in their α chains. They observed all the expected phenotypes and the distribution of phenotypes fitted the Hardy-Weinberg distribution (Tables I and II). Earlier work had described the polymorphism found in *M. fascicularis* but gave rather different phenotype distributions. Thus, Kunkel *et al.* (1957) reported on 6 hemolysates all of which were heterozygous (FS). Eng *et al.* (1960) reported that about half of their 116 animals (origin, Indonesia) were heterozygotes (FS), the remaining were homozygous for the slow component (SS). Tuttle *et al.* (1961) studied 36 animals (origin unknown) and found them distributed as follows: 15 homozygous slow (SS), 20 heterozygous (FS), and 1 heterozygous fast*, slow whose fast band was more anodic than normal FF. Hewitt (1964) examined 175 individuals from Bangkok (34) and ("mostly") Manila (141). He found 11 homozygous

TABLE I. Observed (and expected) numbers of *Macaca fascicularis* distributed according to major hemoglobin phenotype

Source	Number of animals	Major Hb Phenotypes						Reference
		FF	SS	PP	FS	FP	SP	
Thailand								
Batch F	45	0(0·3)	34(34·5)	0(0·1)	6(6·2)	1(0·3)	4(3·5)	Barnicot et al., 1966
Batch G	23	1(1·3)	7(7·9)	1(0·7)	8(6·4)	1(1·9)	5(4·7)	Barnicot et al., 1966
Bangkok	186	0(6·6)	91(94)	2(1·8)	61(50)	9(6·8)	23(25·9)	Ishimoto et al., in press
Malaya	34		10		24			Hewitt, 1964
Singapore	10	0	7	1	0	0	2	Barnicot et al., 1966
	262	0(42·2)	38(74·5)	0(1·1)	190(112)	21(13·2)	13(18·2)	Ishimoto et al., in press
Vietnam								
Batch W	58	1(1·1)	38(37·3)	1(0·2)	13(12·8)	1(1·0)	4(5·6)	Barnicot et al., 1966
Batch Wel	24	0(0·7)	15(15·0)	0(0·1)	7(6·3)	1(0·3)	1(1·6)	Barnicot et al., 1966
Batch WD	40	1(1·4)	23(24·0)	0(0·1)	13(11·6)	0(0·6)	3(2·3)	Barnicot et al., 1966
Cambodia	113		111		2			Ishimoto et al., 1968
Philippines	59				59			Ishimoto et al., 1968
	59	0	0	0	59	0	0	Ishimoto et al., in press
Manila ("mostly")	141	0	1	0	140	0	0	Hewitt, 1964

TABLE II. α-Chain gene frequencies in *Macaca fascicularis*
assuming 1 locus and 3 alleles

Source	Number of genes	F	S	P	Reference
Thailand					
Batch F	90	·08	·88	·04	Barnicot *et al.*, 1966
Batch G	46	·24	·59	·17	Barnicot *et al.*, 1966
	372	·19	·72	·10	Ishimoto *et al.*, in press
Vietnam					
Batch W	116	·14	·80	·06	Barnicot *et al.*, 1966
Batch Wel	48	·17	·79	·04	Barnicot *et al.*, 1966
Batch WD	80	·19	·77	·04	Barnicot *et al.*, 1966
Malaya	524	·40	·53	·07	Ishimoto *et al.*, in press

slow, and 164 heterozygous fast, slow. However, the larger sample from Manila contained only one homozygote while the Bangkok sample contained almost one-third homozygotes. The geographical variation in gene frequencies was also noted by Barnicot *et al.* (1966) in their 2 Thai samples and between the Thai and Vietnam samples (Table II).

Before examining more recent data several technical difficulties which may be affecting phenotypic determinations (especially the earlier work) should be noted. Almost all investigators noted variation in the amount of fast component in FS heterozygotes (a similar situation exists in *M. speciosa*). The amount of F component in FS heterozygotes distributes bimodally with peaks at 30–40 and 60–70% (Barnicot *et al.*, 1966; Ishimoto *et al.*, in press). Hemolysates of animals from Thailand, Malaya and the Philippines (and Vietnam ?) all show the bimodal distribution. When F components from heterozygotes with high and low percentage of F are isolated, mixed, and allowed to dissociate and recombine, no new components are generated (Barnicot *et al.*, 1966). At present it is not known whether 2 types of fast component exist or whether some other factor (presumably at another locus) is controlling the amount of fast component synthesized. The polymerizing component (P) also presents several difficulties. In line with studies on other polymerizing hemoglobins (Riggs *et al.*, 1960; Sullivan and Riggs, 1964; Bonaventura and Riggs, 1967) this component in *M. fascicularis* normally exists *in vivo* as a tetramer. The tetramer has electrophoretic properties similar to those of the F component. When the P component is allowed to oxidize, its migration rate (especially in sieving media) is retarded [note that because of the identical characteristics between the P component in *M. fascicularis* and the polymerizing hemoglobins (Porte Alegre, mouse diffuse, and turtle), I am assuming that the observed polymerization involves disulfide bridging between tetramers; this has not been proved for the P component of *M. fascicularis*]. Aggregates of different sizes apparently form [$S_{20,w}$ of 4·2S and 7·2S (Barnicot *et al.*, 1966), $S_{20,w}$ of 4·2S and 9·2S (Ishimoto *et al.*, in press)] and if molecular sieving is occurring during electrophoresis, this should produce multiple bands. Although the P component was not found by Kunkel *et al.* (1957), Eng *et al.* (1960), Tuttle *et al.* (1961), Hewitt (1964) or Ishomoto *et al.* (1968) (and the latter was aware of the polymerizing component and looked for it in selected samples), it is possible that it was present and interfered with the correct phenotyping.

Ishimoto *et al.* (in press) have examined an additional 507 *M. fascicularis* from Thailand, Malaya and the Philippines. Several important points are apparent from the phenotypic (Table I) and genotypic (Table II) frequencies. First, these workers did not find any FF homozygotes. Second, they did find the polymerizing component (P) in Thai and Malayan samples. Third, their Philippine animals were once again all FS heterozygotes (59 animals had been examined earlier with the same result by Ishimoto *et al.* (1968). Fourth, the distribution of phenotypes does not fit the Hardy-Weinberg equilibrium (which I took the liberty of calculating from their data). The Malayan group deviates very strongly from the expected values.

Can all these data be reconciled? In trying to do so, we must make some assumptions (at least the following and probably others). First, we must assume that the phenotypes have been correctly (almost always) identified; second, that the animals have been correctly (almost always) identified; third, that *M. fascicularis* is not an assemblage of sibling species, nor is it hybridizing with other species in areas of sympatry. The latter may seem to be the weakest assumption since Fooden (1964) has reported that a morphological cline (based on 3 specimens!) exists in Thailand between *M. fascicularis* and *M. mulatta*. He considers them one species. If this is true, then how can the hemoglobin types of *M. mulatta* in Thailand remain monomorphic while those of *M. fascicularis* in the same area are extremely polymorphic?

As is the case with many other non-primate species, the distribution of alleles and phenotypes is geographically dependent (Tables I and II). The most readily explicable situation is in the Philippines where only one of 259 animals examined did not have the FS pattern (it was SS). If we assume that this animal was either misidentified (by origin or by phenotype) or that it was a genetic variant in which both components had the same electrophoretic mobility, then the Philippine data are consistent with the presence of a duplicated gene (presumably α chain). Perhaps F and S are allelic in the Philippine *M. fascicularis* and that the homozygotes are being selected against ? This does not seem likely. If, indeed, the Philippine *M. fascicularis* has a duplicated locus (as in *M. speciosa*), it is pertinent to ask whether or not the duplicated locus exists elsewhere in the range of *M. fascicularis*. The data of Barnicot *et al.* (1966) and Ishimoto *et al.* (in press) indicate that segregation of F and S occurs in Thai populations. The same is true of Vietnamese populations (Barnicot *et al.*, 1966). The Malayan data are most easily explained if the presumed duplication found in the Philippine populations is also present in some of the Malayan animals. In addition there is also an F/S allelic system at one of the loci (F or S locus in Philippine *M. fascicularis*). The only problem with this hypothesis is that it predicts the occurrence of a phenotype with 3 bands (F, S and P). Such a phenotype has not been observed. (I have just received a preprint by Barnicot *et al.* in which they examined 231 *M. fascicularis* hemolysates of Malayan origin. They did find all 3 major components in several animals and conclude, as I have done, that the chain duplication exists in the Malayan *M. fascicularis* populations.) One of the merits of the hypothesis is that it would predict the varying amounts of F and S components in FS heterozygotes which have been observed.

One other complication exists in explaining the hemoglobin phenotypes in *M. fascicularis*. There is a minor component (electrophoretically slower than P) which occurs in some animals but not all (Table III). Since it occurs in heterozygotes (FS, FP, and SP), and since it also has a variant α chain combined with normal β chains (Barnicot *et al.*, 1966; Wade *et al.*, 1967) it involves another α chain duplication. The minor component does not occur in the Philippines (Ishimoto *et al.*, in press) but does occur in Thailand (28·9%-Ishimoto *et al.*, in press; 25·1%-Barnicot *et al.*, 1966) in Vietnam (18·4%-Barnicot *et al.*, 1966) and in Malaya (20·0%-Barnicot *et al.*, 1966; 15·3%-Ishimoto *et al.*, in press). The distribution of the minor component in other parts of the range of *M. fascicularis* is uncertain. Although the distribution of the minor component in Malaya is consistent with linkage to the S locus (or allele), this is not true (unless there has been a crossover) for Thailand where Ishimoto *et al.* (in press) found the minor component in 2 FP heterozygotes.

As difficult as these data are to interpret, they are equally exciting. The problems with *M. fascicularis* can probably be resolved by breeding experiments (for instance a Philippine FS × Thailand PP should be interesting) and by determining the structural equivalence or non-equivalence of the different components in different populations. Interestingly enough, Barnicot *et al.* (1966) suggest from fingerprint data that one of the differences between the α chains of the P and S components may be a glycine to aspartic acid exchange. This is exactly the same exchange found in *M. speciosa* by Oliver and Kitchen (1968), although it does not cause poly-

TABLE III. Observed (and expected) numbers of *Macaca fascicularis* with the minor hemoglobin component classified according to their major hemoglobin phenotype

Source	Frequency of occurence	Major Hb Phenotypes						Reference
		FF	SS	PP	FS	FP	SP	
Thailand								
Batch F	11/45	0(0)	10(8·3)	0(0)	0(1·5)	0(0·2)	1(1·0)	Barnicot et al., 1966
Batch G	6/23	0(0·3)	2(1·8)	0(0·3)	3(2·1)	0(0·3)	1(1·3)	Barnicot et al., 1966
	52/186	0(0)	26(26·3)	0(0·6)	19(17·6)	2(2·6)	5(6·6)	Ishimoto et al., in press
Vietnam								
Batch W	5/58	0(0·1)	4(3·3)	0(0·1)	1(1·1)	0(0·1)	0(0·3)	Barnicot et al., 1966
Batch Wel	8/24	0(0)	3(5·0)	0(0)	5(2·3)	0(0·3)	0(0·3)	Barnicot et al., 1966
Batch WD	10/40	0(0·3)	4(5·8)	0(0)	4(3·3)	0(0)	2(0·8)	Barnicot et al., 1966
Malaya								
Singapore	2/10	0(0)	0(1·4)	0(0·2)	0(0)	0(0)	2(0·4)	Barnicot et al., 1966
	40/262	0(0)	12(5·8)	0(0)	25(29·1)	0(3·2)	3(2·0)	Ishimoto et al., in press

H*

merization of the fast component of *M. speciosa*. The structural comparison of F and S components in different macaque species should prove illuminating and help to solve the questions of possible introgression and gene flow in certain populations of *M. nemestrina* and *M. fascicularis*.

The pattern of hemoglobins observed in *Macaca nemestrina* is also confusing. Hoffman and Gottlieb (1969) report a 2-banded phenotype for the single *M. nemestrina* hemolysate they examined. Crawford (1966) reported extensive polymorphism in *M. nemestrina* hemoglobins. Of 75 animals examined, 12 were homozygous slow, 43 were homozygous fast, 19 had both components, and 1 animal showed 2 components but in the ratio 75:25 (fast/slow). Hadden (personal communication) has not been able to duplicate Crawford's results. After examining hemolysates from a large number of Crawford's experimental animals plus many of their descendents, Hadden reports very little polymorphism. An additional group of *M. nemestrina* (about 100), not available when Crawford examined the University of Washington primate colony, also showed very little polymorphism. Apparently Crawford inadvertently sampled another species (probably *fascicularis*) in addition to *M. nemestrina*. However, to add to the confusion, Ishimoto *et al.* (in press) report on a sample of 85 *M. nemestrina* from Thailand and 160 from Malaya. The Malayan animals were homozygous fast (as were Hadden's which were from Malaya). The Thai sample included 21 fast homozygotes, 62 fast, slow heterozygotes, and 2 slow homozygotes! If the variant chains are assumed to be allelic, the distribution differs markedly from that predicted by Hardy-Weinberg equilibrium (31·6 fast, 40·4 fast, slow, 12·9 slow). One explanation is that the *M. nemestrina* from Thailand also contained some *M. mulatta* (which would be homozygous slow) and some *M. speciosa* (heterozygous fast, slow). Another distinct possibility is that *M. mulatta* genes are being introduced into the *M. nemestrina* gene pool in Thailand by hybridization. Gottlieb and Van Lancker (1963) found 2 hybrids (*M. mulatta* × *M. nemestrina*) in a commercially supplied group of 524 *M. mulatta* and 13 *M. nemestrina*. The hybrids' hemoglobin patterns were 2-banded. A *M. mulatta* × *M. nemestrina* mating in the laboratory produced a viable 2-banded offspring. Introgressive hybridization of *M. nemestrina* by *M. mulatta* in Thailand seems the most likely explanation for the data of Ishimoto *et al.* (in press). On the other hand an additional allele could exist in high frequency in Thailand with strong selection for or against certain phenotypes. The confusion surrounding *M. nemestrina* hemoglobin phenotypes underlines the need for accurate field data, correct identification or deposition of specimens, and adequate sampling of natural populations.

Kitchen *et al.* (1968) reported 2 hemoglobin bands in 47 captive *Macaca speciosa*. They estimated the ratio of fast to slow electrophoretic component (at pH 8·6) to be approximately 65:35. In 3 animals the reverse ratio was found. Oliver and Kitchen (1968) report that the 2 hemoglobins differ by a single α-chain peptide (by peptide mapping) and that 1 α chain contains glycine and the other aspartic acid in tryptic peptide III (probably position 15). Ishimoto *et al.* (1968) reported a single phenotype containing 2 electrophoretic bands from hemolysates of 47 *M. speciosa*. However, Ishimoto *et al.* (in press) report that hemolysates from 2 of 191 *M. speciosa* (from Thailand) had only the slow component, the remainder showing both bands. They determined the amounts of the 2 components in 52 hemolysates and always found a fast/slow ratio of 65:35. One wonders of course whether or not the 2 slow homozygotes might not have been misidentified (perhaps they were *M. mulatta*?) or mislabelled. It is also possible that these 2 specimens were genetic variants in which both bands have the same electrophoretic mobility. One wonders also why unlike earlier workers, they did not observe any quantitative variation in the two components? Kitchen *et al.* (1968) mention that preliminary breeding data support the theory that the 2 α chains are non-allelic. Almost all of the phenotyping also supports the concept of a duplicated α chain in *M. speciosa*.

Papio sp. are commonly imported and a number of species have been examined. With one exception (*P. sphinx*) there is no evidence for a gene duplication. Those species studied are *P. papio* (36-Jolly and Barnicot, 1966), *P. comatus* (*ursinus*) (45-Buettner-Janusch, 1963), *P. cynocephalus* (157-Jolly and Barnicot, 1966; 3-Nute, 1968), *Papio* sp. (pooled *P. anubis*, *P. cynocephalus* and *P. doguera*) (41-Hoffman and Gottlieb, 1969); *P. doguera* (1-Jacob and Tappen, 1958; 539-Buettner-Janusch, 1963), *P. hamadryas* (21-Jolly and Barnicot, 1966; 1-Nute, 1968). *Papio sphinx* has been examined by Nute (1968) Buettner-Janusch and Barnicot (1969b), and Buettner-Janusch *et al.* (1970). Although only 5 animals were involved, all had a 2-banded pattern. One *Papio* (*Theropithecus*) *gelada* had a single-banded pattern.

Colobus sp. examined by Buettner-Janusch and Buettner-Janusch (1964) showed a single-banded hemolysate. Jacob and Tappen (1957) observed single-banded patterns for *Colobus badius* (2) and *C. polykomos* (*abyssinicus*) (3). A single hemoglobin has been reported for *Presbytis cristatus* (21), *P. melanophos* (5), *P. obscurus* (3) and *Pygathrix nemaeus* (1) (Ishimoto and Prychodko, in press).

Thus, the Old World monkeys show an α chain duplication in *Macaca speciosa*, probably 2 α chain duplications in *Macaca fascicularis* (*irus*), a duplication in *Cercocebus albigena* and *C. galeritus*, and perhaps a duplication in *Papio sphinx*.

Hylobates lar (54), *H. l. pileatus* (2), and *H. agillis* (1) all have a single major component, while 4 *H. concolor* showed 2 components in each individual (Ishimoto and Prychodko, in press). Hoffman *et al.* (1967) examined 20 *H. lar* and 2 *H. l. pileatus* and found only a single major band (but species specific) in each hemolysate. Nute (1968) electrophoresed hemolysates from 35 *H. lar* and 3 *H. l. pileatus*. All had single major components, but one *H. lar* sample showed the electrophoretic mobility of *H. l. pileatus* (one wonders again about the possibility of misidentification or mislabelling although a genetic variant could be involved). Hoffman and Gottlieb (1969) have examined 43 *H. lar* (including *H. lar* and *H. l. pileatus*?) and all had a single major hemoglobin component. Barnicot and Jolly (1966) and Sullivan and Nute (1968) investigated the hemolysates from *Pongo pygmaeus* (31) and found extreme polymorphism involving 4 alleles, 2 each at the α and β loci, but no evidence of gene duplication.

Hoffman *et al.* (1967) sampled 109 *Pan troglodytes* (of 4 sub-species) and found only a single band in all but 2 animals. These were both 2-banded, indicating the presence of simple genetic variants. The number examined was extended to 164 (Hoffman and Gottlieb, 1968), yet no additional variation was found. Nute (1968) examined 13 *Pan gorilla* and Hoffman and Gottlieb (1969) examined 1. All were monomorphic for the major component. Additional observations on *Hylobates*, *Pongo* and *Pan* were described in Buettner-Janusch and Buettner-Janusch (1964) and these are similar to the more recent findings.

Thus, among the pongids examined only *Hylobates concolor* shows evidence of a gene duplication.

Minor Components

The minor component (Hb A$_2$) found in *Homo sapiens* is also distributed throughout the Pongidae (*Hylobates*, *Pongo* and *Pan*). It has been observed in all hemolysates. Reports of Hb A$_2$ components in other primates, except Ceboidea, are probably in error. The minor component in *Macaca fascicularis* is an α–β, not an α–δ, chain combination. Minor components reported for loris hemolysates (Buettner-Janusch and Buettner-Janusch, 1964; Stormont and Cooper, 1965) are probably polymers of the major components.

Most investigations of ceboid hemolysates have revealed a minor component migrating more slowly than the major component at alkaline pH values. Boyer *et al.* (1969a, b) have carried out extensive studies on the minor components of New World monkeys including structural studies. The results of their structural investigations (by tryptic peptide compositions) are summarized in Fig. 1. These authors have argued that Hb A$_2$ is essentially functionless in apes and New World monkeys. They have also shown that when the variation in structure (versus a hypothetical archetype) is compared for New World monkey β and δ chains, there is about

Positon	5	6	9	12	16	22	47	50	51	86	116	117	125	126	130	139
Human δ	Pro,	Glu,	Thr,	Asn,	Gly,	Ala,	Asp,	Ser,	Pro,	Ser,	Arg,	Asn,	Gln,	Met,	Tyr,	Asn,
Ateles δ	Gly,	Glu,	Ala,	Ala,	Gly,	Glu,	Ala,	Thr,	Pro,	Ala,	Arg,	Asn,	Gln,	Val,	Phe,	Thr,
Saimiri δ	Gly,	Asp,	Ser,	Ala,	Ser,	Glu,	Ala,	Ser,	Ala,	Ala,	Arg,	Asn,	Gln,	Val,	Phe,	Thr
Saguinus δ	Gly,	Glu,	Ser,	Ala,	Ser,	Glu,	Ala,	Ser,	Pro,	Ala,	Arg,	Asn,	Arg,	Val,	Phe,	Thr
Human β	Pro,	Glu,	Ser,	Thr,	Gly,	Glu,	Asp,	Thr,	Pro,	Ala,	His,	His,	Pro,	Val,	Tyr,	Asn
Ceboid β	Gly,	Glu,	Ser,	Thr,	Gly,	Glu,	Asp,	Thr,	Pro,	Ala,	His,	His,	Gln,	Val,	Tyr,	Asn

FIG. 1. Alignment of the substitutions in human and New World monkey β and δ chains. The ceboid β chain is a precursor derived from *Cebus*, *Ateles*, *Saimiri* and *Saguinus* β chains assuming minimum mutation distances.

twice as much variation in δ chains. This supports their hypothesis that there is little selection on the δ-chain locus.

Since I disagree with their interpretation, I will present a different argument based on their data. They argue that Hb A_2 does not differ in function from Hb A, or that its selective value to the animal is proportional to its percentage composition of total hemoglobin. Since the amount of Hb A_2 present in New World monkey hemolysates is $0\cdot6$–$6\cdot0\%$, depending on the species examined, selection on the δ-chain locus is also reduced to about $0\cdot6$–$6\cdot0\%$ of that on the β-chain locus. The functional studies of Hb A_2 from *Homo sapiens* have not shown any differences between it and Hb A. However, these studies have involved only a few parameters, and the chances that we are simply ignorant of Hb A_2's function are good (remember that it took 50 years to discover that 2,3-diphosphoglycerate, which is approximately equimolar with hemoglobin in the red cell, can drastically affect oxygen binding!). No one has investigated the functional properties of Hb A_2 in other species. Schroeder and Holmquist (1968) have suggested a variety of other functions for minor components; none of these possibilities has been tested. It should be remembered that compared to most enzymes, Hb A_2 is present in large amounts in red cells.

Boyer *et al.* (1969b) have considered the possibility that ape and New World monkey δ chains may not be strictly homologous (that is they did not arise from a single duplication of the β locus, but at 2 different times involving 2 duplications of slightly different β-chain loci). They concluded that the 2 δ chains were homologous. There is evidence both for and against strict homology, but because we are so unaware of the power of convergent evolution at the molecular level, I find it difficult to arrive at either conclusion using the available data. When comparing δ chains (of New World monkeys and apes) with their β chains, the identities at positions 116 and 117 are the strongest evidence for homology (Fig. 1). However, if both types of minor components are doing the same job in each group of animals, then residues 116 and 117 might form a positively charged binding site which has been selected in both groups. This site is already present in β chains and selection might act to increase the strength of binding. The glycine at position 5 in New World monkey δ and β chains versus the proline in the δ and β chains of *Homo sapiens* argues strongly for separate origins of the 2 δ loci. Glycine to proline requires 2 mutational events and it must have occurred twice in 2 separate lineages for the New World monkey and ape δ chains to be homologous. If one compares the variation within New World monkey δ chains and within New World monkey β chains versus their counterparts in *Homo sapiens* another anomaly is apparent (Nute and Sullivan, submitted for publication). Mutations common to New World monkey δ chains occur at positions 12, 47, 130 and

139. There are no such mutations uniting the New World monkey β chains! If we assume homology of the 2 δ chain types, then one must postulate that 4 mutations occurred in δ chains prior to New World monkey speciation, but none occurred in β chains (or they have since been obscured by further variation). It seems unlikely that this would happen. Furthermore, comparative gross morphology and amino acid sequences both indicate a closer relationship between apes and Old World monkeys than between apes and New World monkeys. Since Old World monkeys do not have δ chains, they must have lost it prior to their differentiation, but after differentiation of the ape line. One can postulate that the δ-chain locus is silent in Old World monkeys but active in apes. This, of course, argues against Hb A_2 as a functionless component, and I find it difficult to believe that a gene can remain silent for millions of years and maintain its structural integrity.

As we have seen, there is ample evidence that gene duplication commonly occurs. If we assume independent origins for the New World monkey and ape δ-chain loci, many of the problems discussed above disappear. The variation in sequence among New World monkey δ chains is now less than that among New World monkey β chains (because the common mutations in each line are archetypal). The lack of common mutations for New World monkey δ and β chains weakens this hypothesis. As more sequence data become available, perhaps these archetypal chains will converge. A more rational choice can probably be made when additional pongid δ chains have been sequenced. If the archetypes from each δ chain line converge, they are probably homologous. If they diverge from each other but converge toward their respective β-chain archetypes, then separate duplications probably occurred. If the identities at positions 116 and 117 have evolved in a convergent fashion, this would provide a striking example of the power of convergence at the molecular level!

In summary the minor component, containing δ-like chains, is restricted to the New World monkeys and apes. δ Chains in the 2 groups may have arisen from 1 or 2 duplications of the β chain locus.

Fetal Components

The distribution of the fetal component (Hb F) in the primates is unclear. Certainly fetal hemoglobins occur in Old World monkeys and apes. In the other groups of the primates, either fetuses have not been available, or it is unclear whether or not they have a fetal hemoglobin containing γ chains.

One distinctive property of human fetal hemoglobin is that it denatures more slowly than adult hemoglobin in alkali solutions. Adult hemoglobins from prosimians also denature more slowly than hemoglobins from adult monkeys and apes. This aroused suspicion that perhaps the γ chain is found in the major hemoglobin fraction from adult prosimians. Amino acid sequence studies have shown that prosimian non-α chains are β-like and not γ-like. Similarities in denaturation rates are coincidental or based on some common structural property. One hemolysate from a newborn *Galago c. crassicaudatus* showed 2 electrophoretic bands identical in mobility to those found in the adult (Nute, 1968). Since the β-chain locus is the duplicated locus in *Galago*, one would expect only a single band if a distinct fetal hemoglobin exists in this species. One is assuming of course that the γ-chain locus isn't duplicated and that fetal hemoglobin is the predominant component at birth as in *Homo sapiens*.

Beaven and Gratzer (1959), Sen *et al.* (1960) and Buettner-Janusch *et al.* (1961) report a fetal hemoglobin from *Macaca mulatta*. Differences in oxygen binding of whole blood also indicate that a fetal hemoglobin occurs in *Macaca mulatta*, *M. fascicularis*, *M. nemestrina* (Novy *et al.*, 1969) and *M. speciosa* (Blechner and Stenger, 1970). Kitchen *et al.* (1968) reported the time-dependent disappearance of electrophoretically distinct fetal hemoglobins in *Macaca speciosa* (the α locus in duplicated in this species). Transition from fetal to adult hemoglobins takes place in 35–70 days in macaques and 270–300 days in *Homo sapiens*. Similar results were obtained for *Macaca fascicularis* and a *M. mulatta* × *M. fascicularis* hybrid (Hanly and Hoffman, in press). Nute and Stamatoyannopulos (in press) report 2 fetal hemoglobins in fetal and newborn *Macaca nemestrina*. The hemoglobins, which differ in their γ chains, do not segregate (in 44 animals). Amino acid compositions and peptide mapping indicates the existence of multiple differences between the 2 γ chains. Buettner-Janusch *et al.* (1968) found a distinct electrophoretic component in a newborn and 15-day-old *Papio cynocephalus*. Huisman *et al.* (1960) and Hanly and Hoffman (in press) report a distinct fetal hemoglobin in *Papio* sp. and *Pan troglodytes*.

So far γ chains have been found only in Old World monkeys and apes, but they have not been looked for in any genera other than *Galago* among the prosimians or New World monkeys. At least 2 γ-chain loci exist in *Macaca nemestrina* and in *Homo sapiens*. Probably, 2 different duplications of the γ-chain locus are involved since the human loci produce such similar chains.

Embryonic Components

Embryonic hemoglobins have been reported for *Homo sapiens*. Among the primates, no embryonic hemoglobins have been reported, but Nute *et al.* (in preparation) have discovered a component in a 60-day-old fetus of *Macaca nemestrina* which electrophoretically resembles human Gower-2. Laboratory colonies of breeding primates should offer many opportunities for the study of embryonic hemoglobins.

In summary, the overall distribution of normal components is similar to that found in *Homo sapiens*. That no other species seems to have as many loci probably reflects the small effort put into studies of non-human primates as compared to the effort put into the study of human hemoglobins. Certain components may be restricted to particular primate groups. The distribution of gene duplications indicates that duplication is a common occurrence. This in itself makes it hazardous to assume strict homology for the hemoglobin components throughout the primates.

ABNORMAL OR POLYMORPHIC PRIMATE HEMOGLOBIN COMPONENTS

Hemoglobin variation occurs in many species of primates. Variation of both major and minor components has been reported, but neither fetal nor embryonic hemoglobin variants have been found (with the exception of fetal variants in *Homo sapiens*). As most amino acids are neutral in charge, less than one-fourth of the amino acid exchanges (assuming randomness) will involve a charge change. Since we depend so heavily on electrophoresis as the differentiating technique, most changes in structure probably go unnoticed. Our studies of population structure and gene flow are quite limited (especially among primates). Studies of other species indicate that an unexpected amount of variation exists at most loci (Hubby and Lewontin, 1966; Lewontin and Hubby, 1966). Although we do not know yet how much variation is normal, the interesting primate species will be those with excessive or very little variation. In some species the distribution of hemoglobin variants may reveal patterns of gene flow and population structure.

No electrophoretic variants of the major hemoglobin have been described for tree shrews, lemurs, lorises or New World monkeys. By sequence analysis Nute and Sullivan (submitted for publication) and Boyer *et al.* (1969b) have reported heterogeneity in β chains from *Cebus albifrons* (1 animal examined) and *Saimiri* (1 of 4 animals examined). In both species one neutral amino acid replaces another. The oxygen binding properties of 2 different *Tupaia glis* hemolysates showed distinct differences, but there were no further investigations (Sullivan, unpublished).

Jacob and Tappen (1957) report a variant among 6 *Cercopithecus mitis*. More than 1000 *Macaca mulatta* have been examined but none had variant hemoglobins. In *M. fascicularis* the Hb P component appears to segregate and should be allelic with the slow component. The fast component, Hb F, seems to segregate in parts of the range, but in other parts it may be fixed by gene duplication. The minor component shows one variant (Barnicot *et al.*, 1966). The minor component(s) is not distributed throughout the population and may only exist as a duplication. There is no evidence that it segregates with any of the other α-chain alleles. As discussed in the section on normal components, the variants of *M. speciosa* and *M. nemestrina* described by Ishimoto *et al.* (in press) may result from misidentifications or hybridization. Quantitative variation in some of the macaque components may indicate a genetic polymorphism. The most polymorphic species of Old World monkeys (excepting the system in *M. fascicularis*) appears to be *Macaca (Cynopithecus) niger*. Only a few individuals have been examined (Jolly and Barnicot, 1966; Nute, 1968), and yet 3 alleles appear to segregate. Buettner-Janusch (1963) described a variant phenotype in low frequency in a colony of captive *Papio ursinus*.

No variation in major components has been reported for *Hylobates* if one recognizes *H. lar* and *H. pileatus* as full species. Extensive variation occurs in *Pongo pygmaeus* (Buettner-Janusch and Buettner-Janusch, 1964; Barnicot and Jolly, 1966; Sullivan and Nute, 1968). Single variants of both α and β chains exist in the small population sampled (gene frequencies: α A-0·47; α B-0·53; β A-0·90; β B-0·10). Nine phenotypes are possible, 1 of which is a double heterozygote with 4 major hemoglobins. Sullivan and Nute (1968) observed 5 phenotypes. Isolation of the major components and measurement of their oxygen binding properties failed to disclose any differences (Sullivan and Nute, 1968). The reason for the extreme heterogeneity in *Pongo* is unknown. No variation in *Pan gorilla* hemolysates has been recorded, but Hoffman *et al.* (1967) report 2 variants in *Pan troglodytes*. One α and 1 β chain variant were found once each among members of the sub-species *P. t. schweinfurthi*.

Boyer *et al.* (1969b) reported several variant minor components in New World monkeys. Seven of 100 *Saimiri sciureus* were heterozygous for the minor component. One of 65 *Aotus trivirgatus* contained 2 minor components. In addition to the normal minor component 2 variant components were found among 106 *Ateles*. One, designated W_2, had a gene frequency of 0·35 in *A. geoffroyi*. The second, designated Y_2, was found both in *A. geoffroyi* and *A. fusiceps* but was more common in the latter, where its gene frequency was 0·17. The only other variant minor component described occurs in *Hylobates lar lar*, where it appears to segregate (Nute, 1968). Concomitant variation did not occur in the major component of the animals listed above as having variant minor components, indicating that the variant chain was the δ chain.

It is apparent that many variants are present in populations of *Homo sapiens*. Of course the number of individuals examined is very large compared to the number of non-human primates sampled. Within the primates some species are highly polymorphic, others are remarkably monomorphic. *Homo sapiens* appears to be among the more polymorphic species. Whether the degree of polymorphism in non-human primates can be related to population size (Sullivan and Nute, 1968), breeding behavior (Nute, 1968), malarial infection (Barnicot, 1969a, b), or other factors is unknown. Solutions to these problems require that further studies of the amount of variation in different primate species be completed. The examination of animals of unknown origin in captive colonies can give a distorted pattern of variation.

FUNCTIONAL PROPERTIES OF PRIMATE HEMOGLOBINS

We have seen that each species contains a variety of hemoglobin components. Even by the criteria of electrophoresis most animal hemoglobins are species specific. As we will see later, studies of the amino acid sequences support the hypothesis that almost every species has structurally distinct hemoglobins. Once we appreciate the enormous variation that exists, it is natural to ask whether or not these structural differences are translated into functional differences.

The prime function of hemoglobin is to transport molecular oxygen from the lungs to the tissues. The process is regulated or affected by many factors which, when taken together, produce a rather complicated picture. When studying the binding of oxygen to hemoglobin, one can use either whole blood or hemoglobin solutions. Each approach has its advantages and disadvantages. Sullivan and Riggs (in press) have summarized the oxygen binding properties of vertebrate and invertebrate bloods and oxygen transporting proteins, and the functional properties of hemoglobins are reviewed in Antonini (1965), Bartels (1964), Riggs (1965) and Rossi-Fanelli et al. (1964). Although reversible oxygen binding appears to be the prime function of hemoglobin, it is not its only function. Hemoglobin also binds and transports carbon dioxide and hydrogen ions. It may function in red cell metabolism by binding 2,3-diphosphoglycerate. Undoubtedly there are many other functions. Every protein must be synthesized, ordered in the cellular envelope or in the interstitial fluids, perform its function, interact favorably with its neighboring molecules and be catabolized. Most attempts fail to relate some unique structural feature of a hemoglobin (usually a single amino acid residue) to the morphological integrity of the macromolecule or to the primary function. This is hardly a good argument that selection is not acting on that structural feature. If 2 hemoglobins have similar oxygen binding properties in the physiological pH range they may still differ in many other functional properties.

The advantage of using whole blood is that it most closely duplicates conditions encountered *in vivo*. When making species comparisons, however, it is hazardous to assume that any differences observed are due to differences in hemoglobin structure. Red cell ionic constituents, metabolites and morphology vary widely from species to species. Within a single species these factors may vary with the physiological or nutritional state of the animal. In order to correlate functional changes with structural changes one must assume all other factors are equal. They probably are not.

The oxygen binding properties of whole blood from a number of primates have been studied. These include *Saimiri sciureus* (Lenfant and Aucutt, 1969), *Macaca mulatta* (Novy et al., 1969; Parer, 1967; Lenfant and Aucutt, 1969), *M. nemestrina* (Lenfant and Aucutt, 1969), *M. speciosa* (Blechner and Stenger, 1970), *Papio anubis* (*cynocephalus*) (Parer and Moore, 1968; Lenfant and Aucutt, 1969), *Hylobates lar* (Parer and Moore, 1968), *Pongo pygmaeus* and *Pan gorilla* (Riegel et al., 1966), and *Pan troglodytes* (Riegel et al., 1966; Parer and Moore, 1968; Lenfant and Aucutt, 1969). These workers have stressed the overall similarities of these studies to work on human whole blood, but each has found some differences in oxygen binding between most species. Under standard conditions (pH 7·4, 38°C) whole bloods from *Pongo*, *Pan* and *Homo* are half-saturated with oxygen at about 25–28 mmHg. *Hylobates* requires about 30 mmHg and the New and Old World monkey bloods are 50% saturated at 33–36 mmHg. The Bohr effects (defined here as pH dependence of oxygen binding) are similar, but there seem to be real differences. Fetal bloods from macaques (*M. nemestrina*, *M. mulatta*, *M. fascicularis* and *M. speciosa*) all show increased oxygen affinities (Novy et al., 1969a, b; Blechner and Stenger, 1970). This is also true of fetal blood from *Homo sapiens*. In solution human fetal and adult hemoglobins have very similar properties; inside the red cell fetal hemoglobin has a higher affinity for oxygen. The difference in behavior is caused by the presence of 2,3-diphosphoglycerate (Tyuma and Shimizu, 1969). This compound lowers the affinity of most hemoglobins for oxygen. Apparently, fetal hemoglobin has a weaker affinity for 2,3-diphosphoglycerate than does adult hemoglobin, hence its affinity for oxygen remains high. It is not known whether or not a similar explanation is true for other primate fetal and adult hemoglobins.

The most extensive studies of oxygen binding have been carried out by Sullivan (1968, and in preparation). In these studies dialyzed hemolysates from more than 20 species of primates were examined. Oxygen binding properties were measured under a defined set of conditions (hemoglobin concentration 0·15%, 0·1M phosphate buffer, pH 5·0–8·5, at 15 and 20°C). Under these conditions any differences in function should be related to differences in structure. At the conclusion of these studies the effect of 2,3-diphosphoglycerate on the oxygen binding properties of hemoglobins was reported by Benesch and Benesch (1967) and Chanutin and Curnish (1967). The fact that this metabolite was not eliminated from the hemoglobin solutions that I examined (it cannot be completely removed by dialysis) should affect some of the absolute values obtained, but does not affect the conclusions. Dialyzed hemoglobin solutions from different individuals of the same species gave reproducible results.

The relationships between temperature and oxygen binding were the same for all hemoglobins examined. Values for heme–heme interactions (positive cooperativity) ranged between 2·7 and 3·0. Both the Bohr effect and the oxygen affinity varied considerably from species to species.

The binding of oxygen causes hemoglobin to undergo a change in conformation. These two processes also affect binding sites for carbon dioxide, 2,3-diphosphoglycerate, hydrogen ions, and probably other ligands. Since oxygen binding affects the binding of these ligands, it follows that the concentrations of all ligands affect the binding of the oxygen ligand. This is a mixed blessing. Although it allows a variety of functions to be linked to oxygenation, it also means that mutations which affect one function may affect others as well. For instance, it is not possible to change (by mutation) the Bohr effect without also changing the oxygen affinity. The concept of linked functions is very important for understanding variation in protein molecules. Linked functions are the basis of what might be called molecular pleiotropism. As an example, assume that a particular species "needs" a hemoglobin with a slightly greater Bohr effect. A single nucleotide change produces a hemoglobin with a larger Bohr effect. However, this will also cause an increase in the oxygen affinity which before the mutation, was optimal. If the advantage of an increased Bohr effect more than offsets the disadvantage of an increased oxygen affinity then presumably the mutant allele will be selected for. A second mutation affecting only the oxygen affinity could restore the original oxygen affinity (oxygen affinity can be affected directly or by changes in linked functions). As a result, several changes in structure may be reflected as a single change in function (of course if we could add or subtract the mutations one by one then we could determine the effect of each). Because functions are linked and because many intermolecular interactions must also occur, the changes in protein structure that are selected will probably be those that cause small functional changes. A large change would have too many detrimental pleiotropic effects. Closely related species should have hemoglobins with similar properties not only because their hemoglobins are structurally similar, but also because the animals' physiological limits of tolerance are similar. The potential of each species is largely determined by its past.

Wyman (1948) showed that the pH dependence of oxygen binding could be fitted by the following equation,

$$\log P_{50} = C + \log \frac{(H^+ + K_1')(H^+ + K_2')}{(H^+ + K_1)(H^+ + K_2)}$$

The equation assumes that there are two non-interacting ionizing groups and that one (K_2') becomes a weaker acid and the other (K_1') a stronger acid upon oxygenation (K_2 and K_1). This equation fits the data accumulated for vertebrate hemoglobins. The fact that Perutz *et al.* (1969) have implicated 2 residues in the positive Bohr effect (usually between pH 6·5 and 9·0) does not affect Wyman's model because the pK's of the 2 proton donors are so similar. If one assigns values to 1 group (say K_1' and K_1) and holds these constant while varying K_2' and K_2 several things become apparent. If the mid-point is held constant

$$\frac{K_2' + K_2}{2}$$

and the difference ($K_2 - K_2'$) varied, then the oxygen affinity in the physiological range shifts greatly. This is shown in Fig. 2(a). From an adaptive standpoint, this would be a very effective way to vary the oxygen affinity. This gives a function to the negative Bohr effect (usually between pH 6·5 and 5·0). If the difference between K_2 and K_2' is held constant but the midpoint shifted, very little change in oxygen affinity occurs in the physiological range [Fig. 2(b)]. If we now assign values to the other groups (K_2'

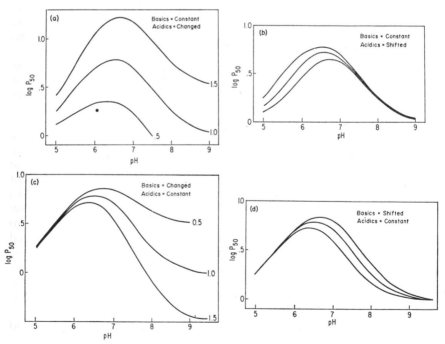

FIG. 2. Hypothetical variation in oxygen binding curves.

and K_2) and vary the magnitude of the pK shift ($K_1' - K_1$), the Bohr effect in the physiological range changes drastically [Fig. 2(c)]. Note that an increase in the slope also causes an increase in the oxygen affinity (lower values of log P_{50}). Shifting the mid-point instead of changing the magnitude [Fig. 2(d)] slightly affects the oxygen affinity and Bohr effect in the physiological range. Now let us compare the oxygen binding properties of primate hemoglobins to see if these mechanisms are used to give different Bohr effects and oxygen affinities.

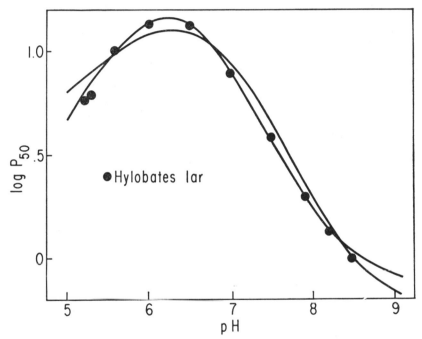

FIG. 3. Oxygen-binding properties of hemoglobin from *Hylobates lar*.

The pH dependence of oxygen binding was determined for dialyzed hemolysates from *Tupaia glis, Perodicticus potto, Galago c. crassicaudatus, G. c. argentatus, G. senegalensis, Propithecus verrauxi, Lemur f. fulvus, L. variegatus, L. catta, Saguinus oedipus, Cebus apella, Cercopithecus aethiops, C. talapoin, C. patas, Papio gelada, P. sphinx, Hylobates lar, H. pileatus, Pongo pygmaeus, Pan gorilla, P. troglodytes* and *Homo sapiens*. Except in the case of *Pongo pygmaeus* (Sullivan and Nute, 1968) no attempt was made to separate and separately test multiple components in those species possessing them. Using Wyman's equation, a theoretical curve was fitted to the experimental data. In Figs 3–5 the solid line represents the theoretical curve. In each instance the data are compared to the theoretical curve for human hemoglobin (Sullivan, 1967) which appears without data points. The values used to generate the theoretical curves are summarized in Table IV.

TABLE IV. The pK differences and oxygenation constants
for primate hemoglobins

	$pK_1'-pK_1$	pK_2-pK_2'	C	$P_{50(pH\ 7\cdot5)}$
Homo sapiens	1·57	·65	·654	4·38
Pan, Pongo	1·4–1·6	·60–·70	·48–·65	4·0–4·4
Hylobates	1·7–1·8	1·2	·24–·32	3·3–3·7
Cercopithecoidea	1·6–1·7	·60–·70	·72–·80	4·6–5·8
Ceboidea	1·7–2·0	·90–1·0	·53–·69	4·9–5·4
Lemuriformes	1·2–1·5	·8 –1·3	·26–·65	8·3–9·5
Lorisiformes	1·8–1·9	·9 –1·0	·75–·84	6·0–9·2
Tupaiiformes	1·9	·8	·59–·63	6·8–7·2

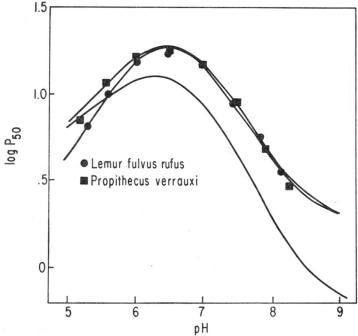

FIG. 4. Oxygen-binding properties of hemoglobins from lemurs.

The pH dependence of oxygen binding of *Hylobates lar* hemoglobin is shown in Fig. 3. Almost all the oxygen binding parameters of gibbon hemoglobin fall outside the range of values including *Homo*, *Pan* and *Pongo* hemoglobins (Table IV). Taxonomically and behaviorally *Hylobates* is the most divergent of the Pongidae, and so it is not surprising that the oxygen-binding properties of its hemoglobin differ significantly from those of other hominoid hemoglobins. Although only 7 species of Old and New World monkey hemoglobins have been examined, they do appear to be distinct. Both the positive and negative Bohr effects are larger in New World monkeys. In line with studies on whole blood, monkey hemoglobins have a lower affinity for oxygen in the physiological range than do ape hemoglobins.

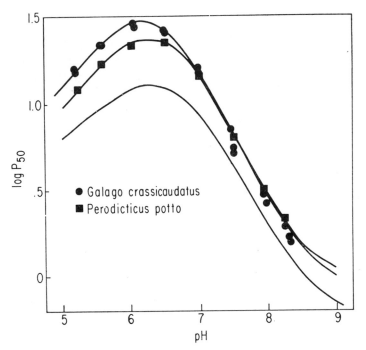

FIG. 5. Oxygen-binding properties of hemoglobins from lorises.

Prosimian hemoglobins are quite different from anthropoid hemoglobins. Note that they all have lowered oxygen affinities at pH 7·5 (Table IV), when compared to monkey and ape hemoglobins. The pH dependence of oxygen binding for loris and lemur hemoglobins is compared to human hemoglobin in Figs 4 and 5. Loris hemoglobins have much larger positive and negative Bohr effects. Lemur hemoglobins have a smaller positive Bohr effect but much larger negative Bohr effect.

It might be instructive to trace a theoretical path by which these hemoglobins might have evolved. As a starting point let us invent a "primitive" hemoglobin with general properties $(pK_1' - pK_1) = 1.7$; $(pK_2 - pK_2') = 0.9$; $C = .533$; $P_{50(pH\ 7.5)} = 5.4$ mmHg). Lemur hemoglobins have lower oxygen affinities. Assume this was accomplished by increasing the negative Bohr effect. Then there was selection for a decreased positive Bohr effect. As this was accomplished it also lowered the affinity for oxygen [Fig. 2(c)]. In order to restore the normal affinity, either the base-level oxygen affinity (affinity irrespective of linked effects) would have to be increased (by lowering the value of C) or the negative Bohr effect would have to be decreased. Since the negative Bohr effect was increased in the first place, perhaps multiple effects were being selected. Thus alteration of the negative Bohr effect might no longer be practical, and the base-level oxygen affinity was changed. Loris hemoglobin also shows a lower affinity for oxygen than our theoretical hemoglobin. Since the negative Bohr effect is unchanged, the base-level oxygen affinity must change. The value of C increases in magnitude indicating a lower affinity. Now there is selection for a greater Bohr effect. As this change is incorporated it increases the oxygen affinity. Decrease in the oxygen affinity again occurs by lowering the base-level affinity. Compared to monkey hemoglobins the oxygen affinities of both lemur and loris hemoglobins are lower in the physiological range. But for the lemur the base level oxygen affinity has actually increased! The Bohr effects have gone in opposite directions for lemurs and lorises. Thus we see that the similarities are really superficial and there is no reason to expect structural similarity. The functional properties in the physiological range may be rather similar but there need be no underlying structural similarity.

The data in Table IV indicate that in spite of the general similarities of oxygen binding in the physiological range, there is a lot of underlying variation. The data are also consistent with the hypothesis that oxygen binding properties are usually species specific and *in general* similar in proportion to the length of time since divergence of 2 given species. Structure and function are probably qualitatively and quantitatively correlated. These data do not support the hypothesis that most substitutions do not affect the functional properties and are selectively neutral.

When we hold the positive Bohr effect constant, the oxygen affinity may be changed by at least 2 mechanisms (varying the negative Bohr effect and varying the base-level oxygen affinity). Varying the affinity of hemoglobin for 2,3-diphosphoglycerate or varying the amount of 2,3-diphosphoglycerate in the red cell are 2 extramolecular means of varying the oxygen affinity. There are so many possible ways to modify the functional properties of a hemoglobin that it is not surprising that it is difficult to correlate structure with function or function *in vitro* with function *in vivo*.

STRUCTURAL STUDIES OF PRIMATE HEMOGLOBINS

Structural studies of primate hemoglobins began with the work of Zuckerkandl *et al.* (1960). Since then much additional work has been done in a number of laboratories resulting in the elucidation of the amino acid sequences of several primate hemoglobins. Yearly summaries of the published data are presented by Dayhoff (1969).

Amino acid sequence data may be viewed from several angles. Initially, one is curious about the location of unusual or functionally significant amino acids such as histidine, proline, cysteine, methionine, tyrosine and tryptophan. If other protein sequences are available, then one can look for regions of sequence homology among different proteins. Elaborate computer programs are available for locating and establishing inter- and intramolecular homologies (Fitch, 1970). If the 3-dimensional structure of a homologous protein is known, then it is possible to attempt to fit the new sequence into the 3-dimensional coordinates of the homologue. As the 3-dimensional structure of the newly sequenced protein is elucidated by X-ray crystallography, then knowledge about the active site may be acquired and sites of co-enzyme attachment may be located. The amount and location of α helices and β structures may be determined and regularities or irregularities in sequence may be correlated with structural features. These features would include α helices, β structures, bends in the polypeptide chain, hydrogen bonding, hydrophobic bonding, ionic bonding and so forth. The residues that occur on the outside of the molecule are compared with those on the inside. When sequences of homologous proteins from closely (or distantly) related animals become available, then common structural regions can be charted and their functional significance investigated. The types of amino acid substitutions (from species to species) are recorded. Finally, differences in sequence can be used to relate species in a phylogenetic pattern. Many of these aspects have already been described for hemoglobins. The 3-dimensional structure has recently been reviewed by Perutz (1969). Rather than present all of these data I will summarize the conclusions concerning protein variation, discuss more recent sequence work, and comment on the use of amino acid sequences as phylogenetic indicators.

The common structural features of globin chains have been discussed in detail by Perutz *et al.* (1965). Myoglobin was included in their comparisons. They found that residues in only 9 positions remained invariant, the remaining residues differ in globins from various species. Four of these invariant residues are involved in the binding of heme to the globin chain. The remaining invariant residues seem to lie at the corners of the molecule or at regions of contact among the polypeptide chains. Although more cytochromes c than hemoglobins have been sequenced, the former are in-variant throughout almost half their sequences. This underlines how differ-ent 2 proteins can be in their patterns of substitution. Internal positions are occupied by non-polar amino acids. Polar amino acids are externally located although some external sites are also occupied by non-polar resi-dues. Proline residues are located either in non-helical regions or at the ends of a helix. This distributional pattern of proline, polar and non-polar residues is common to all proteins whose 3-dimensional structures have been determined.

The amino acid sequences used in the following comparisons vary in the quality of their proofs. Partial sequences have not been used (see Buettner-Janusch et al., 1969; Hill et al., 1963 and Tachikawa, 1969). Hemoglobin sequences from *Pan troglodytes* (Rifkin and Konigsberg, 1965) and *Pan gorilla* (Zuckerkandl in Dayhoff, 1969) are based on peptide compositions. Sequences of the hemoglobin chains of New World monkeys were deduced from tryptic peptide compositions. Certain assumptions were made and these are outlined in the original paper (Boyer et al., 1969b). The tentative sequence of the β chain from *Cebus albifrons* was deduced from composi-tions of the tryptic and thermolytic peptides (Nute and Sullivan, submitted for publication). Complete proof of sequence exists for the globin chains of *Macaca mulatta* (Matsuda et al., 1970a, b, c, d). Proof is available for the sequences of α and β chains from *Lemur fulvus* and *Propithecus verrauxi* and the α chains of *Galago crassicaudatus* and *Tupaia glis* (Nishizaki and Hill, in preparation). Sequence determination by peptide composition is a treacherous art and complete proof of structure is very desirable.

Perutz et al. (1968) have implicated a number of amino acid sites in the binding of heme and in the regions of contact between α and β chains. Substitutions in primate hemoglobins seldom involve these positions. Vari-ations found in abnormal human hemoglobins often occur at these positions and involve changes in function (Perutz and Lehmann, 1968). Figure 6 compares the α chain abnormal human variants with the sequence substi-tutions found among primate α chains. Thirty-four human variants are known (Dayhoff, 1969) and 51 substitutions are known among primate α chains. Because most abnormal human hemoglobins are discovered by electrophoresis, 32 out of 34 variants involve changes in charge. Only 10 out of 57 primate substitutions involve changes in charge. Since the dis-covery of human variants is usually based on changes in charge, it is diffi-cult to calculate how often a variant and substitution should involve the same mutation and occur in both groups. However, of 85 amino acid exchanges, only 2 are common (one each at positions 15 and 68). The dis-tribution of human variants along the polypeptide chain is less systematic than is the distribution of substitutions in the chains of non-human

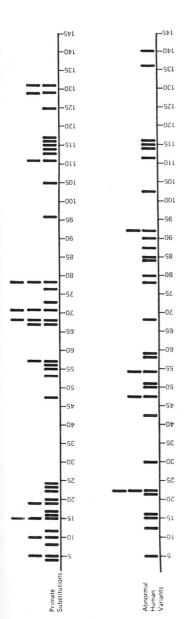

FIG. 6. Distribution of human variants and substitutions along the α polypeptide chains of non-human primates.

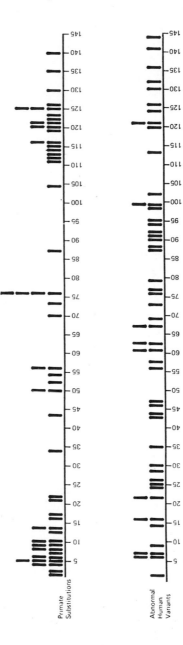

FIG. 7. Distribution of human variants and substitutions along the β polypeptide chains of non-human primates.

primates. There are several regions where substitutions seldom occur (positions 25–55 and 79–110) in non-human primates. Human variants occur commonly in these regions. The distribution of human β chain variants versus non-human primate substitutions in the β chain is shown in Fig. 7. Forty-seven of 56 human β chain variants involve charge changes while 24 of 58 non-human primate substitutions affected the charge. Of the 114 amino acid interchanges only 3 are common to both groups (at positions 6, 22 and 113). The distribution of human variants along the polypeptide chain again appears to be less systematic. The regions of polypeptide chain showing little variation among non-human primates (residues 23–49 and 77–110) are again commonly substituted by human variants. Note also that the distribution of unmodified regions in non-human primate chains is similar for the α and β chains. This emphasizes the homology between them. These distributional data seem to argue against similarity in the 2 types of variation. The pattern of human variants must resemble more closely the pattern of mutation without subsequent selection (except for lethals!). The pattern of primate substitutions must resemble the pattern of mutation after subsequent selection for adaptive changes has occurred. These data together with the data on oxygen binding argue against the hypothesis that only a small part of the amino acid sequence is affected by natural selection. They support the hypothesis that natural selection acts on the entire sequence and that variation is not randomly accumulated even in areas where variation is allowed.

The potential use of amino acid sequences for the construction of animal phylogenies was recognized soon after the elucidation of the genetic code. Progress in this area has been reviewed from time to time (Leone, 1964; Bryson and Vogel, 1965; Hawkes, 1968 and Sibley, 1969). Use of this technique requires some knowledge of 2 well-established disciplines, protein chemistry and systematic biology. Unfortunately, few people are trained in both areas and misunderstandings on both sides have led to some rather heated arguments. One problem seems to be that theory is far ahead of fact. Few sequences are known and these have been interpreted to show that most substitutions are of neutral selective value. If most substitutions (mutations) are selectively neutral, their accumulation should be time-dependent. Thus, the length of time that has elapsed since the members of 2 given species last shared a common ancestor should be roughly proportional to the number of differences in the sequences of homologous proteins found in these species.

Clarke (1970) has argued effectively against the concept that most substitutions are neutral. Differences in function among primate hemoglobins and the approximately quantitative correlation between structural and functional differences argues strongly that the substitutions in the proteins of primates are neither neutral nor functionless. Comparison of the patterns of substitution with the patterns of abnormal variants also suggests that substitutions are being subjected to selective forces. Furthermore, the data presented below indicate that polypeptide chains, like animal populations, evolve at different rates.

A summary of substitutions in primate α chains is shown in Fig. 8. Blank spaces in the sequences indicate that the residue is identical to that found in the hypothetical ancestral chain. Sequence positions not shown are invariant in all primate species examined. The sequence of the ancestral chain was derived from these data plus the data on other mammalian α chains (Dayhoff, 1969). An ancestral residue is generally that one from which other substitutions at that position could be derived with the fewest nucleotide changes. Use was also made of currently accepted phylogenetic relationships. Positions 17 and 71 were omitted since no single ancestral residue could be selected. It is hypothesized that the ancestral chain existed in the lineage leading to the primates. It should be clearly understood that the selection of ancestral residues depends upon the data available. As more sequences are determined the structure of the archetypal molecule will undoubtedly be changed.

Several positions in the α chains should be discussed. The histidine at position 113 is the only substitution that clearly links *Tupaia* with the other primates. The leucine in anthropoid chains is presumed to be a back mutation. A substitution of glycine by serine at position 78 also links *Tupaia* and the rest of the primates in the phylogenetic scheme shown in Fig. 9, but the exact chronological sequence of substitutions is quite uncertain. Note that this would make glycine 78 in *Lemur* a back mutation. Methionine at position 76 links the prosimians (except *Tupaia*) with the anthropoids. Glycine to glutamic acid at position 15 also unites these groups in Fig. 9, but this pathway is very tentative. Positions 76 and 113 are really the only ones that clearly link prosimians with each other and with the higher primates. Of course histidine at position 113 may link primates with insectivores! Five positions involving 6 substitutions link *Propithecus* and *Lemur* Fig. 9). The α chain of *Propithecus* has accumulated substitutions at about twice the rate as that of *Lemur* since the 2 genera diverged. The α chain of *Galago* appear to have undergone little differentiation when compared to that of *Propithecus*. The anthropoids are a much more coherent group than are the prosimians. However, preliminary sequence data for the α chain from *Papio anubis* (=*cynocephalus*) (Sullivan and Nute, unpublished) indicate that it is very divergent also. It differs from macaque α chains by at least 13 substitutions. When compared to the human α chain those of *Macaca* and *Papio* both share the substitutions at positions 68 and 71. The former is a 2-step mutation. Since *Macaca* and *Papio* are closely

Position	4	5	8	10	12	15	16	17	19	20	22	23	24	48	53	55	56	57
Precursor	Pro,	Ala,	Ser,	Val,	Ala,	Gly,	Lys,	Ile,	Gly,	His,	Gly,	Glu,	Tyr,	Leu,	Ala,	Val,	Lys,	Ala,
Homo			Thr,				Val,	Ala,										Gly,
Macaca							Val,	Val,	Ala,									Gly,
Lemur	Ala,		Thr,			Asp,	Ala,	Val,		Glu,			His,		Gly,			
Propithecus	Ala,		Leu,	Thr,	Lys,	Ala,		Ile,	Ser,				His,					Thr,
Galago		Thr,				Glu,						Asp,	His,					
Tupaia		Gly,	Ile,					Ile,		Glu,	Pro,			Met,		Ile,	Glu,	

Position	67	68	71	73	76	78	96	105	111	113	114	115	116	117	125	129	131
Precursor	Thr,	Lys,	?,	Leu,	Leu,	Gly,	Val,	Leu,	Ser,	Leu,	Pro,	Ala,	Glu,	Phe,	Leu,	Leu,	Ser,
Homo	Asn,	Ala,	Val,	Met,	Asn,												
Macaca	Leu,	Gly,	Val,	Met,	Asn,												
Lemur		Ser,		Met,						His,						Phe,	Ala,
Propithecus	Gly,	Asp,		Met,	Thr,					His,	Ser,					Phe,	Ala,
Galago	Ser,	Val,		Met,	Ser,				Cys,	His,							
Tupaia	Thr,	Gly,	Thr,	Ala,	Ile,				Cys,	His,	Gly,	Asp,	Leu,	Phe,	Met,		Asp,

Fig. 8. Alignment of primate α chains.

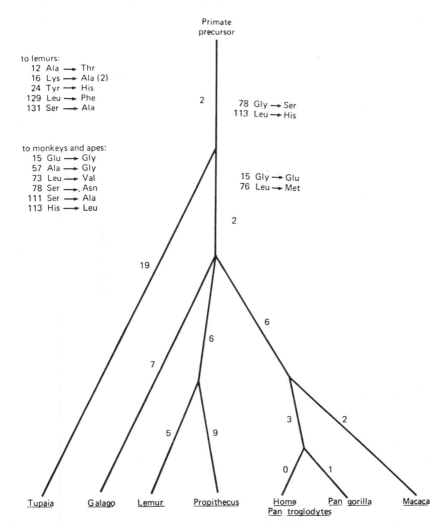

FIG. 9. Primate phylogeny based on the amino acid sequences of the α chains from primate hemoglobins. The numbers along the axes refer to minimal nucleotide changes.

related (some consider them congeneric) these data emphasize the qualitative aspect of sequence comparisons and de-emphasize the quantitative aspect. The location of common mutations determines phylogenies, the number of substitutions may or may not be related to the time of divergence.

The sequence comparisons of primate β chains are shown in Fig. 10.

I

Alignment of primate β chains — positions 1–56

Position	1	2	4	5	6	7	8	9	10	12	13	16	17	21	22	33	43	50	52	54	56
Homo	Val,	His,	Thr,	Pro,	Glu,	Glu,	Lys,	Ser,	Ala,	Thr,	Ala,	Gly,	Lys,	Asp,	Glu,	Val,	Glu,	Thr,	Asp,	Val,	Gly,
P. gorilla																					
Macaca								Asn,								Leu,					Asn,
Cebus				Ala,							Thr,										Asn,
Saimiri				Gly,	Asp,			Ala,			Thr,						Asp,				Ser,
Ateles				Gly,						Ala,											Ser,
Saguinus																					Ser,
Lemur	Thr,	Leu,	Ser,	Ala,		Asp,	Ala,	His,			Thr,	Ser,	Glu,	Glu,				Asn,	Ser,	Ser,	Ser,
Propithecus	Thr,	Leu,	Gly,	Asp,	Val,	Leu,	Ser,	Ala,	Glu,		Ser,	Ala,	Asp,	Glu,	Lys,			Ser,	Ser,	Ile,	Ser,

Alignment of primate β chains — positions 69–139

Position	69	70	73	76	87	104	111	112	113	114	115	116	119	120	121	122	123	125	126	130	135	139
Homo	Gly,	Ala,	Asp,	Ala,	Thr,	Arg,	Val,	Cys,	Val,	Leu,	Ala,	His,	Gly,	Lys,	Glu,	Phe,	Thr,	Pro,	Val,	Tyr,	Ala,	Asn,
Gorilla						Lys,																
Macaca				Asn,														Gln,				
Cebus																		Gln,				
Saimiri				Thr,														Gln,				Thr,
Ateles																		Gln,	Leu,			
Saguinus																		Gln,				
Lemur	Ser,	Glu,	His,	Gln,	Lys,	Ser,	Ala,	Glu,	Ser,	Glu,	Leu,	His,	Asp,	Lys,	Ser,	Ala,				Phe,	Ala,	
Propithecus	Ser,	Glu,	Asp,	Gln,	Ala,					Glu,		His,	Asp,	Ala,	Ser,	Glu,				Phe,	Thr,	

FIG. 10. Alignment of primate β chains.

Again it is clear that the prosimians and anthropoids are separate lineages. A precursor molecule has not been hypothesized nor a phylogeny drawn from the β chain data because so few of the critical species have been examined. However, substitutions at several positions are of interest. At position 87 glutamine is the obvious precursor. The substitution of glutamine by threonine in *Pan gorilla* and *Homo sapiens* serves to link those species. It will be interesting to see if *Hylobates* has threonine. Glutamine 125 appears to unite the Old and New World monkeys. Its distribution as well as the distribution of proline at this position should be important. Lysine 104 links Old World monkeys and apes. The arginine in *Homo sapiens*, ceboids and *Propithecus* does not delineate a natural group. The substitutions at this position illustrate nicely the difficulties involved when too few data are available. There are only 2 possible choices for a precursor. Unless one considers these data in the light of existing classifications very improbable groupings will result. Homologies at positions 5 (proline) and 56 (glycine) also unite Old World monkeys, apes and Man. The latter position may indicate a relationship between New World monkeys and lemurs or it may mean only that two groups of anthropoids have differentiated from the ancestral line. One interesting feature of the New World monkey β chains is that they appear to have differentiated after divergence of the present day species and not before.

In systematics the species is the only taxon that can be biologically defined. The assignment of genera, families and higher categories is rather arbitrary as long as they are not made polyphyletic. As a result of different taxonomic practices the degree of differentiation that defines a genus in one group may serve to distinguish families in another. Sequence data offer a nice framework for making comparisons between groups. Quantitation should involve the averaging of several protein sequences in order to reduce the effects of divergence in any one of them. Quantitation of this type is not similar to determining the point of divergence in time of two lineages using the number of substitutions and a hypothetical mutation rate. Thus, so many substitutions per 100 residues of sequence would constitute the generic, familial or ordinal taxonomic levels. A 2-step mutation represents 2 substitutions. Treatment of data in this way could result in some clarification in primate classification. Note for example that *Pan troglodytes* and *Homo sapiens* have identical sequences while *Pan gorilla* and *Homo sapiens* differ by about 0·7 substitutions per 100 residues. *Pan* and *Homo* are put in separate families. *Lemur* and *Propithecus* are also in different families and yet they differ by 10 substitutions per 100 residues of α chain. This type of analysis, which does not apply at the level of species, stresses the degree of divergence rather than the point in time of divergence.

No specific changes in oxygen-binding properties can be related to particular substitutions at present. One can speculate that interactions between glutamine at position 87, the invariant leucine at position 88, and the carboxyl at 94 in the β chains of monkeys and prosimians may be related to the decrease in oxygen affinity or increase in negative Bohr effect observed in these hemoglobins (Nute and Sullivan, submitted for publication). By hybridizing chains of known structure from different species, it should be possible to map out the areas of the molecule involved with each specific function or functional modification.

Studies on the synthesis of hemoglobin in New World monkeys by Boyer *et al.* (1969a) are really the only experiments in which primates have been used as a model in the study of human hemoglobins. However, as outlined in this review, the study of primate hemoglobins is putting human hemoglobins in their proper perspective. This opens the way for many additional experimental approaches. The similarities and differences between human and non-human primate hemoglobins are numerous. One cannot properly study hemoglobin structure without a knowledge of function. One cannot properly study *Homo sapiens* without knowledge of non-human primates. The co-evolution of hemoglobin structure and function throughout the history of the primates is an exciting story. At present we have only the barest outline of that story.

Acknowledgements

This work was supported by grants from the National Institutes of Health. I am very grateful to those persons who informed me of results prior to their publication. Peter E. Nute read the manuscript and offered many suggestions. Naturally, the errors which remain are mine. I would also like to thank Robert L. Hill for his support and encouragement.

REFERENCES

Abramson, R. K., Rucknagel, D. L., Shreffler, D. C. and Saave, J. J. (1970). Homozygous HB J Tongariki: evidence for only one alpha chain structural locus in Melanesians. *Science, N.Y.* **169**, 194–196.

Antonini, E. (1964). Structure and function of haemoglobin and myoglobin. *In* "Oxygen in the Animal Organism" (F. Dickens and E. Neil, eds), pp. 121–139. Academic Press, New York.

Antonini, E. (1965). Interrelationship between structure and function in hemoglobin and myoglobin. *Physiol. Rev.* **45**, 123–170.

Antonini, E. (1967). Hemoglobin and its reaction with ligands. *Science, N.Y.* **158**, 1417–1425.

Antonini, E. and Brunori, M. (1970). Hemoglobin. *Ann. Rev. Biochem.* **39**, 977–1042.

Barnicot, N. A. (1969a). Comparative molecular biology of primates: a review. *Ann. N.Y. Acad. Sci.* **162**, 25–36.

Barnicot, N. A. (1969b). Some biochemical and serological aspects of primate evolution. *Sci. Prog.* **57**, 459–493.

Barnicot, N. A., Huehns, E. R. and Jolly, C. J. (1966). Biochemical studies on hemoglobin variants of the irus macaque. *Proc. Royal Soc.* Series B **165**, 224–244.

Barnicot, N. A. and Jolly, C. J. (1966). Hemoglobin polymorphism in the orang-utan and an animal with four major hemoglobins. *Nature, Lond.* **210**, 640–642.

Barnicot, N. A., Jolly, C. J., Huehns, E. R. and Dance, N. (1965). Red cell and serum protein variants in baboons. "The Baboon in Medical Research" (H. Vagtborg, ed.), Vol. 1, pp. 323–338.

Bartels, H. (1964). Comparative physiology of oxygen transport in mammals. *Lancet* **2**, 599–604.

Beaven, G. H. and Gratzer, W. B. (1959). Foetal haemoglobin in the monkey. *Nature, Lond.* **184**, 1730.

Benesch, R. and Benesch, R. E. (1967). The effect of organic phosphates from the human erythrocyte on the allosteric properties of hemoglobin. *Biochem. Biophys. Res. Comm.* **26**, 162–167.

Blechner, J. N. and Stenger, V. G. (1970). Oxygen dissociation curves of *Macaca speciosa*. *Amer. J. Obstet. Gynecol.* **107**, 38–43.

Bonaventura, J. and Riggs, A. (1967). Polymerization of hemoglobins of mouse and man: structural basis. *Science, N.Y.* **158**, 800–802.

Boyer, S. H., Crosby, E. F., Fuller, G. L., Noyes, A. N. and Adams, J. G. (1969). The structure and biosynthesis of hemoglobins A and A₂ in the new world primate *Ateles paniscus*: A preliminary account. *Ann. N.Y. Acad. Sci.* **165**, 360–377.

Boyer, S. H., Crosby, E. F., Thurmon, T. F., Noyes, A. N., Fuller, G. F., Leslie, S. E., Shepard, M. K. and Herndon, C. N. (1969). Hemoglobins A and A₂ in New World Primates: comparative variation and its evolutionary implications. *Science, N.Y.* **166**, 1428–1431.

Braunitzer, G. (1966). Phylogenetic variation in the primary structure of hemo-globins. *J. Cell. Physiol.* **67**, Suppl. 1, 1–19.

Braunitzer, G., Hilse, K., Rudloff, V. and Hilschmann, N. (1964). The hemo-globins. *Adv. Protein Chem.* **19**, 1–72.

Brimhall, R., Hollan, S., Jones, R. T., Koler, R. D. and Szelenyi, J. G. (1970). Multiple alpha-chain loci for human hemoglobin. *Clin. Res.* **18**, 184.

Bryson, V. and Vogel, H. J. (editors). (1965). "Evolving Genes and Proteins". Academic Press, New York.

Buettner-Janusch, J. (1963). Hemoglobins and transferrins of baboons. *Folia primat.* **1**, 73–87.

Buettner-Janusch, J. and Buettner-Janusch, V. (1964). Hemoglobins of primates. *In* "Evolutionary and Genetic Biology of Primates" (J. Buettner-Janusch, ed.), Vol. 11, pp. 75–90. Academic Press, New York.

Buettner-Janusch, J., Buettner-Janusch, V. and Mason, G. A. (1969). Amino acid compositions and amino-terminal end groups of α and β chains from polymorphic hemoglobins of *Pongo pygmaeus*. *Arch. Biochem. Biophys.* **133**, 164–170.

Buettner-Janusch, J., Nute, P. E. and Buettner-Janusch, V. (1968). Fetal hemoglobin in newborn baboons *Papio cynocephalus*. *Amer. J. Phys. Anthropol.* **28**, 101–104

Buettner-Janusch, J., Twichell, J. B., Wong, B. Y-S. and van Wagenen, G. (1961). Multiple haemoglobins and transferrins in a macaque sibship. *Nature, Lond.* **192**, 948–950.

Buettner-Janusch, V., Buettner-Janusch, J. and Mason, G. A. (1970). Multiple haemoglobins of mandrills, *Papio sphinx*. *Int. J. Biochem.* **1**, 322–326.

Capp, G. L., Rigas, D. A. and Jones, R. T. (1967). Hemoglobin Portland 1: a new human hemoglobin unique in structure. *Science, N.Y.* **157**, 65–66.

Chanutin, A. and Curnish, R. R. (1967). Effect of organic and inorganic phosphates on the oxygen equilibrium of human erythrocytes. *Arch. Biochem. Biophys.* **121**, 96–102.

Clarke, B. (1970). Darwinian evolution of proteins. *Science, N.Y.* **168**, 1009–1011.

Crawford, M. H. (1966). Hemoglobin polymorphism in *Macaca nemestrina*. *Science, N.Y.* **154**, 398–399.

Dayhoff, M. O. (1969). Atlas of protein sequence and structure 1969. National Biomedical Research Foundation, Silver Spring, Maryland.

Dixon, G. H. (1966). Mechanisms of protein evolution. "Essays in Biochemistry" (P. N. Campbell and G. D. Greville, eds), Vol. 2, pp. 147–204. Academic Press, New York.

Eng, L-I. L., Mansjoer, M. and Donhuysen, H. W. A. (1960). Two types of haemoglobin in Macaca irus mordax monkeys (Thos. and Wrought). *Communs. vet.* **4**, 59–64.

Fitch, W. M. (1970). Further improvements in the method of testing for evolutionary homology among proteins. *J. Mol. Biol.* **49**, 1–14.

Fooden, J. (1964). Rhesus and crab-eating macaques: Intergradation in Thailand. *Science, N.Y.* **143**, 363–365.

Gottlieb, L. I. and Van Lancker, J. L. (1963). Electrophoretic patterns of *Macaca mulatta* and *Macaca nemestrina* hybrid hemoglobin. *Nature, Lond.* **200**, 900–901.

Gratzer, W. B. and Allison, A. C. (1960). Multiple hemoglobins. *Biol. Rev.* **35**, 459–506.

Hanly, W. C. and Hoffman, H. A. Investigations of non-human primate hemoglobin. Fetal hemoglobin. In press.

Hawkes, J. G. (editor) (1968). "Chemotaxonomy and Serotaxonomy". Academic Press, New York.

Hecht, F., Jones, R. T. and Koler, R. D. (1967). Newborn infants with Hb Portland 1, an indicator of α-chain deficiency. *Ann. Human Genet.* **31**, 215–218.

Hewitt, L. F. (1964). Proteins in the erythrocytes of monkeys. *Proc. Roy. Soc.*, Series B **159**, 536–543.

Hill, R. L., Buettner-Janusch, J. and Buettner-Janusch, V. (1963). Evolution of hemoglobin in Primates. *Proc. Natl. Acad. Sci.* **50**, 885–893.

Hoffman, H. A. and Gottlieb, A. J. (1968). Hemoglobins of chimpanzees and gibbons. *Primates Med.* **1**, 27–34.

Hoffman, H. A. and Gottlieb, A. J. (1969). Investigations of nonhuman primate hemoglobin-electrophoretic variation. *Ann. N.Y. Acad. Sci.* **162**, 205–210.

Hoffman, H. A., Gottlieb, A. J. and Wisecup, W. G. (1967). Hemoglobin polymorphism in chimpanzees and gibbons. *Science, N.Y.* **156**, 944.

Hubby, J. L. and Lewontin, R. C. (1966). A molecular approach to the study of genic heterozygosity in natural populations. I. The number of alleles at different loci in *Drosophila pseudoobscura. Genetics* **54**, 577–594.

Huehns, E. R., Dance, N., Beaven, G. H., Hecht, F. and Motulsky, A. G. (1964). Human embryonic hemoglobins. *Cold Spring Harbor Symp. Quant. Biol.* **29**, 327–331.

Huisman, T. H. J., Brande, J. and Meyering, C. A. (1960). Studies on the heterogeneity of hemoglobin. III. The heterogeneity of some animal hemoglobins. *Clin. Chim. Acta* **5**, 375–382.

Huisman, T. H. J., Schroeder, W. A., Dozy, A. M., Shelton, J. R., Shelton, J. B., Boyd, E. M. and Apell, G. (1969). Evidence for multiple structural genes for the gamma-chain of human fetal hemoglobin in hereditary persistence of fetal hemoglobin. *Ann. N.Y. Acad. Sci.* **165**, 320–331.

Ingram, V. M. (1961). Gene evolution and the haemoglobins. *Nature, Lond.* **189**, 704–708.

Ishimoto, G. and Prychodko, W. Hemoglobin types of gibbons and leaf-monkeys. In press.

Ishimoto, G., Tanaka, T. and Prychodko, W. Hemoglobin variation in macaques. In press.

Ishimoto, G., Toyomasu, T. and Uemura, K. (1968). Intraspecies variations of red cell enzymes and hemoglobin in *Macaca irus. Primates* **9**, 395–408.

Jacob, G. F. and Tappen, N. C. (1957). Abnormal haemoglobins in monkeys. *Nature, Lond.* **180**, 241–242.

Jacob, G. F. and Tappen, N. C. (1958). Haemoglobins in monkeys. *Nature, Lond.* **181**, 197–198.

Jolly, C. J. and Barnicot, N. A. (1966). Serum and red-cell protein variations of the Celebes black ape. *Folia primat.* **4**, 206–220.

Kitchen, H., Eaton, J. W. and Stenger, V. G. (1968). Hemoglobin types of adult, fetal, and newborn subhuman primates: *Macaca speciosa. Arch. Biochem. Biophys.* **123**, 227–234.

Kunkel, H. G., Ceppellini, R., Muller-Eberhard, V. and Wolf, J. (1957). Observations on the minor basic hemoglobin component in the blood of normal individuals and patients with thalassemia. *J. Clin. Invest.* **36**, 1615–1625.

Lenfant, C. and Aucutt, C. (1969). Respiratory properties of the blood of five species of monkeys. *Resp. Physiol.* **6**, 284–291.

Leone, C. A. (editor) (1964). Taxonomic biochemistry and serology. Ronald Press, New York.

Lewontin, R. C. and Hubby, J. L. (1966). A molecular approach to the study of genic heterozygosity in natural populations. II. Amount of variation and degree of heterozygosity in natural populations of *Drosophila pseudoobscura*. *Genetics* **54**, 595–609.

Matsuda, G., Maita, T., Igawa, N., Ota, H. and Miyauchi, T. (1970). Biochemical studies on hemoglobins and myoglobins. II. Amino acid sequences of all the tryptic peptides from the α polypeptide chain of adult hemoglobin of the Rhesus monkey (*Macaca mulatta*). *Int. J. Prot. Res.* **2**, 13–26.

Matsuda, G., Maita, T., Ota, H. and Takei, H. (1970). Biochemical studies on hemoglobins and myoglobins. IV. The primary structure of the α and β polypeptide chains of adult hemoglobin of the Rhesus monkey (*Macaca mulatta*). *Int. J. Prot. Res.* **2**, 99–108.

Matsuda, G., Maita, T. and Takei, H. (1970). Biochemical studies on hemoglobins and myoglobins. I. Amino acid compositions and N-terminal amino acids of all the tryptic peptides from the α and β polypeptide chains of adult hemoglobin of the Rhesus monkey (*Macaca mulatta*). *Int. J. Prot. Res.* **2**, 1–12.

Matsuda, G., Maita, T., Yamaguchi, M. and Migita, M. (1970). Biochemical studies on hemoglobins and myoglobins. III. Amino acid sequence in all the tryptic peptides from the β polypeptide chain of adult hemoglobin of the Rhesus monkey (*Macaca mulatta*). *Int. J. Prot. Res.* **2**, 83–97.

Novy, M. J., Parer, J. T. and Behrman, R. E. (1969). Studies of blood-oxygen affinity in adult and fetal macaque monkeys. *Ann. N.Y. Acad. Sci.* **162**, 240–241.

Nute, P. E. (1968). Genetic and evolutionary studies of hemoglobins and transferrins in the Primates. Dissertation, Duke University, Durham, North Carolina. 203 pp.

Nute, P. E., Buettner-Janusch, V. and Buettner-Janusch, J. (1969). Genetic and biochemical studies of transferrins and hemoglobins of *Galago*. *Folia primat.* **10**, 276–287.

Nute, P. E. and Stamatoyannopoulos, G. Evidence for duplication of the haemoglobin chain locus in *Macaca nemestrina*. 15 page manuscript. In press.

Nute, P. E. and Sullivan, B. (1971). Primate hemoglobins: their structure, function and evolution. I. Tryptic peptide compositions of *Cebus albifrons* β chain. Submitted for publication.

Oliver, E. and Kitchen, H. (1968). Hemoglobins of adult *Macaca speciosa*: an amino acid interchange (α 15(gly → asp). *Biochem. Biophys. Res. Commun.* **31**, 749–754.

Parer, J. T. (1967). The O_2 dissociation curve of blood of the Rhesus monkey (*Macaca mulatta*). *Resp. Physiol.* **2**, 168–172.

Parer, J. T. and Moore, C. P. (1968). Respiratory characteristics of the blood of the baboon, gibbon and chimpanzee. *Folia primat.* **9**, 154–159.

Perutz, M. F. (1969). The haemoglobin molecule. *Proc. Royal Soc.*, Series B **173**, 113–140.

Perutz, M. F., Kendrew, J. C. and Watson, H. C. (1965). Structure and function of haemoglobin. II. Some relations between polypeptide chain configuration and amino acid sequence. *J. Mol. Biol.* **13**, 669–678.

Perutz, M. F. and Lehmann, H. (1968). Molecular pathology of human haemoglobin. *Nature, Lond.* **219**, 902–909.

Perutz, M. F., Muirhead, H., Cox, J. M. and Goaman, L. C. G. (1968). Three-dimensional Fourier synthesis of horse oxyhaemoglobin at 2·8 Å resolution: the atomic model. *Nature, Lond.* **219**, 131–139.

Perutz, M. F., Muirhead, H., Mazzarella, L., Crowther, R. A., Greer, J. and Kilmartin, J. V. (1969). Identification of residues responsible for the alkaline Bohr effect in haemoglobin. *Nature, Lond.* **222**, 1240–1243.

Riegel, K., Bartels, H., Kleihaver, E., Lang, E. M. and Metcalfe, J. (1966). Comparative studies of the respiratory functions of mammalian blood. I. Gorilla, chimpanzee, and orangutan. *Resp. Physiol.* **1**, 138–144.

Rifkin, D. and Konigsberg, W. (1965). The characterization of the tryptic peptides from the hemoglobin of the chimpanzee (*Pan troglodytes*). *Biochim. Biophys. Acta.* **104**, 457–461.

Riggs, A. (1965). Functional properties of hemoglobins. *Physiol. Rev.* **45**, 619–673.

Riggs, A., Sullivan, B. and Agee, J. R. (1964). Polymerization of frog and turtle hemoglobins. *Proc. Natl. Acad. Sci.* **51**, 1127–1134.

Rossi-Fanelli, A., Antonini, E. and Caputo, A. (1964). Hemoglobin and myoglobin. *Adv. Protein Chem.* **19**, 73–222.

Schroeder, W. A., Huisman, T. H. J., Sheldon, J. R., Sheldon, J. B., Kleihaver, E. F., Dozy, A. M. and Robberson, B. (1968). Evidence for multiple structural genes for the α chain of human fetal hemoglobin. *Proc. Natl. Acad. Sci.* **60**, 537–544.

Schroeder, W. A. and Jones, R. T. (1965). Some aspects of the chemistry and function of human and animal hemoglobins. *Prog. Chem. Organic Nat. Prod.* **23**, 113–194.

Schroeder, W. A. and Holmquist, W. R. (1968). "A Function for Hemoglobin A_{Ic}? in Structural Chemistry and Molecular Biology" (A. Rich and N. Davidson, eds), pp. 238–255. W. H. Freeman and Co., San Francisco.

Sen, N. N., Das, D. C. and Aikit, B. R. (1960). Foetal haemoglobin in the monkey. *Nature, Lond.* **186**, 977.

Sibley, C. G. (editor) (1969). "Systematic Biology". National Acad. of Sciences Publication 1962. Washington, D.C.

Stormont, C. and Cooper, R. (1965). Hemoglobins and serum proteins in *Galago crassicaudatus panganiensis*. *Fed. Proc.* **24**, 532.

Sullivan, B. (1967). The effect of dilution on the oxygenation properties of cat and human hemoglobins. *Biochem. Biophys. Res. Commun.* **28**, 407–414.

Sullivan, B. (1968). Oxygenation properties of Primate hemoglobins. *Fed. Proc.* **27**, 779.

Sullivan, B. and Nute, P. E. (1968). Structural and functional properties of polymorphic hemoglobins from orangutans. *Genetics* **58**, 113–124.

Sullivan, B. and Riggs, A. (1964). Haemoglobin: reversal of oxidation and polymerization in turtle red cells. *Nature, Lond.* **204**, 1098–1099.

Sullivan, B. and Riggs, A. (1970). Data for constructing oxygen dissociation curves: Vertebrates. *In* "Respiration and Circulation" (P. L. Altman, ed.). FASEB Office of Biological Handbooks, Bethesda, Maryland.

Tachikawa, I. (1969). Isolation of all tryptic peptides from the aminoethylated β polypeptide chain of adult hemoglobin from Japanese monkey (*Macaca fuscata fuscata*) and their amino acid compositions. *Acta med. Nagasaki* **13**, 157–169.

Tuttle, A. H., Newsome, F. E., Jackson, C. H. and Overman, R. R. (1961). Hemoglobin types of *Macaca irus* and *Macaca mulatta* monkeys. *Science, N.Y.* **133**, 578–579.

Tyuma, I. and Shimizu, K. (1969). Different response to organic phosphates of human fetal and adult hemoglobins. *Arch. Biochem. Biophys.* **129**, 404–405.

Wade, P. T., Barnicot, N. A. and Huehns, E. R. (1967). Possible duplication of haemoglobin α-chain locus in the Irus Macaque. *Nature, Lond.* **215**, 1485–1487.

Watts, R. L. and Watts, D. C. (1968). The implications for molecular evolution of possible mechanisms of primary gene duplication. *J. Theoret. Biol.* **20**, 227–244.

Wyman, J. (1948). Heme proteins. *Adv. Protein Chem.* **4**, 407–531.

Wyman, J. (1964). Linked functions and reciprocal effects in hemoglobin: a second look. *Adv. Protein Chem.* **19**, 223–286.

Zuckerkandl, E., Jones, R. T. and Pauling, L. (1960). A comparison of animal hemoglobins by tryptic peptide pattern analysis. *Proc. Natl. Acad. Sci.* **46**, 1349–1360.

Phylogenesis of Immunoglobulins in Primates*

A. O. CARBONARA

Istituto Genetica Medica, University of Turin, Italy

The present communication is divided into three parts: the first is an introduction concerning the chemical structure of the Ig molecules; the second is devoted to the available information concerning the evolution of the antibody molecules from the first vertebrate in which the immunological function was developed and the third deals especially with the last part of this evolution, namely the data about non-human primates and Man.

STRUCTURE OF IMMUNOGLOBULINS

The immunoglobulins (Ig) have been defined (Ceppellini *et al.*, 1965) as a family of serum proteins characterized by possessing an antibody activity and a common structural plan. The molecules of all Ig consist of 2 heavy polypeptide chains (H chains) and two light chains (L chains) held together by covalent and non-covalent bonds (Fig. 1).

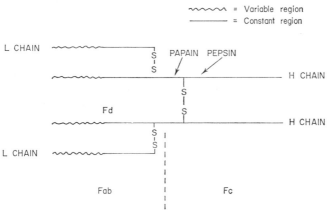

FIG. 1. Diagrammatic representation of the Immunoglobulin molecule.

* The preparation of this review has been assisted by NIH grant AI-08896 and by CNR contract 14.216.2. "Centro di Studio per l'Immunogenetica e l'Istocompatibilità".

The treatment with papain in presence of cysteine splits the molecules in 3 fragments: 2 identical named Fab and one named Fc. Each Fab fragment possesses an M.W. of about 52,000 and consists of one L chain and the NH_2-terminal part of the H chain. The Fc consists of the C-terminal part of H chain. The treatment with pepsin splits the molecules in a single 5 S fragment, named $F(ab')_2$. It consists of the 2 Fab fragments held together by a single disulphide bridge.

The antibody activity is localized exclusively in Fab fragment and is due to the N-terminal part of both H and L chains. The ability to bind different antigens is due to differences in the amino acid sequences. Therefore, the part of chains in which the antibody activity is localized possesses a large variety of primary structures, and is thus called "variable part". The rest of the molecule which does not participate in the combining site is instead called "costant part".

Three different levels of heterogeneity are usually recognized: a first level of heterogeneity consists of, in each individual of the species, several populations of Ig with different physical, antigenic and biological properties. It is possible, as far as we know, to distinguish in Man 9 different variants of H chain (γ_1, γ_2, γ_3, γ_4, α_1, α_2, μ, δ, ϵ) and 2 variants of L chain (κ, λ). The association of each H chain with the 2 types of L chain yields in the serum at least 18 different Ig molecules. The studies on the primary structure have shown that these different H and L polypeptide chains possess in the costant part a given sequence of amino acids with a variable degree of homology. The degree of homology is higher between the polypeptides γ_1, γ_2, γ_3 (hence called sub-classes of IgG) and is lower between the γ, α, μ (hence called classes of IgG). On the basis of the known mechanism of protein synthesis it is possible to argue the existence of several structural genes, one for each variant of H and L chains, derived probably by repeated gene duplications with subsequent divergence.

Some of these structural genes have been also identified by formal genetic analysis based on the presence in the populations of alternative forms or variants of the polypeptide chains. These variants segregate as simple mendelian traits and are the expression of genetic polymorphism. These genetic markers or allotypic specificities (called in Man Gm and InV) are detected by immunological methods using specific allo- or isoantisera.

The Gm (a), (f), (non-a), (z) factors are located in Fc part of the γ_1 chain; the Gm (g), (b), (c), (s), (t) in the Fc part of the γ_3 chain; the Gm (n) in the Fc part of the γ_2 chain (Kunkel et al., 1964; Natvig et al., 1967); the InV (a), (b) and (l) in the κ chains. The differences between the allelic markers consist of 2 amino acids in the case of Gm (a) and Gm (non-a) (Thorpe and Deutsch, 1966; Vyas and Fudenberg, 1969; Wang and Fudenberg, 1969), of 1 residue in the case of Gm (f) and Gm (z) (Edelman *et al.*, 1969) and also of 1 residue in the case of InV (a) and (b) (see Table I and Fig. 2) (Baglioni *et al.*, 1966; Milstein, 1965).

TABLE I. Nomenclature of the Gm factors

Original:	a	x	b²f	b¹	c	r	e	p	b³	b⁴	s	t	b⁰	b⁵	c³	c⁵	z	g	y	n
Numerical:	1	2	3	4	5	6	7	8	9		13	14					17	21	22	

Family studies have shown that the different factors are inherited as mendelian units in fixed gametic combinations or "haplotypes" (Ceppellini *et al.*, 1967) not (or only rarely) separated by crossing over. Population

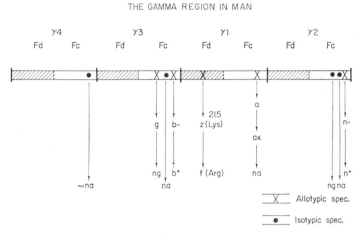

FIG. 2. The γ region in Man. The allotypic specificities are indicated by original nomenclature. For the numerical nomenclature see Table I.

studies have demonstrated significant differences in frequency of these haplotypes in different ethnic groups (Fig. 3). For example, in Caucasians the 3 haplotypes "azg" and "non-a f b¹ n" prevail, although with a different frequency in different populations. In Negro populations the haplotype "a b¹ b³" prevails. The association in the same haplotype of Gm factors carried by different chains is proof that the structural genes controlling these chains are closely linked. Also, the gene which controls the α chain synthesis has been mapped in the same gene cluster (Kunkel *et al.*, 1969) and this is a suggestion that all H chain genes, for which no genetic variants are known, are in fact located in the same chromosomal region.

A third level of heterogeneity is due to structural variability in the part of molecules which contains the combining site. It is very difficult to define the primary structure of a part of the chain which is by definition variable. However, in some malignant diseases of the cell, which normally produce Ig, a homogeneous kind of Ig appears in the serum. These myeloma proteins can be isolated from the serum and their sequences analysed.

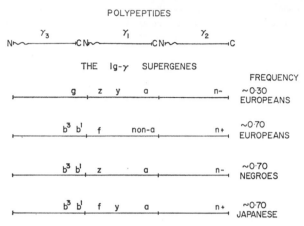

FIG. 3. Phenogroups or haplotypes of the Gm system in different populations.

The sequence analysis performed in the myeloma proteins has shown that the L chain and the H chain can be divided in 2 parts: a costant region (C region) and a variable region (V region) (Hilschmann and Craig, 1965; Putnam *et al.* 1966; Titani and Putnam 1965). The variable region has been found to include the first 105 residues for the L chain and the first 120 residues for the H chain from the NH_2-terminal part. Until now about 300 polypeptide chains have been examined and no identical variable regions have been found.

By comparing the structure of a large number of myeloma proteins it has been found that the amino acid substitutions in variable parts must be considered at 2 distinct levels (Baglioni and Cioli, 1966; Hood and Ein, 1968; Milstein, 1967). One series of residues, which are clustered near the single disulphide bridge of the variant part, have been proved to exhibit an almost infinite diversity of amino acid substitutions; probably this part of the chain corresponds to the combining site. The variability in the rest of the sequence appeared much more restricted and in fact has allowed the description of 3 or more characteristic patterns for each kind of chains. In the case of kappa chains it has been possible to describe three sub-groups (Milstein, 1967) (named κI, κII and κIII) each of which can be distinguished from the others by their characteristic sequences in these areas of the variable part.

Several problems arise trying to explain, on the basis of knowledge of molecular biology, the origin of the variability of the part of Ig involved in the combining site (Ceppellini, 1966; Gally and Edelman, 1970; Milstein and Munro, 1970). One theory suggests that the variable part is controlled by few germ line genes which by somatic mutations give rise to antibody diversity. In contrast, the germ line theory proposes that the vertebrate possesses a structural gene for each functional specific sequence of Ig chains.

This third level of heterogeneity, which is called idyotipic, involves the most basic problems concerning antibody function. No further discussion of this level of heterogeneity will be attempted in this paper, as very little information, and no information at all on non-human primates, is available on the evolutionary aspect of it. Thus we will deal exclusively with the allotypic and isotypic diversities, i.e. with the structural peculiarities of the costant part of the Ig molecules.

PHYLOGENESIS OF IMMUNOGLOBULINS

Regarding generally the evolutionary problem, we have with the Ig another example, besides hemoglobin and cytochrome C, in which a basic outline of chemical evolution can be traced. In fact, due to the combined efforts of many groups of workers (Baglioni and Cioli, 1966; Edelman et al., 1969; Hilschmann and Craig, 1965; Hood et al., 1966; Putnam et al., 1966; Titani and Putnam, 1965) the amino acid sequences of different polypeptide chains from various species is now available. The sequences are analysed on the basis that the differences in amino acids, comparing two polypeptide chains, are directly related to differences in genetic material and are proportional to evolutionary distances.

The results of this kind of comparison are shown in phylogenetic tree as shown in Fig. 4 (Ceppellini, 1966; Grey, 1969; Kunkel et al., 1969; Litwin, 1967; Milstein and Pink, 1970). The general assumption is that all genes present today, which participate in synthesis of Ig, derive from a single original gene coding for a polypeptide chain about 100 amino acids long. This is not directly verifiable since no animal has been discovered in which the Ig possess a monomeric structure.

From a duplication of this primordial gene derive on one side the L chain and on the other side the precursor of all heavy chains (Kunkel et al., 1969). The most recent species in evolutionary sense in which the Ig molecules have been found are the hagfish and the lamprey. The Ig of these animals has a type of H chain which has some peculiarity of the μ chain present in higher organism (Markalonis and Edelman, 1966a, b).

From further duplications of this primordial H gene the other kinds of heavy chain, namely α, δ, ϵ arose. These Ig molecules are present in practically all mammals. The last duplication event to take place was that of the γ gene leading to the appearance of different sub-classes of IgG, i.e. γ_1, γ_2, γ_3, γ_4 among which an homology of very high degree seems to exist.

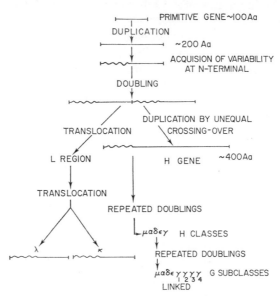

FIG. 4. Scheme of the possible evolution of the Ig genes.

These studies were performed studying the Ig of several species from primitive chordata (lamprey and hagfish) to mammals. Several interesting data concerning the more recent evolutionary events were obtained studying the Ig in non-human primates. These data are particularly useful in several aspects: to establish the last steps of the Ig phylogenesis, to confirm taxonomic relationship in simian population and to evaluate the possibility of using non-human primates in medical research.

The phylogenetic studies in primates cannot be performed on the basis of comparison of amino acid sequences due to the almost complete lack of information. Sequence comparisons are of unique value for 2 reasons. First they concern a phenotipic trait which is very close to the primary product of the gene and is apparently not influenced by environmental conditions. Second, due to the knowledge of the genetic code, differences in sequences can be objectively quantitate.

As data on Ig sequences was not available (due mainly to the absence of myeloma proteins), information concerning non-human primates was obtained by more indirect immunological methods. All methods used comprise an immune reaction between human Ig and specific antisera produced in rabbit or other species (eteroantisera) or in Man (alloantisera). The ability of the Ig of primates to interfere with this reaction is taken as a measure of the degree of relatedness (cross-reactivity) to a human Ig. The general validity of these methods is proved by the fact that whenever applied they have always given phylogenetic relationships between species corresponding to the taxonomic classification established by classical criteria.

However, several points have to be taken into consideration for a correct interpretation of the results. By these methods only the determinants (and maybe not all of them) which are recognized by the antiserum used can be analysed. Moreover, similarities between Man and non-human primates are possible in only one direction using serum against human Ig. Completely different results could be obtained using an antiserum against simian Ig with human Ig acting as inhibitor. It is therefore clear that these immunological reactions give quantitative information only on a relative scale.

Another point concerns the fact that by immunological methods what is realized is an antigenic determinant based on a given tertiary structure, related to primary structure by rules which are not completely known. For instance it may be conceived that the change of a single amino acid, produced by a point mutation, will completely change the tertiary structure abolishing several antigenic determinants. Conversely, the same basic conformation could be maintained despite more than one change which involves "permissive" or "conservative" amino acid substitutions.

PHYLOGENESIS OF Ig IN PRIMATES

The comparative data obtained from studying the Ig molecules in human and non-human primates can be divided into 2 main categories: (i) data concerning the presence in the sera of non-human primates of a given class and type of H and L chain (using antisera which in human detect isotypic variants); (ii) data concerning the genetic and structural relationship of serological markers between human and non-human primates (using antisera which in humans detects allotypic variants).

The data on the physico-chemical structure of IgG in non-human primates show that the structural organizations of these molecules is very similar to that of Man. The γ globulins show in electrophoresis the same migration pattern observed in all mammals: a broad electrophoretic zone extending from the α_2 until the γ region. The heterogeneity of this fraction is evident in immunoelectrophoresis, where the precipitation line of γ globulins extends from the α_2 to the γ_2 region (Picard *et al.*, 1963).

The treatment of baboon and gibbon IgG with papain and pepsin splits the molecules into fragments identifiable on the basis of immunological reactions as Fab, Fc and F(ab′)$_2$ (Litwin, 1969). The reduction and alkylation of the IgG molecules isolated from several primates species yields L and H polypeptide chains. Recombination studies show that, by mixing together H chains from non-human primates and L chains from humans, they reassociate to form a complete Ig molecule. The same result is obtained in the reverse situation: H chains from humans mixed with L chains from non-human primates (Litwin, 1969).

All together, these data suggest that the general structural conformation of the Ig molecules is very similar in different primates species. These results were expected on the basis of very high degree of similarity in this respect between human and even more distant species (Hill *et al.*, 1966).

The degree of immunological cross-reactivity was first studied by qualitative techniques (Milstein, 1965, 1967; Milstein and Munro, 1970); recently the same investigation has been performed by quantitative and more sensitive techniques employing radioactive labelled antigen (Shuster *et al.*, 1969; Wang *et al.*, 1968). The results obtained making use of different antisera against the different human H and L chains are shown in Table II. The following conclusions can be drawn:

TABLE II. Percentage of cross reactivity of non-human primates Ig compared with the human Ig

		Heavy chain						Light chain	
		$\gamma1$	$\gamma2$	$\gamma3$	$\gamma4$	α	μ	κ	λ
Hominoidea	Chimpanzee	90	83	90	0	100	90	100	76
	Gorilla	91	68	85	0	100	28	100	89
	Orang utang	78	73	71	0	100	29	95	70
	Gibbon	60	0	0	0	99	6	87	70
Cercopithecoidea	*M. radiata*	—	0	0	0	100	63	89	36
	M. speciosa	87	0	0	0	96	81	77	37
	M. mulatta	95	0	0	0	92	50	64	36
	Baboon	99	0	0	0	95	76	79	40
Prosimii	*Lemur fulvus*	82	0	0	0	88	0	28	51
	Nycticebus conc.	70	0	0	0	94	9	22	9

From Wang *et al.*, 1968.

1. In some non-human primates the serum contains the same kind of molecules as in Man. For example, in chimpanzee the γ_1, γ_2, γ_3, α, μ chains are very similar to the corresponding human Ig molecules.

2. Some sub-classes of Ig, namely the γ_2 and the γ_3 are present only in Hominoidea and are absent in other species. The γ_4 seems to be absent in all non-human primate species examined.

3. The degree of cross-reaction decreases from the Hominoidea to Ceboidea and Prosimii. This arrangement of various species according to the antigenic relation is in agreement with the currently accepted taxonomic classification of the primates.

4. Comparing the data obtained for the different polypeptide chains, it is evident that the degree of cross-reaction does not fall at the same rate from Hominoidea to Prosimii. In the case of α chain, for example, it is very high and decreases slowly and the same is true for κ and γ_1 chain. In the case of μ and λ chains the degree of cross-reaction decreases rapidly starting from Hominoidea and reaching a very low level in the Prosimii. If the degree of cross-reaction is indicative of similarity in primary structure, this would be an indication that molecular evolution occurs at a different rate in different proteins and in related species.

However, we have already discussed how many objections can be raised against an absolute quantitative significance of these data. The same objection can probably explain the figures (for instance, μ chain in gibbon, λ chain in lemur), that do not fit in with the general behaviour.

The data concerning the absence of the γ_2, γ_3, γ_4 sub-classes also need comment. This result could be interpreted as evidence that some of the gene duplications involving H chain gene took place relatively recently, namely during the evolution of primates. However, the presence of the sub-classes of IgG in all mammals (dog, guinea pig, mouse, horse, cow) (Grey, 1969; Milstein and Pink, 1970) suggests that the duplication of H chain gene must surely have occurred also in all primates. The absence of immunological reaction with the antisera anti-γ_2 and anti-γ_3 means probably that the homologous chain in some species of primates possess different antigenic determinants not shared by humans.

Another possible explanation could be found in a different quantitative expression of the different genes in different species. For example the γ_2 and the γ_3 polypeptide chains could be present in the serum but at a very low level not detectable by the immunological techniques used. A similar situation is known for light chain genes (Hood et al., 1967). The duplication events which produced the κ and λ genes have been located on the basis of chemical paleogenetics before the mammals radiation. This means that all mammals should possess the κ and λ genes. However, the ratio κ/λ in the

A. O. CARBONARA

serum is very different in various species: very high in rabbit and mouse; intermediate in primates; very low in horse, sheep and mink.

The comparative studies of the serological factors (Gm and InV) in human and non-human primates (Alepa, 1969b; Alepa and Terry, 1965; Litwin, 1967, 1969; Schmitt et al., 1965; Shuster et al., 1970; Van Loghem et al., 1968, 1969; Vyas and Fudenberg, 1969) are shown in Table III. It is evident that the order of appearance of these antigenic determinants agrees with the general taxonomic classification of primates. Gm (z), (s), (c) and (b⁰) appear first in Cercopithecoidea, while Gm (a), (b³), (b⁴) are present only in African and Asiatic Apes. Some Gm as the (g), (b¹), (x) and (n) are detected only in human and probably represent recent mutational events occurred during the speciation in Hominoidea.

TABLE III. Gm factors in non-human primates

	IgG1					IgG3									IgG2
	a	x	z	y	f	g	b^1	s	c^5	b^0	b^3	b^4	c^3	t	n
Hominoidea	+	−	+		−	−	−	+	+	+	+	+	+	−	−
Cercopitecoidea	−	−	+		−	−	−	+	+	+	−	−	−	−	−
Ceboidea	−	−	−		−	−	−	−	−	−	−	−	−	−	−
Prosimii	−	−	−		−	−	−	−	−	−	−	−	−	−	−

From Alepa, 1969; Litwin, 1969; Van Loghem et al., 1968.

The results concerning the presence of InV (1) (a) and (b) factors are discordant, some investigators (Alepa, 1969b; Alepa and Terry, 1965) detected these antigens in chimpanzee and orang utang sera, while in other laboratories this finding has not been confirmed (Van Loghem et al., 1968, 1969).

These and many other discrepancies could be explained by the fact that the reagents used in different laboratories do not recognize the same antigenic determinants. For instance, the anti-Gm (b), when obtained by immunization of monkeys, detects this determinant only in Man but not in monkey; if it is derived from a patient with rheumatoid arthritis it reacts also with the gorilla and chimpanzee and if it is obtained from a normal individual (SNagg) it reacts also with some Cercopithecoidea (Milstein, 1965). This means that the Gm (b) factor is a mosaic of determinants some of them peculiar to "human Gm (b)", some in common with the "non-human primate Gm (b)". The same finding has been observed with the anti-Gm (a) and with the anti-Gm (f) reagents (Litwin, 1967).

Whereas Table III gives qualitative information on the presence or absence of different markers in the primate species, in Table IV the same markers are examined in a quantitative fashion, i.e. as proportion of the individuals which possess them in every species (Alepa, 1969a; Van Loghem *et al.*, 1968, 1969).

TABLE IV. Percentage frequency of Gm factors in non-human primates and in various ethnic groups of Man

	No. tested	s	c	b¹	b⁰	b³	b⁴	z	a	f
Hominoidea										
Gorilla	14	100	0	0	100	100	0	100	100	0
Chimpanzee	78	82	82	0	82	14	94	100	100	0
Orang utang	25	100	100	0	70	70	0	0	0	0
Gibbon	14	100	100	0	100	0	0	100	100	0
Man										
Europeans		0	0	90	90	90	90	50	50	90
Japanese		80	0	20	60	60	60	99	100	20
Negroes		19	30–100	99	100	85	85	100	100	0
Cercopithecoidea	90	100	100	100	0	0	0	0	0	0
Ceboidea	3	0	0	0	0	0	0	0	0	0
Prosimii	3	0	0	0	0	0	0	0	0	0

From Van Loghem *et al.*, 1968.

With the reagents used the γ_1 gene seems not to be polymorphic in all non-human primate species examined. All animals tested are positive either for both Gm (z) and Gm (a) or for none. The situation is very similar to that found in some human ethnic groups in which the frequency of the gene Gm^{az} reaches 100%.

A clear genetic polymorphism has been found in orang utang for the gene(s) which controls the expression of Gm (b⁰) and (b³) (Van Loghem *et al.*, 1968, 1969). A genetic polymorphism was found also in chimpanzee sera for several γ_2 factors including Gm (c) and Gm (b⁰). The more frequent phenotype in chimpanzee is (Gm a+ f− b¹− b³+ b⁴+ c+) different in several aspects from that observed in 3 ethnic groups in Man. The Gm (c) factor in human appears only in the Negro populations but is always associated with the Gm (b¹) which is absent in chimpanzee. In chimpanzee the factor Gm (c) has been studied in 4 morphologically and geographically defined different ethnic groups. The frequency of Gm (c) trait has been found to vary from 16 to 31% in the ethnic groups examined.

The presence of polymorphism of the same factor can be interpreted as a similarity at a higher level beside pure structural homology. If it is assumed that genetic polymorphism is substained by selected advantage, we can assume that the species most recently related to Man and facing similar factors of internal and external environmental conditions reaches a higher fitness by following the same polymorphism as in Man. Other species which are more distantly related to Man probably increase their fitness by varying other parts of the molecules.

In Table III each factor detected in the sera of primates has been located on a given polypeptide chain on the basis of known localization of the same factors in Man. At present there is no direct evidence that this is also true in non-human primates. Some γ_1 factors may be shared by different polypeptide chains; in humans, for instance, the antigenic determinant Gm (non-a) is common to γ_1, γ_2 and γ_3. Moreover, some Gm factors, which are located in Man in a given polypeptide chain, can conceivably be located in different molecules in non-human primates. It is the case of Gm (b⁰), (c) and (s) which are present in the gibbon despite the failure to detect the γ_3 polypeptide chain in the serum.

These limitations derive from the above mentioned lack of sequence information. There is, however, one case in which some information of sequences is available (Thorpe and Deutsch, 1966; Vyas and Fudenberg, 1969; Wang and Fudenberg, 1969). This refers to the part of the molecule which is related to Gm (a) and Gm (non-a) factors. In the human, these 2 markers are recognized by specific antisera. The Gm (non-a) is found on γ_1, γ_2, γ_3 polypeptide chains. The Gm (a) is found only in γ_1 and behaves as a true allele of the Gm (non-a). The difference between Gm (a) and (non-a) is due to 2 amino acid substitutions in a specific part of the sequence (Table V). The same sequences are found in primates in which Gm (a) and (non-a) are present. In Cercopithecoidea neither Gm (a) nor (non-a) are present, and when the same sequence stretches are analysed they have been found to be intermediate between (a) and (non-a) (Table V).

TABLE V. Polymorphism — evolution — speciation
as illustrated by the *Gm a* peptide

Arg–*Asp*–Glu–*Leu*–Thr–Lys	"a" γ_1 Man, Chimpanzee, Gorilla
Art–*Glu*–Glu–*Met*–Thr–Lys	"non-a" γ_1, γ_2, γ_3 Man; γ_2, γ_3 Chimpanzee
Glu–Glu–Leu–Thr–Lys	"OWM" Baboon, Thesus

From Wang and Fudenberg, 1969.

Different evolutionary conclusions have been derived from these sequence data. All these interpretations can be subjected to criticism based on the fact that short stretches of sequence do not allow correct establishment of evolutionary events.

However, it is possible to draw an important conclusion from these data.

The fact that in Cercopithecoidea the Gm (non-a) is absent and is present in other primates species simultaneously in different genes suggests that the duplication events which give rise to different sub-classes took place during the primate evolution, after the separation of Old World monkeys from the Antropoidea. On the other hand, as we have already mentioned, IgG sub-classes are present in other species which are relatively distant from Man, for instance cow, dog, guinea pig and horse. It can therefore be suggested that the duplication of the IgG gene takes place and independently in different species or groups of species. Following this suggestion the presence of IgG sub-classes in different species can be interpreted as an example of convergent evolution.

It is clear that the advances in the knowledge on evolution of Ig in non-human primates is linked to the possibility of performing sequences analysis. A substantial advance in this direction would be the isolation of homogeneous populations of Ig molecules. This will be possible either from recognition of pathological conditions like multiple myeloma in Man, or by induction of myeloma by external agents as in mice or obtaining a homogeneous antibody response by repeated injections of some polysaccharide antigens as in the case of rabbits.

REFERENCES

Alepa, F. P. (1969a). Antigenic factors characteristics of human immunoglobulin G, detected in the sera of non-human primates. *In* "Primates in Med", Vol. I, p. 1. Karger, Basle.

Alepa, F. P. (1969b). Antigenic factors characteristics of human immunoglobulin G, detected in the sera of non-human primates. *Ann. N.Y. Acad. Sci.* **162**, 171.

Alepa, P. L. and Terry, W. D. (1965). Genetic factors and polypeptide chain subclasses of human immunoglobulin G, detected in chimpanzee serums. *Science, N.Y.* **150**, 1293.

Allen, J. C. (1966). Gm (b) factor in the γ globulin of non-human primates. *Nature, Lond.* **209**, 520.

Baglioni, C. and Cioli, D. (1966). A study of immunoglobulin structure. II. The comparison of Bence Jones proteins by peptide mapping. *J. Exp. Med.* **124**, 307.

Baglioni, C., Cioli, D., Alescio-Zonta, L. and Carbonara, A. O. (1966). Allelic antigen factor Inv (a) of the light chain of human immunoglobulins. *Science, N.Y.* **152**, 1517.

Ceppellini, R. (1966). Genetica delle Immunoglobuline. *Atti A.G.I.* **12**, 1.

Ceppellini, R. *et al.* (1965). Notation for genetic factors of human immunoglobulins. *Bull. Wld. Hlth. Org.* **33**, no. 5.

Ceppellini, R., Curtoni, S., Mattiuz, P. L., Miggiano, V., Scudeller, G. and Serra, A. (1967). Genetics of leukocyte antigens: a family study of segregation and linkage. *In* "Histocompatibility Testing", p. 149. Muksgaard, Copenhagen.

Edelman, G. M. *et al.* (1969). The covalent structure of an entire γG immunoglobulin molecule. *Proc. Nat. Acad. Sci.* **63**, 78.

Gally, J. A. and Edelman, G. M. (1970). Somatic translocation of antibody genes. *Nature, Lond.* **227**, 341.

Grey, H. M. (1969). Phylogeny of Immunoglobulins. *Adv. Immunol.* **10**, 51.

Hill, R. L., Delaney, R., Lebovitz, A. E. and Fellows, R. E. (1966). Studies on the aminoacid sequence of the H chain from rabbit immunoglobulin. *Proc. Roy Soc. B.* **166**, 159.

Hilschmann, N. and Craig, L. C. (1965). Amino acid sequence studies with Bence Jones proteins. *Proc. Natl. Acad. Sci.* **53**, 1403.

Hood, L. and Ein, D. (1968). Immunoglobulin lambda chain structure: two genes, one polypeptide chain. *Nature, Lond.* **220**, 764.

Hood, L. E., Gray, W. R. and Dreyer, W. J. (1966). On the mechanism of antibody synthesis: a species comparison of L chains. *Proc. Natl. Acad. Sci.* **55**, 826.

Hood, L., Gray, W. R., Sanders, B. G. and Dreyer, W. J. (1967). Light chain evolution. *Cold Spring. Harbor Symp.* **32**, 133.

Kunkel, H. G., Allen, J. C. and Grey, H. M. (1964). Genetic characters and the polypeptide chains of various types of gamma globulin. *Cold Spring Harbor Symp.* **29**, 443.

Kunkel, H. G., Smith, W. K., Joslin, F. G., Natvig, J. B. and Litwin, S. D. (1969). Genetic marker of the γA2 subgroup of γA immunoglobulins. *Nature, Lond.* **223**, 1247.

Litwin, S. D. (1967). Phylogenetic differences among the Gm factors of non-human primates. *Nature. Lond.* **216**, 268.

Litwin, S. D. (1969). The expression of human genetics factors (Gm) on primates immunoglobulins. *Ann. N.Y. Acad. Sci.* **162**, 177.

Markalonis, L. E. and Edelman, G. M. (1966a). Polypeptide chains of immunoglobulins from the smooth dogfish (*Mustelus canis*). *Science, N.Y.* **154**, 1567.

Markalonis, J. and Edelman, G. M. (1966b). Phylogenetic origins of antibody structure. II. Immunoglobulins in the primary immune response of the bullfrog, *Rana catesbiana. J. Exp. Med.* **124**, 901.

Milstein, C. (1965). Interchain disulphide bridge in Bence Jones proteins and in gamma-globulins B chains. *Nature, Lond.* **205**, 203.

Milstein, C. (1967). Linked groups of residues in immunoglobulins k chains. *Nature, Lond.* **216**, 330.

Milstein, C. and Munro, A. J. (1970). The genetic basis of antibody specificity. *Ann. Rev. Microbiol.* In press.

Milstein, C. and Pink, J. R. L. (1970). Structure and evolution of immunoglobulins. *Prog. Biophys. Molec. Biol.* **21**, 209.

Natvig, J. B., Kunkel, H. G. and Litwin, S. D. (1967). Genetic markers of the heavy chain subgroups of human γG globulin. *Cold Spring Harbor Symp.* **32**, 173.

Natvig, J. B., Kunkel, H. G. and Joslin, F. G. (1969). Delineation of two antigenic markers "Non-a" and "Non-g" related to the genetic antigens of human γ globulin. *J. Immunol.* **102**, 611.

Picard, J., Heremans, J. F. and Vandebroek, G. (1963). *Mammalia* **27**, 285.

Putnam, F. W., Titani, K. and Whitley, Jr. (1966). Chemical structure of light chains: amino acid sequence of type K. *Proc. Roy. Soc.* Ser. B. **166**, 124.

Schmitt, J., Krupe, M. and Deicher, H. (1965). Untersuchungen uber Immuno-globulin-Erbmerkmale (Gm- und InV- gruppen) bei subhumanen primaten. *Humangenetik* **1**, 571.

Shuster, J., Wang, A. C. and Fudenberg, H. H. (1970). Evolutionary dissociation of allotypes and other antigenic determinants of immunoglobulins in non-human primates. *Immunochemistry* **7**, 91.

Shuster, J., Warner, N. L. and Fudenberg, H. H. (1969). Cross-reactivity of primate immunoglobulins. *Ann. N.Y. Acad. Sci.* **162**, 195.

Thorpe, N. O. and Deutsch, H. F. (1966). Studies on papain produced subunits of γ-globulins. II. Structures of peptide related to the genetic Gm-activity of γG globulin Fc fragment. *Immunochemistry* **3**, 329.

Titani, K. and Putnam, F. W. (1965). Immunoglobulin structure: amino and carboxyl-terminal peptides of type I Bence Jones proteins. *Science, N.Y.* **147**, 1304.

Van Loghem, E., Shuster, J. and Fudenberg, H. H. (1968). Gm factors in non-human primates. *Vox Sang.* **14**, 81.

Van Loghem, E., Shuster, J., Fudenberg, H. H. and Franklin, E. C. (1969). Phylogenetic studies of immunoglobulins evolution of Gm factors in primates. *Ann. N.Y. Acad. Sci.* **162**, 161.

Vyas, G. N. and Fudenberg, H. H. (1969). Evolutionary dissociation of γG2 and Gm (n) antigens. *Vox sang* **16**, 233.

Wang, A. C. and Fudenberg, H. H. (1969). Genetic control of gamma chain synthesis: A chemical and evolutionary study of the Gm (a) factor of immuno-globulins. *J. Mol. Biol.* **45**, 493.

Wang, A. C., Shuster, J., Epstein, A. and Fudenberg, H. H. (1968). Evolution of antigenic determinants of transferrin and other serum proteins in Primates. *Biochemical Genetics* **1**, 347.

Comparative Cytogenetics in Primates and its Relevance for Human Cytogenetics[*]

A. B. CHIARELLI

Institute of Anthropology, Primatology Centre, University of Turin, Italy

Prerequisite for the comparison of chromosomes from different species is (i) the correspondence between morphologic homology and the homology in genetic content and (ii) the constancy of chromosome number and morphology for each species. The first of these prerequisites is not demonstrable at present, and is the object of prospective research. Support for the second, however, can be found in extensive data in the literature on many species of plants and animals. Therefore, if we consider chromosomes as organic structures, the criteria of homology, analogy and convergence can be applied to chromosome comparison, as is done for many other organic structures in comparative anatomy.

Comparative research on chromosomes, however, deserves special interest as these structures are the direct carriers of genetic information, and this information has a very strict and stable organization. The actual structural affinity between homologous chromosomes for each species of Eukaryota is consistently controlled by their pairing at meiosis in each individual from a chromosome population. Incomplete chromosome pairing at meiosis suggests the presence of structural changes in the organization of the transfer of genetic information which is linearly distributed along the chromosomes. However, structural changes must be restricted within limits beyond which they are no longer compatible with the functional organization of genetic information. The result is the impossibility of homologues pairing, thereby arresting them at meiosis.

In higher organisms, homologous chromosome pairing at meiosis plays the role of a filter through which only a functionally patterned genetic system can pass. Furthermore, in native populations meiosis represents a barrier which prevents exchanges between diverging genetic systems. Its function in the organization of living forms is therefore to stabilize the

* Research supported by a grant from the Italian National Council of Research (C.N.R.), (contract no. 69.02293.115.216) and completed while the author was appointed as Visiting Professor to the Department of Anthropology, University of Toronto, Canada.

genetic information defining a given species by permitting those variations which characterize the adaptive plasticity of each good species. The constancy of the karyotypic features within the species is therefore a consequence of the selective work of meiosis.

In order to ascertain chromosome homology between different individuals the behaviour of the chromosomes at meiosis in the individuals derived from them (F_1) should be inspected. This inquiry is not always an easy task; indeed, inherent technical problems which existed have only recently been solved. For the time being we must therefore tackle the problem of chromosome homology from an essentially comparative morphological standpoint. However, some structurally different chromosomes do exist which show actual functional equivalence. These should therefore be referred to as analogous chromosomes.

Finally, between undoubtedly unrelated species correspondence or similarities exist concerning chromosome number, morphology and even structure, although the organization of their genetic information is absolutely different. These are cases of convergence, almost always casual, but occasionally due to selective factors. Evidence of various kinds is available on the relationship between chromosome number and the genetic variability necessary for the ecological adaptability of a given species. Chromosome morphology itself may be conditioned by selective factors affecting gene organization (John and Lewis, 1968; White, 1970). These instances should be referred to as cases of convergence.

Hence, in the investigation on chromosome homologies, a quite obvious point *a posteriori*, essential to their comparison, should be taken into consideration: namely, the degree of phylogenetic affinity between the species under study. In fact, in the case of morphological chromosome affinity between closely related species, a structural homology rather than convergence is more likely to occur. The closer the kinship between the species under investigation, the higher the probability. In fact, as generations follow one upon another, karyologic variations tend to accumulate, thereby differentiating more and more the chromosome structural organization at the level of individuals and consequently of the whole species. Therefore, in comparing chromosomes between different species, it is necessary to know how these structures can vary and to know the laws controlling these variations.

Leaving aside variations in chromosome numbers such as polyploidy and polysomy, few variations at the level of individual chromosomes are known to result in karyotype patterns or configurations differing from the original. These possible mechanisms are synthesized in Fig. 1. Opinions as to why they succeed in passing the meiosis filter are also given. So far, cytogeneticists have drawn comparisons between somatic chromosomes, at the level of differences that can be detected with the light microscope at a given moment of the cell biological cycle, in which the chromosomes are particularly condensed. It is as if we made a description of all the furniture of a house when it is packed up to be moved into a new house. Hence, the

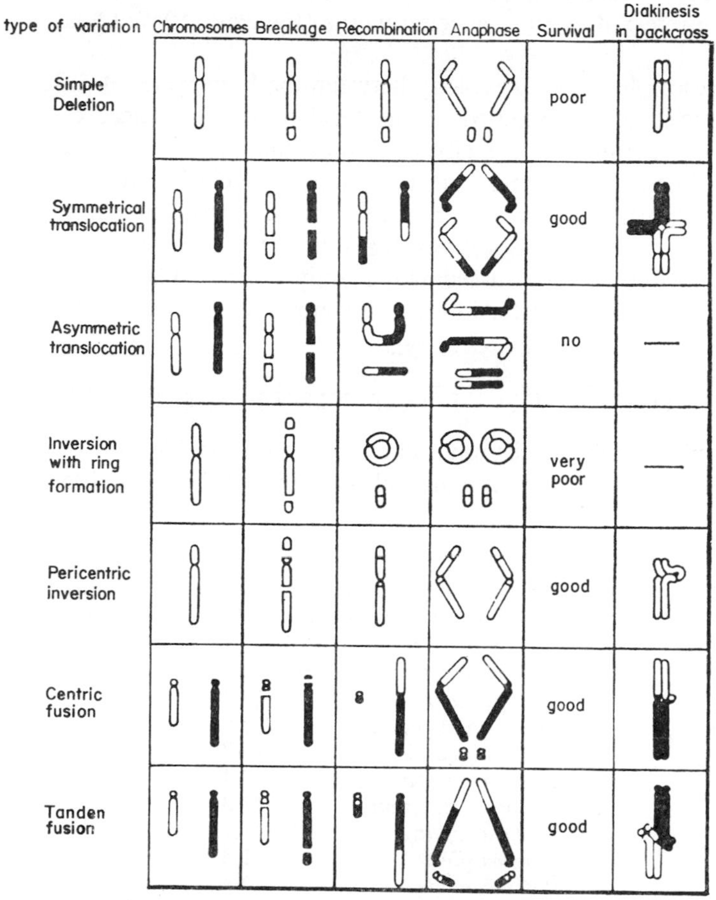

type of variation	Chromosomes	Breakage	Recombination	Anaphase	Survival	Diakinesis in backcross
Simple Deletion					poor	
Symmetrical translocation					good	
Asymmetric translocation					no	—
Inversion with ring formation					very poor	—
Pericentric inversion					good	
Centric fusion					good	
Tanden fusion					good	

FIG. 1. Graphical representation of the possible structural variations of chromosomes.

most accurate observations at this level will fail to let us appreciate the actual organization of genetic information of a given species. Observations of this nature are therefore rather superficial. Furthermore, this comparison is often limited by changes in individual chromosomes due to technical procedures and to the different degrees of chromosome allocyclicity in different tissues. Such limitations, however, do not interfere with the synthesis of information known for various groups of mammals and with its elaboration.

Primate chromosomes have recently aroused the attention of a number of workers and new findings are being published each year. Analytical data pertaining to each species are not pertinent to this synthesis. The reader is therefore referred to more specific works (Egozcue, 1967; Chiarelli, 1966, 1968).

The research on the homology between the karyotypes of different primate species can be studied at the level of DNA nuclear content, the chromosome number, the size and morphology of individual somatic chromosomes and chromosome behaviour at meiosis. In this paper the available information for various primate species will be examined at each level. Then future perspectives for our better understanding will be evaluated. The data discussed will mainly concern Catarrhine monkeys because they are of greater interest in comparison with the human karyotype.

COMPARISON OF DNA NUCLEAR CONTENT

As is widely known, histophotometric methods enable us to measure the nuclear DNA content of any animal or vegetal cell (Vialli, 1957). In this

TABLE I. DNA content in arbitrary units using the Deeley histophotometer with human $12\cdot4 \pm 1$ as reference (simplified from Manfredi-Romanini, 1970)

Simiae Catarrhinae	
Cercopithecus aethiops	10·4
Cercopithecus cephus	12·5
Cercopithecus diana	16·7
Erythrocebus patas	12·4
Cercocebus torquatus	17·3
Macaca mulatta	11·1
Macaca silenus	11·2
Papio hamadryas	12·5
Colobus polykomos	12·7
Nasalis larvatus	15·3
Hylobates lar	10·6
Hylobates agilis	10·4
Symphalangus syndactylus	11·1
Pongo pygmaeus	14·5 ± 8
Gorilla gorilla	12·6 ± 7
Pan troglodytes	12·6 ± 5
Homo sapiens	12·4 ± 1

field extensive investigations on primates are now being carried out by M. G. Manfredi-Romanini and her co-workers at the Institute of Anthropology of Pavia University (Manfredi-Romanini, 1968; Manfredi-Romanini and Fontana, 1968; Fontana, 1969). Despite their incompleteness, these findings can lead us to some remarkably interesting considerations. In Table I the values for individual species in reference to Man are given.

COMPARISON OF CHROMOSOME NUMBER

The chromosome number is known for 110 of the 180 living primate species. Numerical variations within species are rarely found and are almost always due to centric fusion or fission, although in some instances polyploidy (*Galago senegalensis* and *G. crassicaudatus*) or polysomy (*Cercopithecus, Nasalis, Symphalangus*) could be suggested as being responsible for these variations.

Within all the primates the chromosome number is found to range between 20 and 82 with a sharply asymmetric distribution. Within the Old World primates the chromosome number varies from 42 to 72 and the distribution curve exhibits 2 modes: a pronounced mode for number 42 and a lower one for number 66 (Fig. 2). The tail of the curve from 54

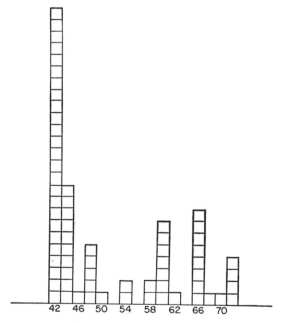

FIG. 2. Numerical distribution of the chromosomes in the Catarrhine monkeys.

onward is made up entirely of the chromosome numbers of species and subspecies of *Cercopithecus*. The variability of the chromosome number in the various species of this genus is rather peculiar and the mechanisms responsible for it remain to be clarified (Chiarelli, 1964, 1968a, b). The remaining Catarrhine species, including Man, occupy the numbers from 42 to 50.

The distribution of the various species of the Catarrhines with chromosome numbers between 42 and 50 is peculiar. Number 2n = 46, for instance, is characteristic of Man alone, 2n = 48 of the 3 large anthropoid apes and of *Nasalis*; 2n = 50 of *Symphalangus*; 2n = 44 of the various species of *Hylobates*, *Presbytis* and *Colobus*, with the exception of one species of gibbon (*H. concolor*) which seems to have 52 chromosomes (Wurster and Bernirschke, 1969). Finally the species from the genera *Papio*, *Macaca*, *Theropithecus* and *Cercocebus* all exhibit 2n = 42 chromosomes.

Man therefore is the only species whose karyotype contains 46 chromosomes, with a position midway between the numbers of the *Hylobates*, *Presbytis* and *Colobus*, on one side and those of *Nasalis*, *Pongo*, *Pan* and *Gorilla* on the other. Within these 2 number groups, 44 and 48, our attention should be directed to finding out the karyotypes most similar to that of Man.

COMPARISON OF CHROMOSOME MORPHOLOGY

A remarkable morphological uniformity is seen in the karyotype of the various species having a chromosome number of 42. It almost always consists of 13 pairs of sub-metacentric and 6 pairs of metacentric chromosomes. In addition there is a pair of chromosomes characterized by a large achromatic region, and also a pair of sex chromosomes. Very slight differences have been described within the various species and their consistency is doubtful (Fig. 3). This general karyotype homogeneity is demonstrated by the frequent descriptions of hybrids that often are also fertile, not only among species from the same genus, but also among species belonging to different genera (Chiarelli, 1961) (Fig. 4). The uniformity in chromosome behaviour at meiotic diakinesis further indicates this homogeneity (Fig. 5). Examination of meiotic chromosomes from a hybrid of two *Papio* species has shown the presence of a small translocation in one of the parental species (Fig. 5).

Turning now to the other major distribution of primate chromosomes; the chromosome numbers between 54 and 72 are characteristic of the various species from the *Cercopithecus* genus (Table II). In the various species from this genus the chromosomes may be morphologically divided into sub-metacentric, metacentric and acrocentric (Fig. 6). In these species

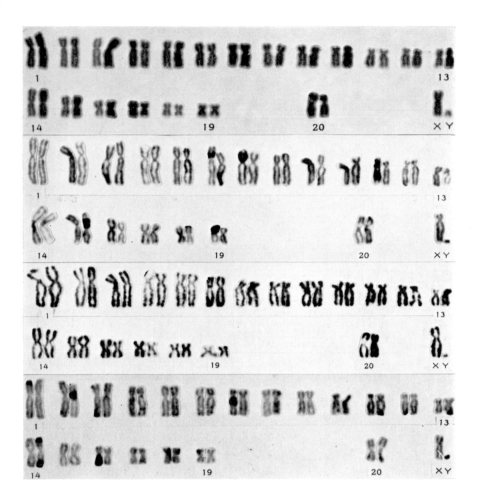

FIG. 3. The karyotypes of (a) *Papio*; (b) *Theropithecus*; (c) *Macaca* and (d) *Cercocebus*.

there also exists a chromosome pair characterized by a wide achromatic region. However, this chromosome appears to be homologous to that of the species having 42 chromosomes only for the arm containing this achromatic region: the other arm is missing or is barely represented. The chromosome number for each morphologic group is not constant and is seen to vary within individual species. Karyological mechanisms involved in the differentiation of this group of species, as we have already said, therefore becomes rather problematic.

K

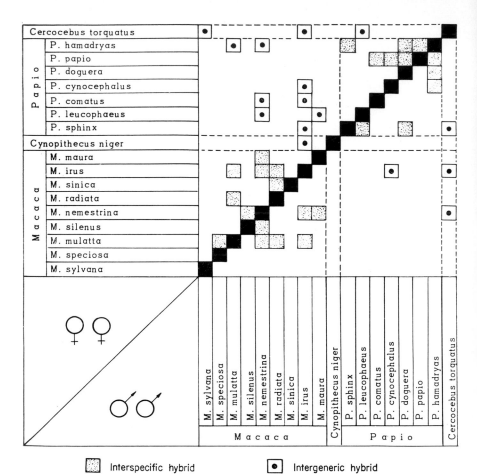

FIG. 4. Known hybrids among the species of the genus *Macaca, Papio, Cynopithecus* and *Cercocebus.*

TABLE II. Chromosome numbers in the different species of *Cercopithecus*

Species	54	56	58	60	62	64	66	68	70	72
C. patas	—									
C. talapoin	—									
C. diana			—							
C. l'hoesti			—	—						
C. neglectus			—		—					
C. nigroviridis			—							
C. aethiops			—							
C. cephus							—			
C. mona							—	—		
C. nictitans							—		—	
C. mitis										—

TABLE II (a) Total Haploid Chromosomal Length of individuals
with different numbers of chromosomes of the genus *Cercopithecus*

Cercopithecus	Mean in μ
2n = 54	94·5 ± 14
2n = 60	101·6 ± 24
2n = 66	112·1 ± 13
2n = 72	125·1 ± 20

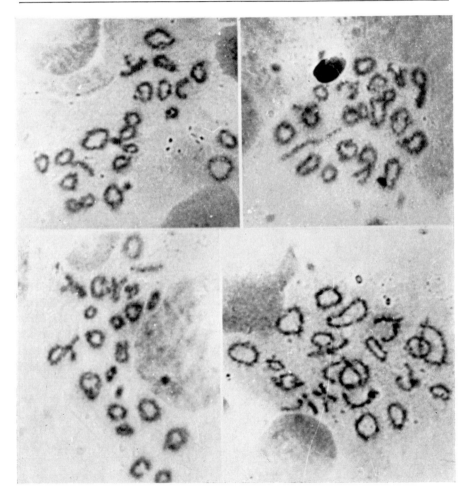

FIG. 5. Diakinesis of *Papio hamadryas* (top left), of *P. papio* (top right), and *P. cynocephalus* (bottom left). In the diakinesis on the bottom right obtained from a possible hybrid between *P. cynocephalus* and *P. ursinus*, only 20 instead of 21 bivalents are counted; note the 8-shaped bivalent in the middle.

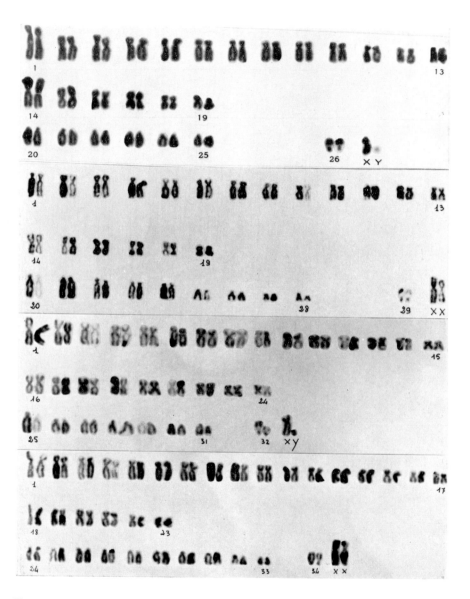

FIG. 6. Karyotypes of several species of genus *Cercopithecus* (a) *C. talapoin*; (b) *C. nigroviridis*; (c) *C. mona*; (d) *C. mitis*.

The karyotypes of *Hylobates*, *Presbytis* and *Colobus* with 2n = 44 chromosomes, plus that of *Nasalis* with 48 chromosomes, and *Sympholangus* with 50 (Fig. 7), are very different from those of the anthropoid apes (*Gorilla*, *Pan* and *Pongo*) with 48 chromosomes (Fig. 8). The *Nasalis* karyotype, although presenting 48 chromosomes, is morphologically very similar to that of *Colobus* and *Hylobates*, and differs from the karyotype of the apes, which is in keeping with traditional systematics. According to a hypothesis advanced some time ago (Chiarelli, 1968), the chromosome number in this

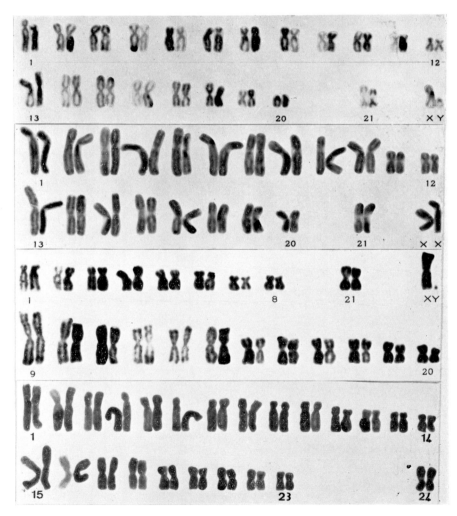

FIG. 7. Karyotypes of (a) *Presbytis*; (b) *Colobus*; (c) *Hylobates*; (d) *Nasalis*.

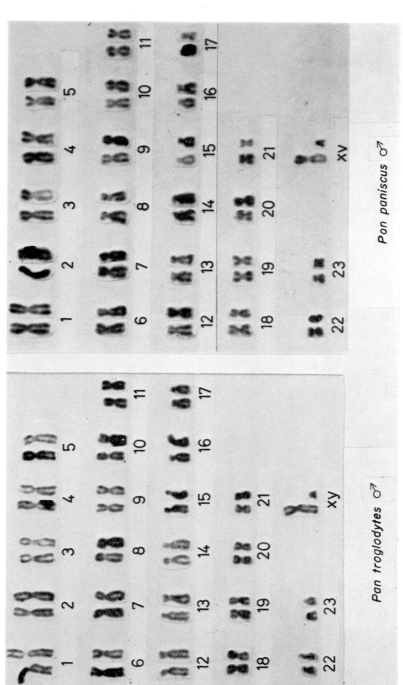

Fig. 8. (a) Karyotypes of *Pan troglodytes* and *Pan paniscus*.

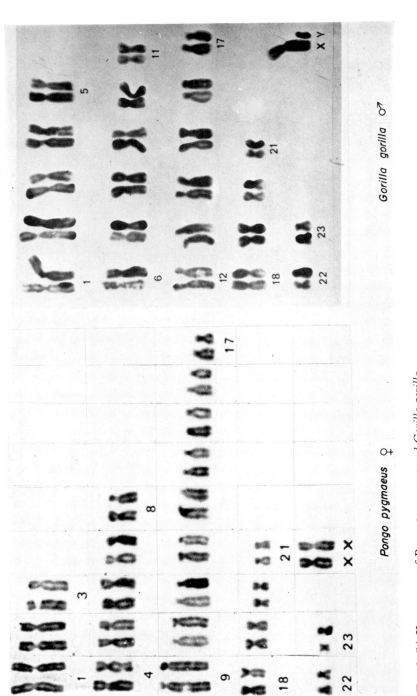

FIG. 8. (b) Karyotypes of *Pongo pygmaeus*; and *Gorilla gorilla*.

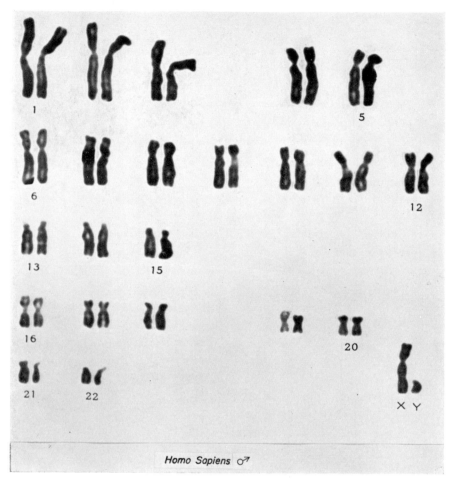

Fig. 8. (c)—Karyotypes of *Homo sapiens*.

species could have arisen through polysomy of some peculiar chromosomes. *Nasalis*, therefore, from the standpoint of chromosome morphology, may be grouped with the species which have 44 chromosomes. The same goes for *Symphalangus*.

The most striking karyological differences between the groups with 44 (Colobidae and Hylobatidae) and 48 (Pongidae) chromosomes are the following: (i) the lack of acrocentric chromosomes in the group with 44 chromosomes (except *Presbytis* which has one pair) whereas in the anthropoid apes 5 or 9 pairs of this type of chromosome are found; (ii) the presence of a pair of chromosomes characterized by a large achromatic region in almost all the Cercopithecinae, while these marker chromosomes are altogether missing in the karyotype of the anthropoid apes.

When the human karyotype is compared with that of the karyotypes of these other 2 primate groups, its greater similarity to that of the species having 48 chromosomes stands out immediately. Indeed, at first sight it is often impossible to distinguish a human metaphase plate from that of the chimpanzee. These data have been phylogenetically interpreted in a previous paper (Chiarelli, 1968); I wish to stress here the comparison between individual chromosomes. This comparison may be conducted by comparing indices and relative measurements (Chiarelli, 1962). However, this type of procedure is marred by the fact that the measurements are carried out on photographic enlargements. Difficulties in accurately measuring the chromosomes in this manner have sometimes led to erroneous results. Moreover, the detected correlation existing between the degree of spiralization and the chromosome morphology in metaphase increase the unreality of the direct comparison of idiograms.

Another, more direct method, is the comparison of chromosomes with identical degrees of spiralization from different karyotypes, and with chromosomes from homologous cells cultured by the same techniques. To do this it must necessarily be postulated that the chromosome material from anthropoid apes is identical to that of Man. This assumption is supported by the measurements of the total length of karyotypes that have shown an actual identical length of the karyotypes, as well as by the histophotometric determinations of nuclear DNA. These results are valid for the African anthropoid apes, at least. The orang utang shows differences at the limit of significance in both types of data. Therefore, the chromosomes from Man and African apes in plates of equal length may be directly compared. In order to facilitate this comparison, each chromosome pair from *Pan troglodytes* was at first compared with those of *Gorilla gorilla*. The results obtained are shown in Fig. 9 in which single chromosomes are reported and compared. Chromosomes are divided according to the Denver classification of the human karyotype to facilitate their identification. In the first group (AB) which consists of 5 large-sized metacentric and sub-metacentric chromosomes, an actual difference is seen to occur in the third chromosome. In *Gorilla* the centromere of this chromosome appears to be in a more terminal position. In the second group (C) which consists of 7 metacentric and sub-metacentric chromosomes, smaller than the ones described above, a difference is seen to occur in the seventh chromosome: in *Gorilla* this chromosome has a more medially located centromere. The third group consists of 6 acrocentric or sub-acrocentric chromosomes which do not show clear cut differences between the 2 species. The same morphological similarity is found in the fourth group consisting of 4 metacentric and sub-metacentric chromosomes, and in the fifth group which consists of two acrocentric chromosomes. On the whole, the sex chromosomes are similar, although the X chromosome seems to be somewhat longer in *Gorilla*. Hence chromosome differences between the 2 African anthropoid apes are displayed in pairs no. 3, nos 6 and 7.

For the time being, comparison of human chromosomes and those of the anthropoid apes has been made by studying the chimpanzee karyotype, using the same procedure just described for chimpanzee and gorilla. At first, without altering the chimpanzee karyotype, all the chromosomes of

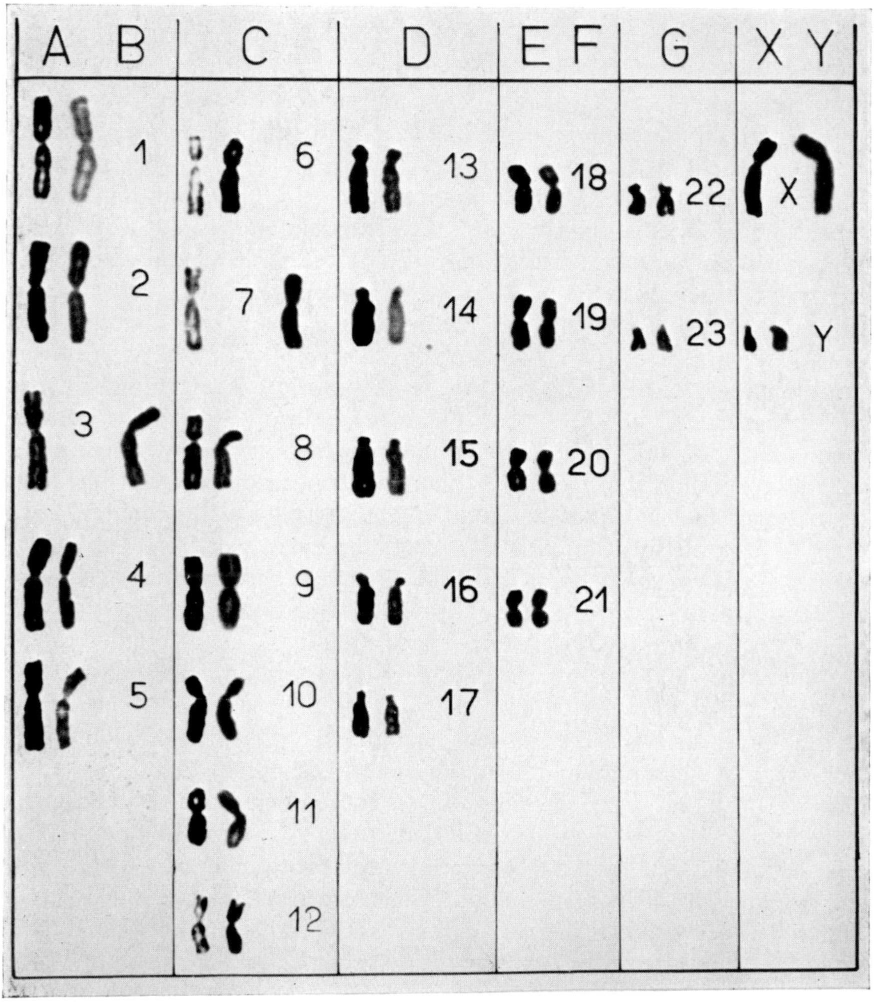

FIG. 9. Direct comparison of chimpanzee chromosomes with those of the gorilla.

the human karyotype which were morphologically similar, if not identical, to those of the chimpanzee (namely, chromosomes 3, 7, 11, 12, 13, 18, 19, 22, X and Y of Man) were selected (Fig. 10). Chromosomes 2, 9, 14, 15, 16, 20 and 21 were seen to display some analogy, or at any rate some possible dislocation, at the level of corresponding chimpanzee chromosomes,

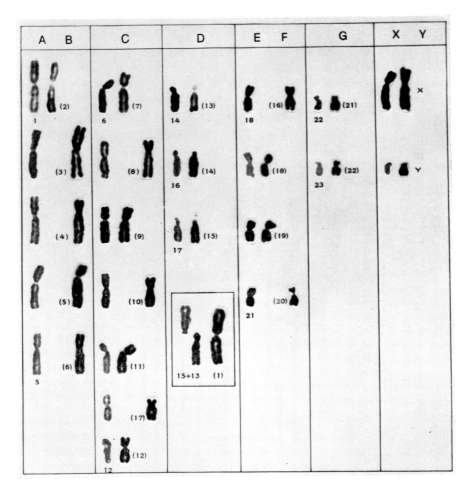

FIG. 10. Direct comparison of chimpanzee chromosomes with those of Man.

although the comparison was difficult to make due to a different centromere position or to their different sizes. Lastly, chromosomes 4, 5, 6, 8, 10 and 17 appeared to be quite different from the corresponding chimpanzee chromosomes and were correlated with those from the chimpanzee solely because of some homology in one of their arms.

In the human karyotype, chromosome 1 was very hard to place among chimpanzee chromosomes, and conversely, for 2 nearly acrocentric chimpanzee chromosomes no partners could be found in the human karyotype. It is reasonable to infer, therefore, that chromosome 1 in Man may be derived from the centric fusion of chromosomes 13 and 15 of a karyotype like that of the chimpanzee. The correspondence between these chromosomes is evident. A centric fusion seems therefore to be responsible for the reduction in chromosome number in Man from 48 to 46, starting from a common ancestor with the apes (Chiarelli, 1962).

As far as the comparison of the karyotype between Man and the orang utang is concerned analogies or homologies are not so exact, because of the reasons mentioned above. However, in Table III the chromosomes of these 2 species which might, according to our classification, be considered homologous, are shown. Some chromosomes of the orang utang, such as chromosome no. 9, are peculiarly characteristic of this species. It is curious also that chromosomes exhibiting a morphology very similar to that of chromosome 9 from the orang utang have been found occasionally in human cell lines stabilized *in vitro* (Nuzzo *et al.*, 1961).

TABLE III. Possible chromosomal homology between Man and Orang utang

Homo ⟷	Pongo
1	1
4	3
6	5
7	6
10	8
19	21
13	13
14	14

Chromosome comparison between Man and the other Cercopithecoidea species is far more complicated. Instead of "homology" the term "similarity" would be more suitable here, and instead of individual chromosomes, it would be better to consider groups of morphologically similar or dissimilar chromosomes for discussion. In almost all Cercopithecoidea, a pair of marked chromosomes is present. The only Cercopithecoid species studied so far without this chromosome pair is *Symphalangus syndactylus*, which, however, exhibits a chromosome of the same shape as the marked chromosome in *Hylobates*, lacking the chromatic portion superimposed on the achromatic region (linear satellite). Perhaps during the complex rearrangement which took place in the *Symphalangus* karyotype this portion was translocated upon another chromosome or in the same on the opposite arm (Chiarelli, 1971).

Comparison of sex chromosomes requires special attention. The actual identification of the X chromosome by means of incorporation of tritiated thymidine has been done only for *Macaca speciosa* (Huang *et al.*, 1969) and *Pan troglodytes* (Low and Benirschke, 1969). For other species the identification of the X chromosome is still a matter of discussion, however, the general consensus is that the X chromosome is almost always medium sized with a subterminal centromere.

On the other hand, Y chromosomes are highly heterogeneous in both morphology and size in the various Catarrhine species. In fact, their size differs from that of *Hylobates* Y chromosomes which appears punctate to that of Man, which is moderate in size (Fig. 11). Significant variation is also found in the centromere position. The Y chromosome of Man is perfectly homologous morphologically only with the chromosomes of the chimpanzee and gorilla. The Y chromosome of the orang utang is different in that it exhibits a wide achromatic process on its shorter arm. These are the available data today on the possible morphological homologies or similarities of human chromosomes with those from other primates at an optical level. Other information will eventually be harvested although much more progress does not seem likely at this level of observation.

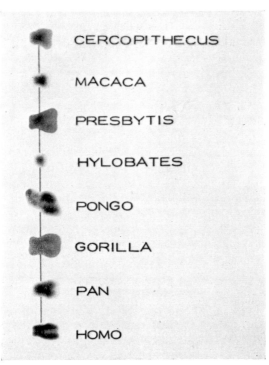

FIG. 11. The Y chromosomes of diverse primates species.

PERSPECTIVES

For a synthesis to be of some value it must succeed in forecasting, or at least in giving a glimpse into future developments. Electron microscopy has so far been of little help in the comparative study of metaphase chromosomes in vertebrates. The scanning electron microscope is likely to be of greater help for the study of some structural details, namely the centromere or the achromatic regions. On the other hand, the electron microscope is providing exceptionally important information for the comparison of chromosome ultrastructure and organization. No doubt this will open new prospects in comparative genetics; but first of all the ultrastructural organization of chromosomes in higher vertebrates should be clarified. Only recently has this problem been decisively tackled (Du Praw, 1965, 1966). The 2-dimensional microdensitometry method devised by Hughes (1968) and the one more recently proposed by Casperson (1970) on the DNA-binding fluorescent agents may also be significant in the objectification of data in investigations of comparative cytogenetics.

The Replication Pattern

Another method which can reasonably be regarded as profitable in yielding useful information for chromosome comparison between different species is the one based on replication patterns. In recent years this type of investigation has been used for individual chromosomes in Man, utilizing tritiated thymidine as a precursor of DNA (German, 1962; Schmidt, 1963; Moorhead and Vittorio, 1963; Kikuchi and Sandberg, 1963; Hsu *et al.*, 1964; Lima de Faria, 1964). As a rule, homologous chromosomes display the same pattern of thymidine uptake, so that this method is of value in recognizing them. Identification of some chromosomes from the human karyotype has been facilitated by this procedure. For instance, in chromosome 1, the distal portion of the shorter arm was seen to synthetize DNA before remaining chromosome regions. Furthermore, in the same chromosome there is a small region proximally located to the small achromatic region of the longer arm that seems to synthetize DNA relatively later than the other regions.

DNA synthesis in the long arms of chromosome 4 is terminated later than in the longer arms of chromosome 5, while in chromosome 5 DNA is consistently synthetized later in the short arms than in the long arms. Within group 6–12, pair 7 and pair 11 show an intense uptake of labelled DNA precursors in the proximal region of their long arms at the final stage, while all the others are unlabelled. Within group 13–15, 3 different patterns of synthesis completion are found, which may be of use in the identification of these 3 chromosome pairs. Chromosome pair 13 carries on DNA synthesis for a long time over extensive segments of the long arms; the arms of pair 14 are seen to achieve DNA synthesis later only in the region next to the centromere, whereas pair 15 terminates DNA synthesis much earlier. Pair 17 generally completes DNA synthesis earlier than pair 18. In both pairs 19 and 20 DNA synthesis is completed quite soon, indeed, they are the first to complete their synthesis of DNA within the whole chromosome complement in Man. The X chromosomes, when paired, are asynchronic in their replication, one of them replicates earlier than the other. Replication of chromosome Y occurs relatively very late especially with respect to chromosomes 20 and 21. On the basis of these premises it might be extrapolated that genetically homologous chromosomes from different species show a tendency to replicate with an identical rhythm.

Investigations on chromosomes replication patterns in the various non-human primates species are still scarce at this point. Besides the painstaking work by Huang and co-workers (1969) on DNA synthesis in *Macaca speciosa*, which will be taken into consideration again later on, we are particularly interested in the investigation by Low and Benirschke (1969) on the pattern of chromosome replication in *Pan troglodytes*. These authors have brought to light various interesting similarities in the replication patterns of whole or parts of chromosomes in chimpanzee and Man. However, a revision of their classification system for chimpanzee chromosomes could provide further and more interesting comparisons.

For example, the non-homology shown by their study in the replication pattern between chromosome 1 in the chimpanzee and chromosome 1 in Man should be re-examined. In fact, the largest chromosome of the chimpanzee karyotype cannot be homologized with the largest chromosome of the human karyotype. According to the hypothesis discussed before, the largest chromosome of the human karyotype (1 in the Denver classification) is derived from the centric fusion of 2 acrocentric chromosomes of the 13–15 type of the chimpanzee karyotype. On comparing the replication of these two chromosomes in the chimpanzee with those of chromosome 1 from Man the homology seems to become very strict indeed (Fig. 12). In our opinion, this provides further evidence supporting the hypothesis of the origin of human chromosome 1 through the fusion of 2 acrocentric chromosomes of types 13 and 15A from a karyotype similar to that of the chimpanzee. This re-examination would lead to the complete reorganization of the karyotype in groups G, A and B proposed by Low and Benirschke for the chimpanzee. In any case, the correspondence of the replication patterns of chimpanzee chromosomes with those from Man is fairly good, especially when the final stage is taken into account.

Although consistent studies of the replication pattern in various species are much needed and desired they are not an absolute test of genetic homology. The study of chromosome replication patterns in related species, however, will undoubtedly provide interesting information on their degree of phylogenetic affinity by bringing to light or confirming chromosome rearrangements such as translocations or pericentric inversions which took place during their evolution and differentiation.

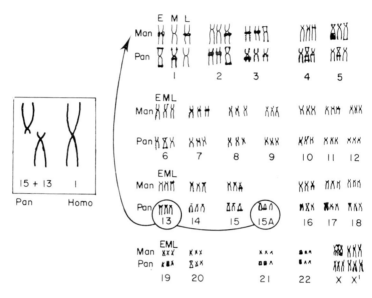

FIG. 12. Reduplication patterns of the chromosomes of the chimpanzee and of Man, at early (E), middle (M) and late (L) stages, according to Benirschke's re-interpretation (modified from Low and Benirschke, 1968).

The Trisomy of Chromosome 21

Substantial information on the genetic homology of a particular chromosome in Man and chimpanzee has recently been obtained by McClure *et al.* (1970). These scientists of the Yerkes Primate Center have identified a chimpanzee exhibiting a small supernumerary acrocentric chromosome. This chimpanzee displays several phenotypic anomalies similar to those of human individuals with Down's syndrome or mongoloid idiocy who have a trisomy of chromosome 21. The morphologic homology between chromosome 21 in Man and the triplet occurring in this Yerkes chimpanzee is very impressive (Fig. 13). Hence the view that the genetic homology of these chromosomes in Man and chimpanzee, which is partial in the least, may reasonably be sponsored.

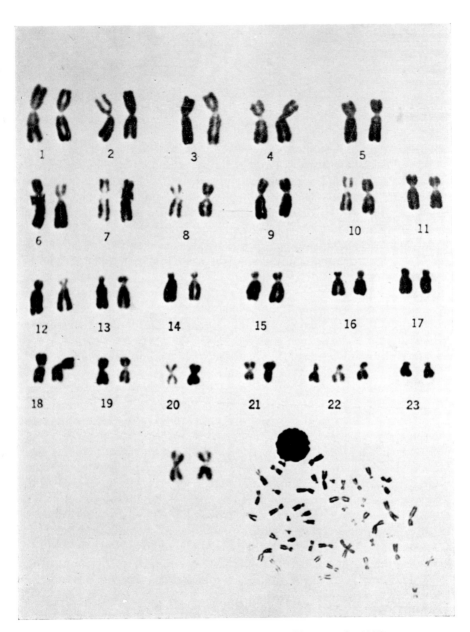

FIG. 13. Trisomy 22 in the chimpanzee. From McClure *et al.*, 1970.

The Sex Chromosomes

Still more reliable information about the genetic homology of the X chromosomes from different species will no doubt be obtained with relative facility in the near future. A variety of markers are already available for these chromosomes in our species (Siniscalco, 1963) and some of them have already been found in some non-human primate species (Gavin *et al.*, 1964).

The Pair of Marked Chromosomes

In our opinion, special attention should be paid to the marked chromosomes, which are characteristic of several species of Old World monkeys. To begin with, these peculiar chromosomes are interesting because of their replication pattern as revealed by labelled thymidine uptake. In fact, Huang *et al.* (1969) have brought out that in *Macaca mulatta* these chromosomes replicate asynchronously with respect to the other chromosomes. In lightly labelled metaphases, this pair was always free of labelling. In heavily labelled cells, it was generally more lightly labelled than most of the other chromosomes. In moderately labelled cells, there was a single labelled block near the centromere opposite the arm carrying the secondary constriction. This observation indicates that in the marker pair DNA synthesis is completed earlier than in most chromosomes in the complement. The fact that this chromosome exhibits this large secondary constriction is regarded as an indication of the sites of nucleoli formation (Swanson, 1960). The nucleolus is the centre of ribosomal RNA synthesis (Darnell, 1958).

Early completion of DNA synthesis in a chromosome associated with nucleolus formation seems to be to the advantage of cell economy. After achieving its synthesis of DNA, this region can start nucleolus reorganization in early telophase, and exercise its function well before chromosomal DNA unravelling in the next interphase. But still another feature lends interest to these chromosomes. In Man, the nucleolar regions of the autosomes are situated in the "stalks" or secondary constrictions, which attach the satellites to the short arms of the acrocentric chromosomes (13–15, 21–22) (Levan and Hsu, 1959; Hamerton, 1961; Slizynski, 1964; Valencia, 1964). Obviously direct comparison of these chromosomes with the nucleolus-organizing pair of the Papinae is speculative. However, the similarity in both shape and size that would result from the association of these 2 chromosome types in Man and true anthropoid apes with the marked chromosome in *Macaca* is remarkable. Aside from its genetical interest, the prospect of this homology would offer a new approach to tackling the problem of the ancestral karyotype from which the various groups of Old World monkeys have arisen. In fact, in the phylogenetic interpretation which we have proposed (Chiarelli, 1968) of the evolution of the karyotypes of this

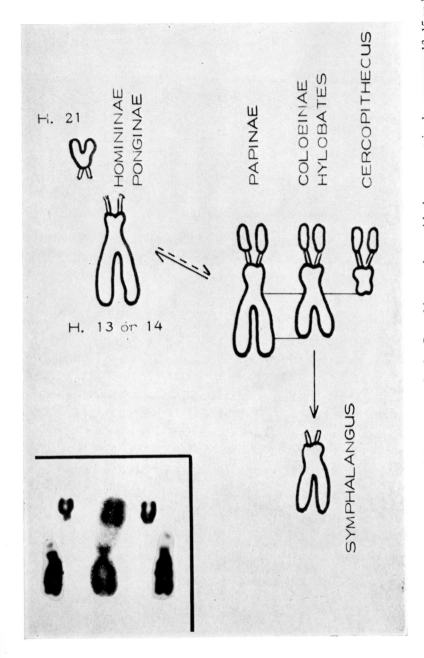

FIG. 14. A possible analogy of marked chromosomes in the Catarrhine monkeys with the acrocentric chromosomes 13, 15 and 12 in Man.

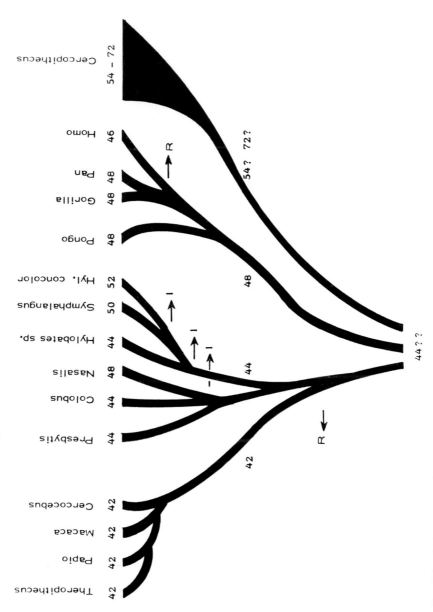

FIG. 15. Possible phylogenetic relationship among the diverse genus of the Catarrhine monkeys.

group of primate, 3 evolutive lines were outlined, but no relationships between them were established (Fig. 15).

The Meiotic Chromosomes

A different way to face the problem of genetic chromosome homogeneity in different though related species is to study the meiotic chromosomes. Two stages in meiosis are particularly important to this end: the pachytene and diakinetic stages. It is obvious from what is known about meiotic prophase chromosomes in different animals and plants (Darlington, 1937; Swanson, 1960) that a detailed analysis of the structure and pairing behaviour of the meiotic prophase chromosomes in Man and other primates should bring forth genetic information of paramount importance.

The study of pachytene chromosomes in Man is, however, far from easy. Since the first attempts at description of chromosomes during these stages by Schultz and St. Lawrence in 1949, some progress has been made (Slizynski, 1964; Yerganian, 1963; Eberle, 1963; Ferguson-Smith, 1964; Anneren et al., 1970). The type and pattern of chromosome distribution in the pachytene stage is generally constant and is used as a means of chromosome identification and therefore can be used for identifying homologies among different species. The diakinetic stage permits the interpretation of the findings on somatic chromosomes on a more strictly genetic level. In fact, the frequency of chiasmata is to a great extent genetically determined (Nicoletti, 1968) and is a reliable index of variability of the species. The observation of chromosome behaviour during meiosis can lead to a sure identification of chromosomal rearrangements such as translocations and inversions, not only in individuals belonging to the same species, but also in fertile and sub-fertile hybrids of different species. A particularly interesting aspect that can be brought to light by the study of meiotic chromosomes is the behaviour of the sex chromosomes.

Compared to the amount of available information about somatic chromosomes, our knowledge of meiotic chromosomes from non-human primates has not progressed very much. Only some 20 individuals belonging to 8–9 different species have been investigated so far. We have already mentioned the results obtained from *Papio* and *Macaca*.

Here, I wish to give an idea of the problems of the comparison of human meiotic chromosomes with those of anthropoid apes and mention the interest in chiasmata. Meiotic chromosomes from Man have been fairly extensively studied. Comparative investigations which were carried out on a group of 8 complete diakineses from a single individual (Chiarelli and Falek, 1967) have shown constant morphological similarities between homologous bivalents of different spreads. This resemblance of particular chromosomes from spread to spread was due, in large measure, to the number and

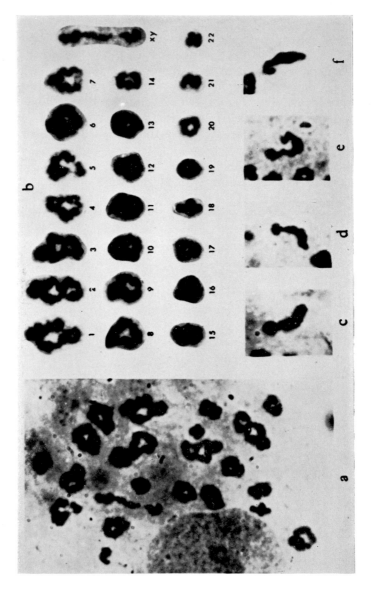

Fig. 16. Meiotic human chromosomes. From Chiarelli and Falek (1967).

position of the chiasma along the chromosome filament (Fig. 16). The mean number of recognizable chiasmata for each bivalent in the diakinetic spread ranges from 3·6 to 1. The first 3 chromosomes have a mean of more than 3 chiasma; chromosomes 4–12 have a mean between 2 and 3 chiasma, and chromosomes 13–20 have less than 2 chiasma per chromosome. In the 8 spreads examined, the total number of chiasmata per cell varied from 49 to 57.

The X and Y chromosomes are linearly attached in nearly all the spreads examined. The Y chromosome appears as a short triangularly-shaped segment terminally attached at the end of the long arms to one of the extremities of the X chromosomes. In some spreads the Y chromosome appears terminally attached to the X chromosome, while in others the Y chromosome is well separated from the X. The evidence of a true chiasma between

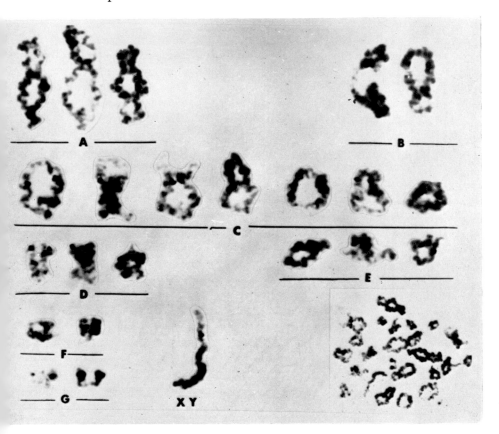

Fig. 17. Meiotic human chromosomes in which the centromere is visible. Courtesy of A. T. Chen and A. Falek.

the two sex chromosomes is, therefore, still open to speculation. It may be that the chiasma between the X and Y chromosomes exists for only a short period at a very early stage of diakinesis. This may be the reason why it is not visible in many of the spreads. Moreover, the fact that the X chromosome often appears bent at a sub-medial point leads us to believe that this may represent the position of the centromere and that the Y chromosome is attached to the terminal part of the shorter arm of the X.

Most recently, Chen and Falek (1969) using special techniques have succeeded in localizing the centromere in meiotic chromosomes (Fig. 17). These findings provide fresh fascinating prospects for a comparative study on the localization and frequency of chiasmata between bivalents of different species whose homology has been postulated. Evidently great interest lies in the prospect of comparative studies at this level among anthropoid apes and Man.

The Hybrids

Obviously, these homologies would be conclusively demonstrated by the study of diakinesis in hybrids. As previously mentioned, meiosis acts as a filter through which only the structures that are compatible with a functional genic organization are allowed to pass. This was suggested as possible for a hybrid of the 2 *Papio* species. It is fairly likely that many, if not all, hybrids deriving from different species and genera of the Papinae group are fertile. Hybridological information so far available concerning this group is considerable. In many of these hybrids, fertility with at least one of the parental species is known (Chiarelli, 1961).

Leaving aside the well known instances of hybrids between the horse and the donkey, as well as between the hare and the rabbit, which are sterile because of the impossibility of chromosomal organization at meiosis, a similar case was accurately described by Low and Benirschke (1968) between 2 species of *Saguinus* (*S. oedipus* and *S. midas*). Despite the close similarity shown by the somatic chromosomes of the hybrid with those of the 2 parental species, complete aspermia was found in histological sections of the testis. The seminiferous tubules were of normal size, patent but empty. Their basement membranes where not thickened in special stains and numerous interstital cells were present. Most tubular lining cells, according to the authors, appeared to be Sertoli cells and toward the tubular lumen many large vacuoles were present. Spermatogonia were also present, they however were sparse and had a clear cytoplasm. A number of mitoses were seen in these cells and early stages of meiosis were observed in some areas but there were no figures of diakinesis. Cellular debris was found in some tubules which presumably were a result of faulty meiosis. In this instance the barrier occurs prior to diakinesis, in the earliest stages of meiosis. The species barrier may, however, manifest itself at other levels.

The hybrid offspring may find difficulties in post-natal development, notably at the stages in which complex biological rearrangements take place. This might result in the frequency of mortality observed in post-natal life stages in hybrids from differing primate species (Chiarelli, 1961). However, this barrier certainly manifests itself more efficiently during the orderly sequence of gene activation and repression occurring during embryonic development. In spite of gestation, many hybrids are never born. The chance of fertilization being achieved, is however the initial limit of the complex mechanism maintaining the gene assortment within specific variability. A number of complex investigations are being conducted in many laboratories in an effort to identify all the barriers limiting gametic coupling within the same species or between closely related species. Some of these studies also concern non-human primates.

The gradual succession of barriers has very often a phylogenetic importance. In fact, the greater the phylogenetic distance between the parental species, the more precocious the barrier. This is clearly demonstrated by the frequency of hybrids, almost all also fertile, described for Papinae. From a cytotaxonomic and evolutionary standpoint, however, the study of meiosis in hybrids produced by *Cercopithecus* species having different chromosome numbers would be of particular interest. Indeed, this type of approach is likely to be the only one that will allow a solution to the problem of the peculiar variability of the chromosome number in this group of species. Obviously, another group of hybrids which would excite special interest not only at a scientific level, is that of hybrids between anthropoid apes and Man. The failure of some attempts of this kind made by Ivanoff at the Pasteur Institute in 1930 was due to inadequate knowledge of fertilization processes and of embryonic implantation. Theoretically, it may reasonably be assumed that fertilization and the earliest developmental stages of a foetus could be attained in a mating between Man and African apes. In fact similarity between chimpanzee or gorilla chromosomes and those from Man is equal, or even greater, than the similarity between chromosomes of the horse and donkey, or the hare and rabbit. From the standpoint of phylogenetic differentiation a real genetic barrier between African apes and Man may be chronologically traced back to 5 and 6 million years. In terms of generation, calculating each generation every 15 years, this would mean 300,000–400,000 generations. This distance in term of genic mutations is responsible for differentiation that took place between the horse and donkey or between the hare and rabbit from their respective common ancestor.

The DNA Hybridization in vitro

But the most up-to-date method used to establish the existence of homogeneity in genetic material from different species is the DNA hybridization *in vitro*. As is known, this method is based on the dissociation of DNA from 2 species into single strands and seeing how much recombination to form hybrid double strands can occur (Kohne, 1968). Different versions of this procedure have been used to investigate DNA homologies in some mammals including a few primates. Hoyer *et al.* (1964) immobilized single-stranded DNA from various vertebrates in agar and determined the amount of recombinations which occurred with fragmented single-stranded DNA from Man and mouse. A more perfected version of this method consists of dissociating DNA from 2 different species and allowing it to compete for combination with agar-embedded DNA and fractions of DNA with varying base ratios prepared by centrifugation.

Using this method Martin and Hoyer (1967) established that the Guanine-Cytosine (G.C.) rich fractions of human DNA had 95% homology with DNA from the chimpanzee, 85% with that from the gibbon and 74% with *Macaca* DNA. Homologies for the Adenine-Thymine (A.T.) rich fraction were not as close. This might suggest that the G.C. rich fraction represents genes that are relatively stable in evolution. More refined methods are becoming increasingly available. The molecular hybridization of radioactive DNA with stationary DNA of cytological preparations, devised by Pardue and Gall (1969), or of RNA and *in situ* DNA in tissue sections (Buongiorno-Nardelli and Amaldi, 1970), or in squash preparations (Gall and Pardue, 1969) or also on single chromosomes (Jones, 1970) are opening new and unexpected possibilities in the research of homology between the chromosomes of different species without causing severe ethical controversies.

REFERENCES

A proposed standard system of Nomenclature of Human Mitotic Chromosomes. Denver, Colorado, 1960. *In* "Birth Defects: Original Article Series", **2**, 2 (The National Foundation, New York 1966).

The London Conference of Normal Human Karyotype. *In* "Birth Defects: Original Article Series", **2**, 2 (The National Foundation, New York, 1966).

Chicago Conference: Standardization in Human Cytogenetics. *In* "Birth Defects: Original Article Series", **2**, 2 (The National Foundation, New York, 1966).

Anneren, G., Berggren, A., Stahl, Y. and Kjessler, B. (1970). Preparation methods for human pachytene chromosomes. *Hereditas* **64**, 211–214.

Buongiorno-Nardelli, M. and Amaldi, F. (1970). Autoradiographic Detection of Molecular Hybrids between RNA and DNA in Tissue Sections. *Nature, Lond.* **225**, 946–948.

Bender, M. A. and Chu, E. H. Y. (1963). The chromosomes of Primates. *In* "Evolutionary and Genetic Biology of the Primates" (Buettner-Janusch, ed.), Vol. I, pp. 261–310. Academic Press, New York.

Caspersson, T. (1970). Identification of the human Chromosomes by DNA-Binding Fluorescent Agents. *Chromosoma* **30**, 215–227.

Chen, A. T. and Falek, A. (1969). Centromeres in Human Meiotic Chromosomes. *Science* **166**, 1008–1010.

Chiarelli, B. (1961). Ibridologia e sistematica nei Primati. 1. Raccolta di dati. *Atti A.G.I.* **6**, 213–220.

Chiarelli, B. (1962a). Comparative morphometric analysis of the Primates chromosomes 1°. The chromosomes of the Anthropoid Apes and of Man. *Caryologia* **15**, 99–121.

Chiarelli, B. (1962b). Comparative morphometric analysis of the Primate chromosomes. 2°. The chromosomes of the genera *Macaca, Papio, Theropithecus* and *Cercocebus. Caryologia* **15**, 401–420.

Chiarelli, B. (1963a). Primi risultati di ricerche di genetica e cariologia comparata in Primati e loro interesse evolutivo. *Riv. Antropol.* **50**, 87–124.

Chiarelli, B. (1963b). Observations on P.T.C.-Tasting and on Hybridization in Primates. *Symp. Zool. Soc. London* **10**, 277–279.

Chiarelli, B. (1964). Genetica comparata in Primati e genetica umana. (Relazione tenuta all'XI Convegno della salute. Ferrara) 53–79.

Chiarelli, B. (1966). Caryology and Taxonomy of the Catarrhine Monkeys. *Am. J. Phys. Anthr.* **24**, 155:170.

Chiarelli, B. (1966a). Marked chromosome in Catarrhine monkeys. *Folia Primat.* **4**, 74–80.

Chiarelli, B. (1966b). Interesse tassonomico e filogenetico dei dati cariologici e ibridologici nei Primati del Vecchio Mondo. *Boll. Zool., Fasc.* **II**, 381–387.

Chiarelli, B. (1967a). La morfologia del cromosoma Y in differenti specie dei Primati. *Riv. Antropol.* **54**, 3–6.

Chiarelli, B. (1967b). Ibridologia e Sistematica in Primati. Deduzione sul cariotipo degli Ibridi. *Riv. Antropol.* **53**, 112–117.

Chiarelli, B. (1968). Dati comparativi preliminari sui cromosomi meiotici in diacinesi di alcuni Primati. *Atti Ass. Gen. Ital.* **13**, 183–192.

Chiarelli, B. (1968a). Caryological and hybridological data for the taxonomy and phylogeny of the Old World Primates. *In* "Taxonomy and Phylogeny of the Old World Primates" (Chiarelli, ed.) Rosenberg and Sellier, Torino.

Chiarelli, B. (1968b). Chromosome polymorphism in the species of the genus *Cercopithecus. Cytologia* **33**, 1–16.

Chiarelli, B. (1968c). From the karyotype of the Apes to human karyotype. *South Afr. J. Sci.* **64**, 72–80.

Chiarelli, B. (1970). The chromosomes of Chimpanzee. *In* "The Chimpanzee" (Bourne, ed.), Vol. 2, pp. 254–264. Karger, Basle, New York.

Chiarelli, B. (1971). The chromosomes of the Gibbons. *In* "The Gibbons" (D. M. Rumbough, ed.), Karger, Basel (in press).

Chiarelli, B. and De Carli, L. (1959). Evoluzione sulle conoscenze sui cromosomi umani e loro significato in Antropologia. *Arch. Antr. ed. Etn.* **89**, 148–167.

Chiarelli, B. and Falek, A. (1967). I cromosomi meiotici dell'Uomo. *Arch. Antr. Etnol.* **97**, 47–52.

Chiarelli, B. and Vaccarino, C. (1964). Cariologia ed evoluzione nel genere *Cercopithecus. Atti Ass. Gen. It.* **9**, 329–339.

Chu, E. H. Y. and Bender, M. A. (1962). Cytogenetics and evolution of Primates. *Ann. N.Y. Acad. Sci.* **102**, 253–266.

Darlington, C. D. (1937). "Recent Advances in Cytology", 2nd edition, p. 671. Churchill, London.

Darnell, J. E., Jr. (1958). Ribonucleic acid from animal cells. *Bact. Rev.* **32**, 262–290.

Du Praw, E. J. (1965). Macromolecular organization of nuclei and chromosomes: a folded fibre model based on whole-mount electron microscopy. *Nature, Lond.* **206**, 338–343.

Du Praw, E. J. (1966). Evidence for a "folded-fibre" organization in human chromosomes. *Nature, Lond.* **209**, 577–581.

Eberle, P. (1963). Meiotische Chromosomen des Mannes. *Klin. Wschr.* **41**, 848–856.

Ferguson-Smith, M. A. (1964). The sites of nucleolus formation in human pachytene chromosomes. *Cytogenetics* **3**, 124–134.

Gall, J. G. and Pardue, M. L. (1969). Formation and detection of RNA-DNA hybrid molecules in cytological preparations. *Proc. Nat. Acad. Sci.* **63**, 378–383.

Gavin, J., Noades, J., Tippett, P. Sanger, R. and Race, R. R. (1964). Blood Group Antigen Xg[a] in Gibbons. *Nature, Lond.* **204**, 1322–1323.

German, J. (1962). III. DNA synthesis in human chromosomes. *Trans. N.Y. Acad. Sci.* **24**, 395–407.

German, J. (1964a). Identification and characterization of human chromosomes by DNA replication sequence. Symp. Int. Soc. Cell Biology, Vol. 3, pp. 191–207. Academic Press, New York.

German, J. (1964b). The pattern of DNA synthesis in the chromosomes of human blood cells. *J. Cell. Biol.* **20**, 37–55.

Hamerton, J. L. (1961). Sex chromatin and human chromosomes. *Int. Rev. Cytol.* **12**. In press.

Hamerton, J. L., Klinger, H. P., Mutton, D. E. and Lang, E. M. (1963). The somatic chromosomes of the Hominoidea. *Cytogenetics* **2**, 240–252.

Hoyer, B. H., McCarthy, B. J. and Bolton, E. T. (1964). A Molecular Approach in the Systematics of Higher Organisms. *Science, N.Y.* **144**, 959–967.

Hsu, T. C., Schmid, W. and Stubblefield, E. (1964). DNA replication sequences in higher animals. *In* "The Role of Chromosomes in Development" (Locke, ed.), pp. 83–112. Academic Press, New York.

Huang, C. C. (1968). Karyologic and autoradiographic studies of the chromosomes of Rattus (*Mastomys*) natalensis. *Cytogenetics* **7**, 97–107.

Huang, C. C., Habbitt, H. and Ambrus, J. L. (1969). Chromosomes and DNA synthesis in the stumptail monkey (*Macaca speciosa*), with special regard to marker and sex chromosomes. *Folia Primat.* **11**, 28–34.

Hughes, D. T. (1968). Density Patterns in Human Chromosomes. *Lab. Pract.* **17**, 800–803.

Ivanoff, E. (1930). L'insemination artificielle des Mammiferes en tant que methode scientifique et zootechnique. *Bull. Acad. Vet. Fr.* **3**, 49.

John, B. and Lewis, K. R. (1968). The chromosome complement. Protoplasmatologia (Springer-Verlag) **4**, 1–206.

Jones, K. W. (1970). Chromosomal and Nuclear Location of Mouse Satellite DNA in Individual Cells. *Nature, Lond.* **225**, 912–915.

Kikuchi, Y. and Sandberg, A. (1963). Chronology and pattern of human chromosome replication. I. Blood leucocytes of normal subjects. *J. Nat. Cancer Inst.* 1109–1143.

Kohne, D. E. (1968). Taxonomic Applications of DNA Hybridization Techniques. *In* "Chemotaxonomy and Serotaxonomy" (Hawkes, ed.), pp. 117–130. Academic Press, London.

Levan, A. and Hsu, T. C. (1959). The Human Idiogram. *Hereditas* **45**, 665.

Lima de Faria, A. (1964). DNA replication in human chromosomes – a review; *In* "Mammalian Cytogenetics and Related Problems in Radiobiology", pp. 31–37. Pergamon Press, London.

Low, R. J. and Benirschke, K. (1968). Chromosome study of a marmoset hybrid. *Folia Primat.* **8**, 180–191.

Low, R. J. and Benirschke, K. (1969). The replication pattern of the chromosomes of *Pan troglodytes*. Proc. 2nd int. Congr. Primat., Atlanta, GA 1968, Vol. 2, pp. 95–102. Karger, Basle, New York.

Manfredi-Romanini, M. G. (1968). Quantitative relative determination of the nuclear DNA in the Old World Primates. *In* "Taxonomy and Phylogeny of the Old World Primates with References to the Origin of Man" (B. Chiarelli, ed.), pp. 139–150. Rosemberg and Sellier, Turin.

Manfredi Romanini, M. G. (in press). Nuclear DNA content and Area of Primate Lymphocytes as a cytotaxonomic tool. *J. Hum. Evol.* **1**

Manfredi-Romanini, M. G. and Fontana, F. (1968). Nuclear DNA content (a.u.) in some species of *Lemuroidea*. *Riv. Antropol.* **55**, 85–94.

Martin, M. A. and Hoyer, B. H. (1967). Adenine plus Thymidine and Guanine plus Cytosine enriched Fractions of Animal DNA's as Indicators of Polynucleotide Homologies. *J. molec. Biol.* **27**, 113–129.

McClure, H. M., Belden, K. H. and Pieper, W. A. (1970). Autosomal Trisomy in a Chimpanzee: Resemblance to Down's Syndrome. *Science, N.Y.* **165**, 1010–1012.

Moorhead, P. S. and Vittorio, D. (1963). Asynchrony of DNA synthesis in chromosomes of human diploid cells. *J. cell Biol.* **16**, 202–209.

Nicoletti, B. (1968). Il controllo genetico della meiosi. *Atti Ass. Genet. It.* **13**, 3–71.

Nuzzo, F., De Carli, L. and Chiarelli, B. (1961). Analisi morfologica degli assetti cromosomici in un gruppo di cellule umane stabilizzate in vitro. *Atti Ass. Gen. Ital.* **6**, 11–124.

Pardue, M. L. and Gall, J. G. (1969). Molecular hybridization of radioactive DNA to the DNA of cytological preparations. *Proc. Nat. Acad. Sci.* **64**, 600–604.

Schmidt, W. (1963). DNA replication patterns of human chromosomes. *Cytogenetics* **2**, 175–193.

Schneider, E. L. and Salzman, N. P. (1970). Isolation and Zonal Fractionation of Metaphase Chromosomes from Human Diploid Cells. *Science, N.Y.* **167**, 1141–1143.

Siniscalco, M. (1963). Localization of genes on human chromosomes. *In* "Genetics Today", Vol. 3, pp. 851–870. Pergamon Press, Oxford.

Slizynski, B. M. (1964). On human pachytene chromosomes. *In* "Mammalian Cytogenetics and Related Problems in Radiobiology" (Pevan, Chagas, Frota-Pessoa, Caldas, eds), pp. 171–186. Pergamon Press, Oxford.

Smith, I. (1968). Indoles, Amino Acids and Imidazoles as an Aid to Primate Taxonomy. *In* "Chemotaxonomy and Serotaxonomy" (Hawkes, ed.), pp. 29–37. Academic Press, London.

Swanson, C. P. (1960). "Cytology and cytogenetics". McMillan, London.

Taylor, J. H. (1959). Autoradiographic studies of the organization and mode of duplication of chromosomes. *In* "Symposium on Molecular Biology" (Zirkle, ed.). University of Chicago Press, Chicago, Ill.

Valencia, J. I. (1964). Discussion on the B. M. Slizynski paper: On human pachytene chromosome, pp. 186–187. *In* "Mammalian Cytogenetics and Related Problems in Radiobiology" (Pevan, Chagas, Frota-Pessoa, Caldas, eds). Pergamon Press, Oxford.

Vialli, M. (1957). Volume et contenu en ADN per noyau. *Exp. Cell Res.* Suppl. **4**, 249–284.

White, M. J. D. (1970). Heterozygosity and genetic polymorphism in parthenogenetic animals. *In* "Essay in Evolution and Genetics" (K. Hecht and W. G. Steere, eds). North Holland Publ. Co., Amsterdam.

Wurster, D. H. and Benirschke, K. (1969). Chromosomes of some Primates. Mammalian chromosome Newsletter 10:3.

Yeager, C. H., Painter, T. S. and Yerkes, R. M. (1940). The chromosomes of the chimpanzee. *Science, N.Y.* **91**, 74–75.

Yerganian, G. (1963). Chromosome cytology of medical anomalies. *In* "Radiation-induced Chromosome Aberrations" (S. Wolff, ed.), p. 259. Columbia University Press, New York.

Young, W. J., Mergt, T., Ferguson-Smith, M. A. and Johnston, A. W. (1960). Chromosome number of the chimpanzee (*Pan troglodytes*). *Science, N.Y.* **131**, 1672–1673.

ADDENDUM

While this paper was in press an interesting note was published on the use of Quinaquine mustard in identifying the Y chromosome (Pearson *et al.*, 1971, in *Nature* **231**, 326–329). The authors are also comparing other chromosomes using the same technique. The results seem to contradict the hypothesis of the origin of chromosome No. 1 in Man by the centric fusion mechanism and point towards the non-homology of chromosome No. 5 in the human karyotype with those of the Apes (personal communication).

These results, although in agreement with our original interpretation (Chiarelli, 1962) are still debatable because of the as yet nuclear physiological meaning of the banding. Research is in progress in our laboratory to demonstrate similar banding patterns for the whole karyotype using a number of different techniques, especially those including NaOH and trypsin treatments.

Moreover within the past few months, thanks to the cooperation of Dr. A. Chen of the Department of Psychiatry of Emory University, I was able to study 5 meiotic diakinesis of *Pan troglodytes*. The bivalents appear to have a shape very similar to those of the human and the number of total chiasmata ranges between 46 and 52.

Concluding Remarks

N. A. BARNICOT

Department of Anthropology, University College London, England

The contributions are by such a wide spectrum of experts that it would be both difficult and presumptuous to attempt a systematic and complete summary. I shall therefore confine myself to some general questions arising from the meeting.

Perhaps it is worth asking why we do comparative work. The first, and most obvious, answer is that we hope that an examination of the resemblances and differences between living primates will help us to reconstruct the phylogeny of the order, a phylogeny of special interest since it includes our own lineage. Here we are following the logic of post-Darwinian comparative anatomy, though often with new, and we hope more powerful, methods of observation and analysis. There are well-known difficulties in tracing historical pathways from the evidence of living forms alone and I shall discuss these later.

However, even if we happen to be less interested in phylogeny than in the workings of organs, tissues or molecules it is still worthwhile examining a wide range of species. By seeing many variations on a theme we may appreciate what is essential to the system and the extent to which it can be modified to cope with special circumstances. For example, evolutionary divergence has produced many versions of haemoglobin differing from one another in various ways and to varying degrees. We can regard them as the results of so many ready-made experiments and by comparing their structure and properties we can learn something about the design of the molecular machine. We find that some parts are seldom changed and this strongly suggests an essential function. Taken together with data on rare deleterious mutants the comparative evidence is complementary to purely physico-chemical studies on the inter-relations of molecular structure and function. A welcome bonus of comparative work is that we may sometimes find that a certain feature of the system is uniquely visible or accessible to experiment in particular species. The dipteran salivary chromosomes, the urodele lampbrush chromosomes and the giant axons of squids are good examples. Some may feel that this type of comparative study is opportunistic and shallow, merely using the abundant variety of nature as an adjunct to anatomical or physiological research and ignoring the wider significance; but historical questions are never far away. If we ask why a system has the form it does and not some other, part of the answer must surely lie in its antecedents.

The third reason for comparative work, already mentioned by Kalmus, is more practical and applies specially to primates or at least to some primates. Certain branches of medical research require experimental animals with responses as close as possible to those of Man. The successful transmission of the slow virus of Kuru (a central nervous disease previously obscure in aetiology) to chimpanzees is a case in point and other examples could be found in transplantation and brain research. Because of their relatively close relationship to Man we are led to exploit the great apes and certain monkey species even to the point of threatening their survival in the wild. Conservation problems were not on our programme but it is worth recording that many of us were worried by this situation and discussed possible remedies unofficially.

This discussion was, however, about comparative *genetics* rather than comparative primatology in general. A comparative geneticist should presumably compare genes in various species, but until quite recently there was no satisfactory way of doing this. If, for example, we found a very similar inherited variation of coat colour in several species we could investigate its genetics in each of them but we could not be sure that the same locus, let alone the same allele, was involved in all the species. We know in fact that the effects of different genes can be very much alike at phenotypic levels remote from their sites of primary action. Unless we happened to be dealing with genera or species that can be successfully crossed we could not pursue the analysis further by breeding experiments and even if this were possible it would be slow and costly work in the case of most primates.

This situation was radically changed by the discovery of the coding mechanism of DNA and its role in protein synthesis and by the development of methods for analysing the amino acid sequences of proteins. Genes can be identified by the polypeptide chain they produce, the sequence of the chain reflecting the base sequence of the gene. By examining the amino acid sequence of a given protein in a wide range of organisms gene homologies can be inferred and the amount of change that has occurred in evolutionary divergence can be estimated. Admittedly code degeneracy limits the accuracy with which gene structure can be inferred from protein structure alone but nevertheless molecular biology has opened the way to a much more precise and wide-ranging comparative anatomy of genes than was conceivable by the classical methods of genetics. Goodman and Koen have already given a lucid account of these matters.

The blood groups were probably the first examples of common human variations that obey mendelian rules with gratifying precision. The chemical basis of these serological traits has been hard to unravel but it now seems clear that ABH specificities depend on the pattern of terminal sugar residues on the carbohydrate chains of these muco- or lipopolysaccharides. The genes are thought to act on enzymes involved in the synthesis of these chains so that the variations we detect serologically are at least one step away from their primary sites of action. In their monumental studies of primate blood-group specificities Wiener and Moor-Jankowsky have shown that some unexpected and interesting complications may emerge when the red cells of apes and monkeys are studied with human grouping reagents. Thus ABH specificities cannot be detected at all on Old World monkey red cells and, strangely enough from a taxonomic viewpoint, a similar phenomenon turns up in the gorilla. Perhaps some of these anomalies would disappear if the effects of the genes on the enzymes themselves could be studied.

These authors have drawn attention to our lack of family data on lower primates. They were obliged to take an alternative approach and to test their genetic hypotheses by examining the population frequencies of phenotypes on the assumption that Hardy-Weinberg equilibrium applies. Often enough this works out quite well both with blood groups and simple electrophoretic variations of red cell and serum proteins; but sometimes it does not and we are then left wondering whether the genetic hypothesis is wrong or whether the assumption of equilibrium is unjustified. Departure from equilibrium due to the mating system or oddities of sampling can be checked by examining several different protein polymorphisms since all should be affected, but this would not be so if disequilibrium were due to strong selection at a particular locus. Many primates live in small social groups that may well be inbred because in some species they contain only one male while in others the genetic contributions of the males may be very unequal due to a strong male dominance hierarchy. In this case samples of the population as a whole would not be in equilibrium. The fact that a reasonable fit to Hardy-Weinberg expectations is quite often found may be partly due to significant gene flow between breeding groups and partly to the insensitivity of this method of detecting deviations from random-mating. Another possible source of confusion, to which Sullivan drew attention, is that taxonomic identification of the animals may sometimes be faulty so that more than one species may be represented in the sample.

Even haemoglobins may present complex genetical problems which would be easier to solve if we had family data. In certain macaque populations, and probably in some other genera, there are evidently gene duplications (Barnicot et al., 1966, 1970 and Sullivan in this volume) so that the frequencies of individuals with 2 types of haemoglobin are far in excess

L

of the frequency expected if they were simple heterozygotes. The wealth of variation in the isoenzyme bands of the red-cell phosphoglucomutase in *Macaca fascicularis* (Ishimoto *et al.*, 1968; Barnicot and Cohen, 1970) is another case where interpretation would be easier given family material.

The immunoglobulins are a class of proteins of exceptional complexity and interest and Carbonara (this volume) gives an expert account of them. The most intense interest is focused on the variable regions of the light and heavy chains which evidently confer antibody specificity. The conformation of these regions and the mechanism by which such high variability arises are still unclear. Identification of different chain types has depended on serology but chemistry is beginning to catch up. Serological comparisons of the heavy chains of the IgG fraction as between Man and other primates shows that the γ_1 chain is quite similar to human in apes, monkeys and even prosimians but that γ_2 and γ_3 can only be detected in Man and apes and γ_4 only in Man. Again the heavy (α) chain of IgA in apes, monkeys and prosimians seems to be much like that of Man but the μ chain of I_gM is substantially different from human in prosimians and even in the gibbon. It appears that the structure of these various chains has been stabilised to very varying degrees in primate evolution. However it must be remembered that the tests are for determinants present in Man and do not necessarily give an accurate picture of the extent of differentiation between non-human primates themselves. Many of the numerous Gm specificities that are found in Man on certain IgG heavy chains have also been detected in apes and some even in Old World monkeys. The Gm (a) specificity is known to depend on substitutions at 2 nearby sites on the γ_1 heavy chain. In gorilla, chimpanzee and Man Gm (a) has aspartic acid and leucine at these 2 sites but in baboons and rhesus monkeys glutamic acid is substituted for aspartic.

We may now turn briefly to a number of characters that for one reason or another are less satisfactory than biochemical traits for comparative genetics.

The chemical specificity of taste-insensitivity to PTC and related compounds suggests that it may be due to some small steric change in a receptor molecule but at present we have no direct chemical test and must rely on tests of perception. Casual observations on the responses of lower primates are not always easy to interpret and time-consuming training routines would be needed to get really convincing results.

The epidermal ridges on the palms and soles of primates are adaptive to arboreal life and they also occur in some tree-living marsupials. Presumably they increase frictional resistance and perhaps also tactile sensitivity. In Man, individual variations of pattern and their varying frequencies in populations have been studied for many years but as Mavalwala emphasized, the subject has tended to become overburdened with a mass of descriptive detail of uncertain relevance. Holt (1968) showed that variations of digital ridge-count are largely due to genetical differences but the effects of individual genes could not be discerned. Individual variations of pattern broadly comparable to those in Man certainly occur in lower primates, at least in apes and some Old World monkeys (Cummins and Midlo, 1943). Interest in dermatoglyphics had quickened recently because the patterns are found to be affected by certain chromosomal abnormalities. Evidently the patterns are determined early in embryonic life. Penrose (1965) sees them as topological consequences of covering curved surfaces with parallel lines, the ridges tending to follow directions of greatest curvature on the early limb rudiments. Possibly an experimental approach might be feasible using cultures of limb buds. If so primate embryos, or perhaps marsupials, which are born in a very immature state, would be needed.

The Berrys told us about a type of skeletal variation that is deceptively simple at the phenotypic level since it can be scored as presence or absence of certain anomalies, but is nevertheless very complex genetically. Threshold effects are superimposed on a multifactorial background and furthermore in pure lines of mice, in which these "quasi-continuous" variations were first analysed, dietary modifications can affect the manifestation of the traits. The fact that population differences in frequency can be demonstrated when more laborious measurements of bones show little or nothing, may be due to an amplification effect resulting from the sensitivity of the phenotype to small genotype changes near the threshold. As they have shown, variations of the kind occurring in mice are also found in men and apes but they wisely refrained from claiming genetical homology. It is interesting that Schultz (1956) reported that digital and other skeletal anomalies that would be expected to be crippling in arboreal life may be equally or even more frequent in various wild primates than in Man. The Berrys have combined their work on complex traits in natural mouse populations with observations on certain blood protein variants and have demonstrated interesting season fluctuations of gene frequencies. Anyone who, like myself, would like to apply biochemical methods to the study of population genetics in non-human primates, can only envy those who work on a species so prolific and easy to trap as the mouse.

With chromosome studies we seem to enter a region somewhere between morphology and genetics. This of course is a distortion of perspective due to method since it is not inconceivable that one day we may be able to recognize the sequence of genes on chromosomal strands by direct observation. As Chiarelli's review showed us, an impressively large proportion of living primate species have now been examined cytogenetically. Once again we note that comparative studies have proliferated as a result of technical advances which were first applied in the human field. The light microscopist sees the chromosomes effectively only at 2 special phases of their activity when they are relatively condensed; at mitosis when they are concerned with the even distribution of genetic material between daughter cells and at meiosis when synapsis, crossing-over and reduction occur. The detailed mechanisms of these essential processes are still largely obscure but, whatever structural changes may occur in evolution the chromosomes must remain capable of performing them, and, as Chiarelli said, in sexual forms, meiosis acts as a filter such that rearrangements leading to defective gametes are subjected to negative selection. Apart from this does it matter how the genes are arranged on the chromosome? Apparently it may, since White (1970) has shown that in parthegenetic species, where no meiotic filter is interposed, selection for certain chromosomal mutations can still occur. The nature of such position effects at the molecular level is obscure.

To what extent changes of karyotype parallel divergence at the gene level is not entirely clear. Certainly some primate species, such as the chimpanzee, gorilla and Man, that are held to be closely related on bio-chemical and other grounds, have quite similar karyotypes. On the other hand the macaques have a very similar karyotype to the baboons, manga-beys and gelada baboon but their haemoglobin appears to differ widely in structure from that of the latter group (Barnicot and Wade, 1970 and un-published). In general one might expect karyotype divergence to increase with time of separation but the strange case of the 2 bush-baby species, *Galago senegalensis* and *G. crassicaudatus* shows (unless taxonomy has gone badly astray) that large changes can sometimes occur in a relatively short period. Comparison of relative lengths of chromosomes and of arm ratios is, of course, a rather crude method of detecting structural changes and apparent similarity may conceal many changes at a level beyond the visi-bility of the light microscope and yet far from the level of single genes.

As Chiarelli pointed out, various methods are now coming into use that may help to determine chromosome homologies in primates. These include radio-labelling, fluorescent staining and cell fusion methods. The study of meiotic material in interspecific or intergeneric crosses is feasible in some primate groups and preliminary results should obviously be followed up. As Goodman remarked during the meeting, the implications of repetitive DNA, which may comprise quite a large proportion of the genome, for studies of the karyotype are at present unexplored.

I want to return now to the problem of reconstructing phylogenies from data on living species. If the fossil record was sufficiently complete we could trace the lineages of living primates back to ancestral forms and obtain estimates of divergence times by geological methods. Even so we should only have direct evidence of the course of evolutionary change in the skeleton and whatever can be inferred about soft parts from the bones.

We are not likely to get fossil haemoglobins, chromosomes or even finger-prints.

In fact, the fossil record is still very incomplete so that inferences about major aspects of primate phylogeny based on comparative anatomy, bio-chemistry or karyology still command attention. There are long time spans with few or no known fossils, for example, between the Miocene apes and modern pongids and between Eocene and living prosimians. We are there-fore obliged to make sizeable extrapolations, often on rather subjective grounds. In addition some fossil forms are represented only by fragments of limited regions of the skeleton.

In general we expect forms derived from a common ancestor to become more dissimilar as time passes; but they will retain some features that betray their common ancestry because not all parts of the body change equally fast. The extent of divergence, assuming that we can measure it satisfactorily will not necessarily have a simple relation to time. Palaeon-tology provides many examples of widely varying rates of change both within and between lineages (Simpson, 1944, 1949, 1953). Furthermore, resemblances are not an infallible guide to relationship because similar features may sometimes be acquired independently in different lineages, usually by adaptive response to similar conditions of life.

It is important to note that evolutionary change includes 2 processes, progressive modification within a lineage and diversification by which new lineages arise. The same forces of selection, drift and mutation are opera-tive in each but the latter involves the additional factor of reproductive isolation. The 2 processes may be intercorrelated as when deployment into a new environment leads to adaptive diversification and rapid divergence of the lineages so formed. A related point is that the term relationship is ambiguous; in Fig. 1 A and B are more closely related in so far as they are

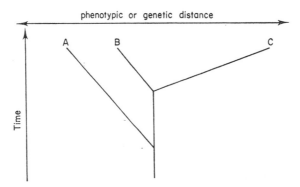

FIG. 1. Remarks and prospectives for comparative genetics between non-human primates and Man.

more alike phenotypically or genetically but B and C are more closely related in that they have the more recent common ancestor.

Change in the genome is, in a sense, the most fundamental aspect of evolutionary divergence, but changes in the form of complex organs tell us little about it. Procedures such as DNA hybridization and protein sequency are much more informative, the first in relation to overall change in the genome and the second in relation to change at specific loci.

Protein sequences lend themselves readily to quantitative comparisons of evolutionary divergence since the number of changed sites can be counted. By referring to the genetic code estimates of the number of mutational changes can be derived; but these estimates are necessarily minimal because we can only see the results of the last change and an indefinite number may have preceded it.

It would be meaningless from a phylogenetic point of view to compare proteins that are not derived from a common ancestral protein. When we find that a functionally similar polypeptide from 2 species has, say, 50% of identical sites it seems intuitively obvious that homology is the best explanation, but what if there are only 5% of identities ? Margoliash *et al.* (1968) suggest a statistical approach to this problem. It depends on examining whether similarities between chains are greater than would be expected by chance resemblance. Very often functionally similar chains show identity of certain sites in a wide range of species and in some cases there is independent evidence that they play some essential functional role. For this reason, once such conformations are achieved, they are likely to persist for long periods due to the stabilizing action of selection and should be good indicators of homology. However, as these authors argue, their adaptive value also makes it more likely that they could arise by convergence. They therefore eliminate such sites and base the test of homology on the remainder. Whether sites are really divided into two sharp categories is open to question; perhaps a gradation with respect to tolerance of change is more likely. Tolerance of change should not be regarded as a fixed property of an individual site as such, since there may be interactions such that change in one is contingent on change in others. Awkward problems of homology are not necessarily confined to proteins of remotely related species. Minor haemoglobins are present in both hominoids and ceboids. Their δ chains are closely similar to β chains and doubtless arose by duplication of these; but compositional evidence as to whether they were derived from the same or different duplications is equivocal (Boyer *et al.*, 1969).

Assuming that problems of homology of chains and sites have been solved we can construct a matrix of mutational distances between species. How can we use such data to make phylogenetic trees ? Various statistical methods have been suggested (Cavalli-Sforza and Edwards, 1967; Fitch and Margoliash, 1967; Dayhoff, 1969). We have seen such trees exemplified at this meeting and have also heard some disapproval of them. Cavalli-Sforza and Edwards, who were perhaps the first to explore this field, were certainly aware of the difficulties and limitations. Common to most of these methods is the assumption that the optimal tree, among many possible ones, is that involving the least evolutionary change. This is difficult to justify *a priori* and may be untrue in particular instances. It is therefore comforting to find that the resulting trees resemble those based on comparative anatomy and palaeontology quite closely. Nevertheless there is an important difference between them. The former have no time scale; they are diagrams of mutual relationship in terms of mutational or phenotypic distance. The branching patterns in the Fitch procedure is generated from comparisons of mutational distance and a temporal phylogeny like the one shown in the figure above would be misrepresented. A large mutational distance can mean either a long time lapse or higher rates of divergence. Dayhoff's method again assumes minimal paths but involves reconstruction of ancestral sequences. In essence the argument is that, at a given site, the amino acid present in the majority of descendants is most likely to have been present in the ancestor. Some bias may result if descendant lines are unequally represented. In some phylogenies many of them may in fact be extinct and they might have given contrary information.

Although the limitations of these methods may sometimes lead to wrong conclusions about the evolution of a particular protein it is reasonable to suppose that such errors will become less important if data from several proteins are combined.

In order to supply a time-scale we need either empirical data from palaeontology or a theoretical basis for relating mutational distance to time. It has been suggested that protein evolution, unlike morphological evolution, may proceed at relative constant rates because it is largely due to random fixation of selectively neutral mutations. Kimura (1968) was led to this conclusion by considering the likely rate of substitution in the genome as a whole. His figure was so high that, in his view, the resulting genetic load would be intolerable if change were due to selection. His argument has been criticised by Maynard Smith (1968) and by King and Jukes (1969) who put forward other reasons for believing in protein evolution by random processes.

A common but weak argument is that many mutations seem not to affect the properties of a protein; but this may merely mean that our methods are insensitive or inappropriate. Extrapolations from test-tube experiments to the much more complex conditions in living cells may be questionable. Boyer *et al.* (1969) suggested that haemoglobin minor components can have little effect on oxygen carriage since they are present in such low concentrations, and they may therefore have experienced little selective pressure. But as Sullivan (this volume) points out, haemoglobin binds other things besides oxygen. In any case a component present in small amounts might still exert significant efforts by interactions. Sullivan's own work on oxygen dissociation parameters in various primates and Goodman and Koen's studies on LDH point the way to a more functional approach without which studies on the comparative anatomy of proteins are apt to become somewhat formalistic.

Both Goodman (this volume) and Sarich and Wilson (1967) have commented on the close similarities between certain proteins of the great apes, especially the chimpanzee, and those of Man; but they drew different conclusions. The latter authors assumed constant rates of change and thus arrived at divergence times much more recent than palaeontologists would accept. Read and Lestrel (1970) have pointed out that their equation relating serological distance to time was based on slender evidence and equations that would yield larger divergence times are certainly not excluded.

Goodman, on the other hand, thinks that protein evolution has slowed down, not only in hominoid but also in other anthropoid lineages and he offers certain explanations. While admiring his ingenuity I propose to assume the role of devil's advocate. He invokes the concept of anagenesis which is understood to mean a process by which more complex, highly integrated systems evolve. Although this concept has meaning when the evolution of animals is viewed on a very wide scale, I must confess that I find it nebulous. It is not easy to believe that anthropoid proteins are really much more delicately adjusted machines than are those of, say, prosimians and that change is inhibited for this reason. Even if evolutionary change in certain hominoid proteins has decelerated it is apparent that the lineage did not lose the capacity for rapid change in other respects. We cannot say how long it took to evolve bipedal locomotion but changes in the face and braincase seem to have taken place remarkably rapidly. The idea that speciation has failed to occur in *Homo sapiens* because anagenesis has reached a culminating point in this species seems superfluous. Man is a very mobile species with unusual powers of communication and the time span is not so long as to make absence of speciation remarkable. There is clearly quite a lot of genetical variability between human populations but reproductive isolation has not supervened.

Goodman has also suggested that deceleration of protein evolution may have been due to increasing length of generations. This would presumably affect genetic change as a whole. If mutation rates *per generation* tend to be roughly constant in various organisms, incorporation of mutational changes by drift should indeed be slower and this should also be true of selection, which is generation–time dependent. Nevertheless, palaeontology seems to provide little or no evidence of any relation between generation–time and evolutionary rates (Simpson, 1949). The American opossum is a good example of a mammal with a short generation-time which, as far as we can tell from the bones, has undergone little change during the Tertiary and the elephants despite large size and presumably long generation-times, were able to produce quite a complex radiation of forms.

Another of Goodman's suggestions is that the presence of a haemochorial placenta in anthropoids has been a factor restricting change in certain proteins. It is assumed, mainly on anatomical grounds, that this type of placenta forms a less effective barrier than the epitheliochorial placenta to the passage of foetal antigens into the mother and of maternal antibodies into the foetus so that immunological incompatibilities may arise and have deleterious effects. Whether such a mechanism would affect cell proteins as well as plasma proteins is not clear. As far as the red cell Rh antigens are concerned it is known that in Man foetal cells enter the maternal blood stream largely at parturition; it is possible that transfer of red cells at this time is more likely to happen with a haemochorial than with an epithiliochorial placenta. It should be noted that the haemochorial type of placenta is not confined to anthropoids and is indeed quite widespread in mammals, for instance in rodents. It would be interesting to know more about rates of protein evolution in this group which, far from showing any evolutionary restriction, is one of the most diverse and successful of mammalian orders.

REFERENCES

Barnicot, N. A., Huehns, E. R. and Jolly, C. J. (1966). Biochemical Studies on haemoglobin variants of the irus macaque. *Proc. Roy. Soc. B.* **165**, 224–2444.
Barnicot, N. A., Wade, P. T. and Cohen, P. (1970). Evidence for a second haemoglobin α-locus duplication in Macaca irus. *Nature, Lond.* **228**, 379–381.
Barnicot, N. A. and Wade, P. T. (1970). Protein structure and the systematics of old world monkeys: *In* "Old World Monkeys, Evolution Systematics and Behaviour" (J. R. and P. H. Napier, eds). Academic Press, New York and London.
Barnicot, N. A. and Cohen, P. (1970). Red cell enzymes of Primates (Anthropoidea). *Biochem. Genet.* **4**, 41–56.
Boyer, S. H., Crosby, E. F., Thurmon, T. F., Noyes, A. N., Fuller, G. F., Leslie, S. E. and Shepard, M. K. (1969). Haemoglobins A and A2 in New World Primates: Comparative variations and its evolutionary implications. *Science* **166**, 1428–1431.
Cavalli-Sforza, L. L. and Edwards, A. W. F. (1967). Phylogenetic analysis Models and Estimation procedures. *Am. J. Hum. Genet.* **19**, 233–257.

Cummins, H. and Midlo, C. (1943). "Fingerprints, Palms and Soles". The Blakiston Co., Philadelphia.

Dayhoff, Margaret O. (1969). Atlas of protein sequence and structure. National Biomedical Research Foundation. Silver Spring, Md., U.S.A.

Fitch, W. M. and Margoliash, E. (1967). Construction of phylogenetic trees. *Science* **155**, 279–284.

Holt, S. (1968). Genetics of Dermal Ridges. C. C. Thomas, Springfield, Illinois.

Ishimoto, G., Toyomasu, T. and Uemara, K. (1968). Intraspecies variations of red cell enzymes and haemoglobins in Macaca irus. *Primates* **9**, 395–408.

King, J. L. and Jukes, T. H. (1969). Non-Darwinian evolution. *Science* **164**, 788–798.

Kimura, M. (1968). Evolutionary rate at the molecular level. *Nature, Lond.* **217**, 624–626.

Margoliash, E., Fitch, W. M. and Dickerson, R. E. (1968). Molecular expression of evolutionary phenomena in the primary and tertiary structure of cytochrome c. *Brookhaven Symp. Biol.* **21**, 259–305.

Maynard-Smith, J. (1968). "Haldane's dilemma" and the rate of evolution. *Nature, Lond.* **219**, 1114–1116.

Penrose, L. S. (1965). Dermatoglyphic topology. *Nature, Lond.* **205**, 844–846.

Read, D. W. and Lestrel, P. E. (1970). Hominid phylogeny and immunology; a critical appraisal. *Science* **168**, 578–580.

Sarich, V. M. and Wilson, A. C. (1967). Immunological time scale for hominid evolution. *Science* **158**, 1200–1203.

Schultz, A. (1956). The occurrence and frequency of pathological and teratological conditions and of twinning in non-human primates. *Primatologia* **1**, 965–1014.

Simpson, G. G. (1944). "Tempo and Mode in Evolution". Columbia University Press, New York.

Simpson, G. G. (1949). "The Meaning of Evolution". Yale University Press.

Simpson, G. G. (1953). "The Major Features of Evolution". Columbia University Press, New York.

White, M. J. D. (1970). Heterozygosity and genetic polymorphism in parthenogenetic animals: *In* "Essays in Evolution and Genetics" (K. Hecht and W. G. Steere, eds). North Holland Publ. Co., Amsterdam.

Author Index

Numbers in *italics* indicate the pages on which the references are listed in full

A

Abel, W., 58, *60*
Abramson, R. K., 214, 250
Achs, R., 59, *60*
Adams, J. G., 226, 250, 251
Agee, J. R., 220, *255*
Aikit, B. R., 230, *255*
Alepa, F. P., 266, 267, *269*
Alescio-Zonta, L., 259, *269*
Allen, F., 121, *124*
Allen, J. C., 258, *269, 270*
Allfrey, V. G., 154, 210
Allison, A. C., 213, *252*
Alter, M., 45, 57, 58, 59
Amaldi, F., 304, *304*
Ambrus, J. L., 291, 293, 296, *306*
Amos, D. B., 104, 124
Anderson, D. R., 109, *127*
Anderson, J. E., 18, *36*
Anderson, P. K., 33, 36
Anneren, G., 299, *304*
Antonini, E., 213, 233, *250, 255*
Apell, G., 215, *253*
Appella, E., 196, *208*
Arnheim, N., 181, *208*
Astwood, E. B., 69, *69*, 133, *149*
Aucutt, C., 234, *253*
Austall, H. B., 7, *12*

B

Baglioni, C., 5, *11*, 259, 260, 261, *269*
Bains, G. S., 29, 38
Baird, H. W., 58, *60*
Baitsch, H., 55, 60
Baker, L. L., 144, 150
Balan, B. B., 17, 18, 28, 30, 31, 32, *36*
Baldwin, M., 76, *95*
Bali, R. S., 45, 57, *60*
Balner, H., 97, 98, 104, 109, 110, 112, 114, 115, 118, 122, *125, 126, 127, 128*

Bandou, K., 59, *63*
Bansal, P., 53, *60*
Bardhan, A., 54, 62
Barge, A., 109, *127*
Barnabas, J., 166, 172, *208*
Barnicot, N. A., 6, *11*, 129, *149*, 201, 203, 204, *208*, 213, 218, 219, 220, 221, 222, 223, 225, 226, 231, 232, 233, *251, 253, 256*, 311, 312, 314, *319*
Barrett, R. J., 133, *149*
Bartels, H., 233, 234, *251, 255*
Barto, E., 196, *211*
Basilio, C., 153, *210, 211*
Bat-Miriam, M., 52, *60*
Baumgarten, A., 144, *149*
Baur, K., 189, *208*
Baur, S., 182, *208*
Beadle, G. W., 7, *11*
Beaven, G. H., 215, 230, *251, 253*
Becker, E., 52, *60*
Behrman, R. E., 230, 234, *254*
Bekkum, D. W. van, 108, 125
Belden, K. H., 294, 295, *307*
Bender, M. A., *305*
Benesch, R., 161, *208*, 234, *251*
Benesch, R. E., 161, *208*, 234, *251*
Benirschke, K., 6, *11*, 278, 293, 294, 302, *306, 307, 308*
Bennett, J. H., 144, *150*
Berg, J. M., 55, *60*
Berggren, A., 299, *304*
Bernstein, F., 72, *92*
Bernstein, I. S., 138, 141, *149*
Berry, A. C., 17, 18, 27, 28, 29, 30, 31, *36*
Berry, R. J., 13, 15, 16, 17, 18, 25, 27, 28, 29, 30, 31, 33, 34, 35, *36, 37, 38*
Bertillon, J., 43, *60*
Beutler, E., 7, *12*, 203, *208, 211*

321

Subject Index